T0264915

Service Delivery Platforms

Developing and Deploying Converged Multimedia Services

Service Delivery Platforms

Developing and Deploying Converged Multimedia Services

Edited by
Syed A. Ahson • Mohammad Ilyas

CRC Press
Taylor & Francis Group
Boca Raton   London   New York

CRC Press is an imprint of the
Taylor & Francis Group, an **Informa** business
AN AUERBACH BOOK

Auerbach Publications
Taylor & Francis Group
6000 Broken Sound Parkway NW, Suite 300
Boca Raton, FL 33487-2742

© 2011 by Taylor and Francis Group, LLC
Auerbach Publications is an imprint of Taylor & Francis Group, an Informa business

No claim to original U.S. Government works

International Standard Book Number: 978-1-4398-0089-8 (Hardback)

Visit the Taylor & Francis Web site at
http://www.taylorandfrancis.com

and the Auerbach Web site at
http://www.auerbach-publications.com

Contents

Contributors

Sofiane Abbar
PR*i*SM Laboratory
University of Versailles-Saint-
    Quentin-en-Yvelines
Versailles, France

Nilanjan Banerjee
IBM Research India
New Delhi, India

Mokrane Bouzeghoub
PR*i*SM Laboratory
CNRS and University of Versailles-
    Saint-Quentin-en-Yvelines
Versailles, France

Teck Yoong Chai
Institute for Infocomm Research
A*STAR (Agency for Science,
    Technology and Research)
Singapore

Dipanjan Chakraborty
IBM Research India
New Delhi, India

Byung-Chul Choi
Department of Engineering and
    Technology Management
Portland State University
Portland, Oregon

Rolan Christian
Centre for Telecommunications Access
    and Services (CeTAS)
School of Electrical and Information
    Engineering
University of the Witwatersrand
Johannesburg, South Africa

C. Costache
Siemens Program and System
    Engineering
Brasov, Romania

Tugrul U. Daim
Department of Engineering
    and Technology
    Management
Portland State University
Portland, Oregon

M. Demeter
Siemens Program and System
    Engineering
Brasov, Romania

Goran Đorđević
Institute for Manufacturing
    Banknotes and Coins NBS
Beograd, Serbia

Richard Good
Centre for Broadband Networks
University of Cape Town
Cape Town, South Africa

Athanasios Karantjias
Department of Computer Science
University of Piraeus
Piraeus, Greece

E. Kayafas
National Technical University of
Athens
Athens, Greece

Helmut Krcmar
Information Systems
Technische Universität München
München, Germany

Teck Kiong Lee
Institute for Infocomm Research
A*STAR (Agency for Science,
Technology and Research)
Singapore

Jan Marco Leimeister
Information Systems
Universität Kassel
Kassel, Germany

Stéphane Lopes
PR*i*SM Laboratory
University of Versailles-Saint-
Quentin-en-Yvelines
Versailles, France

Han-hua Lu
Nanjing University of Posts and
Telecommunications
Nanjing, People's Republic of China

Milan Marković
Security Department
Banca Intesa ad Beograd
Beograd, Serbia

Li-juan Min
Nanjing University of Posts and
Telecommunications
Nanjing, People's Republic of China

Toshiyuki Misu
NEC Corporation
Tokyo, Japan

Sumit Mittal
IBM Research India
New Delhi, India

A. V. Nedelcu
Transilvania University of Brasov
Brasov, Romania

Lek Heng Ngoh
Institute for Infocomm Research
A*STAR (Agency for Science,
Technology and Research)
Singapore

George Pentafronimos
Department of Computer Science
University of Piraeus
Piraeus, Greece

Nineta Polemi
Department of Computer Science
University of Piraeus
Piraeus, Greece

Christoph Riedl
Information Systems
Technische Universität München
München, Germany

D. N. Robu
Siemens Program and System
 Engineering
Brasov, Romania

Joaquín Salvachúa Rodríguez
Telematics Engineering Department
Polytechnic University of Madrid
Madrid, Spain

Judith E. Y. Rossebø
Department of Automation Networks
ABB Corporate Research
Billingstad, Norway

Ragnhild K. Runde
Department of Informatics
University of Oslo
Oslo, Norway

Luis Angel Galindo Sánchez
Innovation and Strategy
Telefonica Spain
Madrid, Spain

F. Sandu
Transilvania University of Brasov
Brasov, Romania

Anett Schülke
NEC Laboratories Europe
NEC Europe Ltd.
Heidelberg, Germany

Xu Shao
Institute for Infocomm Research
A*STAR (Agency for Science,
 Technology and Research)
Singapore

Richard Spiers
Centre for Broadband Networks
University of Cape Town
Cape Town, South Africa

Yan-fei Sun
Nanjing University of Posts and
 Telecommunications
Nanjing, People's Republic of China

Joseph Chee Ming Teo
Institute for Infocomm Research
A*STAR (Agency for Science,
 Technology and Research)
Singapore

Neco Ventura
Centre for Broadband Networks
University of Cape Town
Cape Town, South Africa

Ya-shi Wang
Nanjing University of Posts and
 Telecommunications
Nanjing, People's Republic of China

Jing Zhang
School of Economics and
 Management
Beijing University of Posts and
 Telecommunications
Beijing, People's Republic of China

Shun-yi Zhang
Nanjing University of Posts and
 Telecommunications
Nanjing, People's Republic of China

Yong Zheng
Nanjing University of Posts and
 Telecommunications
Nanjing, People's Republic of China

Luying Zhou
Institute for Infocomm Research
A*STAR (Agency for Science,
 Technology and Research)
Singapore

Chapter 1

Service Composition and Control for Next-Generation Converged Applications

Nilanjan Banerjee, Dipanjan Chakraborty, and Sumit Mittal

Contents

1.1 Introduction

Convergence between the two largest networks (Telecom and IP) is taking place very rapidly and at different levels: (1) network level: unification of IP networks with traditional Telecom networks through evolving standards (Session Initiation Protocol (SIP), Realtime Transfer Protocol (RTP), SS7, 3G) to support interoperability; (2) service level: traditional Telecom services like *voice calls* are being provisioned on the IP backbone (VoIP), while traditional IP services (most data-driven services such as multimedia, browsing, chatting, gaming, etc.) are accessible over the Telecom network.

Significant investment has been made at different layers of the converged stack, toward supporting data-driven enriched services such as multimedia, gaming, browsing, and so on. Such investments range from core network infrastructure changes (2G to 2.5G, leading to 3G, 4G) to adoption experiments with several standards (e.g., Wireless Application Protocol (WAP)) to support data-driven services (e.g., news feeds, mobile commerce, location-based service access), traditionally accessed over the IP network. Most content providers today have a *mobile* version of their information portals, which is accessible using mobile and cellular devices. Different game changing and competing technologies (e.g., Global Positioning System (GPS) versus cellular triangulation-driven location services, VoIP versus traditional telephone calls) in this converged market place are driving the big players to force an evolution in their business models.

With the market reaching saturation in several countries and revenues from voice calls decreasing rapidly, Telecom operators are aggressively looking at newer sources of revenue. So far, the typical model for providing data-driven services over the Telecom infrastructure has been through partnerships with *content providers*. In recent years, however, these partnership-driven services have been facing strong competition from *similar* technologies and applications provided by Internet Content providers. These applications can be accessed through a browser-enabled phone, while paying only for the connectivity charges, and thereby adversely affect revenues from the paid-for services hosted on the Telecom operator portal. Examples of such services range from VoIP and telephony conferencing services (e.g., skype, lycatalk etc.) to various content services (maps, ringtones, etc.). An increasing number of mobile users are now using browser-enabled phones to access these services, bypassing the Telecom portal. For example, it has been estimated that 60% of the mobile content traffic in the United States and 90% in Europe is off portal [1].

Telecom operators, however, have an edge over Internet service providers in terms of their still unmatched core functionalities of location, presence, call control, and so on, characterized further by carrier-grade Quality-of-Service (QoS) and high availability. As an example, imagine a next-generation instant conferencing application that dynamically connects friends who are *in the same location* in real time (e.g., mall) with a single click. Further, each user's device dynamically redirects/rejects calls depending on real-time preferences (e.g., allow access to my location only if I am in a public place).

Enablement of several such enriched converged services requires the core functionalities of the Telecom operator to be easily *accessible* and *composable* with third-party services providing the core application logic. Moreover, the underlying converged infrastructure (IP + telephony) needs to be smart enough to be able to provide and manage such enhanced converged services. Such services and applications require enhanced message routing and control in the core stack.

As we can see, a potential channel for the Telecom operators to increase their revenue is to offer these functionalities as services to developers for creating such new innovative applications. These developers can belong to not only the select partners of the Telecom operator, but also those involved in creating a variety of long-tail applications [2]. Additionally, due to the IP and telephony networks converging, developers can also compose their applications with third-party services available on the IP network. For example, Location and Presence information from Telecom can be clubbed with Google Maps to provide new workforce management solutions for mobile settings [3].

Recognizing the potential, Telecom operators have started investing heavily to redesign their back-end support systems (e.g., billing, provisioning, and network support systems) to address the challenges of providing the user with a unified communication, collaboration and service access experience. However, the core functionalities of location, presence, call control, and so on were so far used only internally by the carrier operator's core services (e.g., calls). As such, they were not accessible outside the network, and not easily integrable and interoperable with third-party services.

Toward alleviating these problems, there has been a concerted effort toward the creation of a blueprint for a common service delivery platform (SDP) over the past few years. Next-generation SDP* is an architectural solution that enables the reuse of service components trapped in "vertical service silos" by adopting a horizontal layered approach. There are capabilities beyond network enablers that need to be exposed through the operators SDP—these include mobility, operation support system–billing support system (OSS–BSS), functions such as billing and provisioning, subscriber profiles, QoS attributes of network elements, as well as profiles of devices supported by the operators' network.

---

* http://en.wikipedia.org/wiki/Service Delivery Platform

The salient components of an SDP are service creation environment, service orchestration environment, service execution environment, and service control and management environment. A number of standard bodies are working to come up with reference architecture for SDP. Examples are open mobile alliance service environment (OSE), Telemanagement Forum's Service Delivery Framework (SDF), Open Service Access (OSA) Parlay, and so on. Most of these reference designs expose the Telecom capabilities through flexible service-oriented architecture (SOA)-based programming interfaces such as Parlay-X [4]. Recently, Telecom operators are also exposing their capabilities through lightweight mashable interfaces or widgets to compete with the Web 2.0 Internet service providers. One such example is BT Web 21C [5]—a service aggregation environment that enables a user to create Telecom mashup applications very easily.

This chapter focuses on two integral aspects of next-generation SDPs—service *creation* and *control* infrastructure during service execution. In particular, it captures some of the inherent challenges in effective composition and control of Telecom capabilities for next-generation converged applications, in an open, collaborative environment—challenges that are inadequately addressed in existing standardized frameworks and protocol specifications.

1.2 Service Composition Framework

As mentioned earlier, core Telecom functionalities need to be accessible and composable with converged IP services to support the evolving business models. This leads to rapid integration and reuse of these services. Web 2.0-oriented mashup-style applications should also be able to effectively use Telecom functionalities (e.g., call services, SMS, location, etc.) in their application creation environments.

To support the basic operations such as voice and SMS, the building blocks of a Telecom infrastructure—location registries (Home Location Register/Visitor Location Register), accounting and billing services, SMS gateways, call servers, and so on—are already in place. However, these are not easy to utilize in new applications because they are not exposed using standardized frameworks and component models. Toward this, Telecom operators are steadily adopting SOA that would let developers access these services without knowledge of the underlying platform implementation. Web services, as an instantiation of SOA, have received much interest in the community due to their potential in facilitating seamless business-to-business or enterprise application integration. The Parlay consortium has defined a standard, called *Parlay-X* [4], that exposes Web service interface for core Telecom functionalities. On similar lines, IP Multimedia Subsystem (IMS) [6] provides a reference framework to expose these functionalities as services to Web-engineered systems using SIP [7].

Although efforts like Parlay-X and IMS are a step in the right direction, the rapid development of applications that utilize Telecom functionality still faces a number of challenges in a realistic setting. First, one needs to provide interfaces that shield the converged service developer from different Telecom protocols (Parlay-X, SIP, etc.), including the legacy ones. Second, one needs to package the Telecom functionalities so that they can be *readily* used in different programming styles (Java, HTML/JavaScript, etc.) other than pure Web service-based composition (e.g., BPEL). Finally, one also needs to encapsulate the invocation of Telecom functionality with various *coordination* rules, for example, those that correspond to managing the usage of a service, including monitoring, metering, and access control.

In this section, we focus on the above challenges and present SewNet—a framework that addresses these to enable rapid composition of Telecom services. In addition, we also provide a survey of related art in this space. SewNet is a service composition environment designed bottom-up to support rapid integration of Telecom functionalities with converged services. Its design is based on its own Telecom Service Reference, Encapsulation, and Coordination (T-Rec) "Proxy" model. This enables developers to seamlessly incorporate Telecom functionality and apply various coordination rules. Further, SewNet provides an eclipse-based* service composition environment based on the T-Rec model. SewNet can be used by different categories of developers, including Java, BPEL, and HTML/JavaScript programmers.

It is important to note that although we focus on Telecom here, the T-Rec proxy model is generic and applicable to third-party services available on the Web.

1.2.1 Problem Illustration and Motivation

We illustrate the problem with respect to the component-oriented diagram of a service that utilizes Telecom functionality, as shown in Figure 1.1. In general, such an application can be broken into two major blocks. First, there are *Telecom blocks* (represented as black rectangular boxes in the figure) that invoke a Telecom network functionality (e.g., invoking the location service or capabilities like SMS, Third-Party Call Control). The others are *non-Telecom blocks*, where the developers can

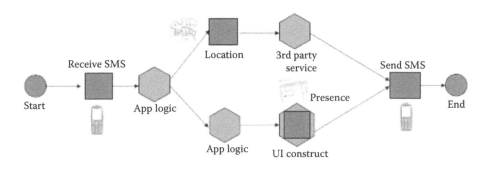

Figure 1.1   Model of a Telecom service.

* http://www.eclipse.org

embed various constructs (depicted by gray hexagonal boxes in the figure). For example, in a workforce management solution, these blocks can contain logic for scheduling agents on the basis of Location and Presence information provided by Telecom operator. Alternatively, these blocks can be user interface (UI) constructs, for instance, those enabled by various Ajax-based platforms. Finally, the non-Telecom blocks can also be invocation points for orchestration with third-party services available over the Web. Also, as the figure shows, to bind these Telecom and non-Telecom blocks, specification of complex control and data flows is also required during the service design.

We believe that for application development utilizing Telecom blocks composed with the non-Telecom ones, the following challenges need to be addressed:

First, functionality available within a Telecom operator is, in general, exposed using multiple protocols. For example, Presence-related information can be accessed via the SIP protocol, and Messaging capabilities using the Short Message Peer-to-peer Protocol (SMPP). As a step toward hiding protocol heterogeneity and complexity, the Parlay consortium has come up with the Parlay and the subsequent Parlay-X standards. Parlay-X exposes a Web Services interface for several Telecom functionalities. However, it does not cover the whole gamut of functionalities that can be offered by the Telecom operator, especially those requiring session control. Further, some of the Telecom functionalities can also be exposed through legacy protocols. Therefore, we need an *abstraction* model that provides interfaces shielding the application developer from the underlying protocols. This model should also allow seamless switching between different protocols, for example, when moving from legacy interfaces to the Parlay-X ones.

Second, developers who want to utilize Telecom functionality in their application can belong to different categories [2]. More specifically, composite applications modeled in Figure 1.1 can be written in Java, BPEL, HTML/JavaScript, and so on. Although editors corresponding to the various programming styles provide the developer with constructs for the non-Telecom blocks, they still require the Telecom functionalities packaged in a format suitable for incorporation. For example, in the case of Java applications, a developer needs a Java interface to invoke these functionalities (while a Java-based programming environment lets her code much of the non-Telecom blocks). Similarly, developers require a Web Service Definition Language (WSDL) interface for a BPEL-based composition, JavaScript for a HTML/JavaScript-based composition, and so on. Therefore, the abstraction model (outlined above) for core Telecom functionalities also needs to be *broad* to cover a range of programming styles.

Third, even though interfaces like WSDL (for Parlay-X) and SIP have tools to generate "clients" for invoking the corresponding functionalities, in real life, however, there is effort required to integrate these clients within the application. For example, code needs to be written to incorporate the client in the application code, while taking care of tertiary library dependencies for this client. It would help the developer immensely if the abstraction model pregenerates the clients corresponding to different programming styles, and packages them in a *structured* manner. Having a well-defined structured format would enable any application development environment (with some extensions to interpret this structure) to integrate these clients seamlessly.

Finally, when Telecom functionality gets used in an application, the Telecom operator, as well as the application developer, wants to coordinate its usage. For instance, one needs mechanisms for embedding logic for charging, specifying access control policies, and so on. Furthermore, with recent trends suggested by Web 2.0, application developers should be able to contribute, implicitly or explicitly, to the enrichment and refinement of the exposed Telecom functionality (and its usage). Therefore, the abstraction model needs to be *rich* enough to enable all of this.

A number of operators are already moving in the direction of making their core functionalities available for application development. For example, British Telecom has released a Software Development Kit [5] that enables its network services to be utilized in Web mashups. However, what is missing is an abstraction model which is broad, structured, and rich, as motivated above. We describe such a model next, and thereafter present a service creation framework on top of this model.

1.2.2 T-Rec Model

Figure 1.2 represents the basic concept of the T-Rec "Proxy" model. In essence, once an application has been broken down into Telecom and non-Telecom blocks, this model is used to realize the Telecom blocks, considering the programming style, while also enabling mechanisms for coordination and enrichment. In practice, these proxies would be created by a Telecom operator and be made available to application developers over the Web (or a converged IP network).

1.2.2.1 T-Rec Structure

The rich, structured T-Rec Proxies consist of the following elements:

- *Proxy representation:* Contains signatures of the methods (Application Programming Interface (API)) exposed by the proxy along with a textual

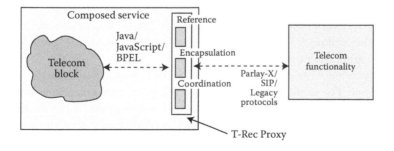

Figure 1.2   T-Rec "Proxy" model.

description of the service it represents. The APIs are designed to hide protocol-specific details and abstract the Telecom functionality to the programming language level. As discussed before, APIs corresponding to multiple styles (Java, BPEL, and JavaScript) should be created to support different environments.

- *Implementation:* This module connects to the Telecom service using the underlying protocol, and is available in different formats. For example, the implementation could be in the form of a *.jar* file for a Java proxy, a *.js* file for JavaScript, or could be encapsulated by visual constructs, such as widgets, and used inside HTML pages.

- *Configuration file:* Proxies come with a default setting but can be further configured by developers. This includes assigning default values for some of the parameters in an API, specifying the access control list, and so on. Such settings could also be functionality specific, for instance, restricting the size of SMS messages.

- *Metadata:* To enable easy look up, keywords and tags related to the proxy functionality are associated with it. New tags can be added to the proxies if required; for example, when a developer utilizes a proxy in a way that was not originally foreseen by its creator.

- *Utility snippets:* The proxies are populated with multiple code snippets on top of the basic functionality. For example, a "Presence" proxy may have a program fragment that parses the returned response (usually an XML document) for different attributes. These utilities can be suggested, in an appropriate manner, to developers who intend to use the proxy.

- *Unit test code:* Proxies contain codes that let different APIs supported in the proxy be tested in isolation. These are very helpful during testing and debugging.

- *Link to blogs:* Each proxy is linked to a blog entry where developers can log their experience of using the proxy. If multiple proxies are suggested during a look up, analyzing the blog entries can help the developers choose the most appropriate one for their task.

1.2.3 T-Rec Benefits

Intuitively, a T-Rec proxy acts as a "wrapper" for Telecom functionality, including its underlying protocol. Using this wrapper, the proxy creator can provide several benefits to application developers.

Encapsulation: APIs defined in a proxy can hide protocol-specific details from the developer. For example, interfaces in Parlay-X throw exceptions with error codes that require knowledge of Parlay-X for interpretation. As an instance, an application developer using Parlay-X would need to know that the error code *SVC0004* stands for invalid addresses in a message. Using the proxy model, we can encapsulate these error codes with higher-level exceptions, such as throwing *InvalidAddressException* whenever error code SVC0004 is returned. Moreover, using proxies, similar APIs can be exposed across different protocols. For example, various APIs in the Location proxy can have similar signatures for Parlay-X and SIP-Presence-based implementations.*

Coordination: When Telecom proxies get used in an application, the Telecom operator as well as the developer can manage and meter its usage. For instance, whenever the proxy corresponding to Location information gets invoked within an application, the Telecom operator can authenticate the developer and also charge some amount. In this case, proxies are configured to collect the relevant information, for example, developer Id, from the developer and send it to the operator. Similarly, the developer can configure the Location proxy to cache the location information locally within itself, and avoid connecting to the operator's infrastructure at each invocation.

The proxy model also provides an easy mechanism to incorporate various business contracts between the operator and the developer. For example, the implementation module in a proxy can be extended to make the proxy display advertisements on behalf of the operator, whenever it is invoked. In this case, logic can be such that the proxy picks what to advertise on a real-time basis.

Collaboration and reuse: Using the proxy model, developers can collaborate, share, and contribute toward enriching Telecom functionality. For instance, the user of the Location proxy in an application can publish a utility to parse the output of this service. This utility can be reused by other developers while incorporating this proxy in their applications. Similarly, the proxies can be configured to provide updates to a developer about new entries on the blog, utilities published recently, bug fixes, and so on. In the case of bug fixes, logic can be embedded in the proxy to automatically download the latest implementation modules.

1.2.4 SewNet Details

Next, we present the SewNet framework that utilizes the rich, structured T-Rec proxy model to enable seamless weaving of Telecom functionality with application logic and other constructs required to develop a service.

1.2.4.1 SewNet Architecture

As Figure 1.3 shows, SewNet has two main architectural components—SewNet Core and Composition Studio(s).

SewNet Core forms the backbone architecture that exposes Telecom functionality to developers through simple, intuitive interfaces for lookup and selection while allowing for developer participation and feedback through publishing and blogging.

---

* In SIP, location information is obtained by subscribing for the presence information, and parsing the returned document. We can wrap this under a *getLocation()* interface.

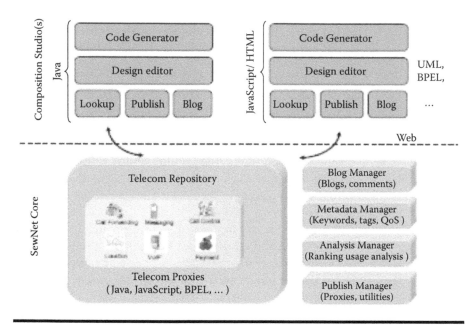

Figure 1.3   SewNet architecture.

Telecom Repository consists of proxies for different functionalities exposed by the Telecom operator. These proxies are available in various implementation styles, for example, Java proxies to be used inside Java applications, JavaScript proxies to run on a Web browser, and so on. As mentioned earlier, each proxy hides the underlying protocol (Parlay-X, SIP, etc.) and offers a rich set of APIs to facilitate integration with the application being developed.

Metadata Manager helps the Telecom Repository organize and maintain relevant metadata (keywords and tags, textual description, etc.) associated with a proxy and the APIs it offers. This information is used to suggest proxies on a lookup.

Blog Manager organizes and stores free-form textual comments associated to a proxy and its APIs. These inputs are presented to the developer while browsing and selecting proxies from the Telecom repository.

Analytics Manager maintains qualitative information about proxies, including a rating and ranking of each proxy. We envision this manager containing tools to analyze blogs by different developers, collect usage statistics, and so on, and making such information available to application developers.

Publish Manager defines the interface to publish new proxies as well as new artifacts associated with an existing proxy; published items become available to other developers.

Composition Studio(s) are developers wishing to use Telecom proxies exposed by SewNet need to integrate their development environments (or composition studios) with SewNet Core. These studios range from programming platforms (e.g., Eclipse environment) to model-driven tools (such as those containing Unified Modeling Language (UML) editors for service design and representation) to work-flow editors allowing services to be composed in a language like BPEL. For integration with SewNet Core, a studio needs extensions along three dimensions. First, its service design (or programming) editor should provide the ability to identify the Telecom blocks from the non-Telecom blocks. Second, it should offer Lookup, Publish, and Blogging interfaces for proxies provided by SewNet core. Third, once proxies have been selected for different Telecom blocks, a Code Generator module should traverse the structured format of each proxy to seamlessly integrate it with the rest of the application code.* It is interesting to note that a particular composition studio may use one or more different types of proxies. For example, while creating a Java Server Pages (JSP), developers can incorporate Java as well as JavaScript proxies.

1.2.4.2 Composing Telecom Services Using SewNet

Once a composition studio has been integrated with SewNet Core, the following steps illustrate the process that application developers follow to compose Telecom services:

1. *Service design:* The developer designs the service using drag-and-drop or other mechanisms supported on the design editor. In this step, Component services, Logic blocks, Control flow (sequencing, fork, join), and so on are defined by the developer (cf. Figure 1.1).
2. *Proxy lookup:* For each Telecom block, the developer obtains a set of matching proxies from repository, based on keywords, input–output, or both.
3. *Proxy selection:* From candidate proxies suggested by SewNet, the developer selects those that best fit the requirement. Proxy selection is based on suitability (e.g., reading more about each proxy) and other metadata (QoS parameters like reputation, etc.).
4. *Proxy configuration:* The developer optionally associates various service coordination rules with each proxy. For example, she specifies the time period after which cached location information is to be refreshed by the Location proxy.
5. *Code generation:* Outputs code for the designed service (BPEL, Java, JavaScript inside HTML, etc.). This step is described in more detail in the next subsection.
6. *Incorporate other constructs:* The developer incorporates appropriate application logic, UI elements, and so on to complete the service. At this point, the developer also takes care of data flow between different constructs.

---

* In practice, a Composition Studio would already have some code generation capabilities. In this case, we just need to extend these capabilities to incorporate the proxy model.

1.2.4.3 Code Generation

From a developer's view, once a service has been designed and relevant proxies for the Telecom blocks selected, she expects the composition environment to generate a skeleton code that not only captures the service flow, but also integrates the code for the selected proxies. Further, this code should provide her the extension points to include application logic and other constructs for the non-Telecom blocks. We divide this process of "Code Generation" into three steps (Figure 1.4):

1. *Capture service design:* Service design is captured in a structured document that we call *processDoc*. Further, for each Telecom block, processDoc stores information about the proxy that was selected, and the API that was chosen under this proxy. It should be noted that the structure of processDoc and the information it stores is dependent on the programming style of the application. More specifically, it should be able to represent each programming construct of that style.

2. *Generate process skeleton:* This step takes processDoc and transforms it to a concrete, fully compiled code. For this purpose, Code Generator parses processDoc and converts each element into the corresponding programming construct. While parsing this document, it creates place holders for the non-Telecom blocks and adds comments to aid the developer when she examines the generated code. For the Telecom blocks, it populates the proxy code, as described next.

3. *Populate Telecom proxy code:* In this step, the Code Generator imports the relevant implementations of the proxies from the Telecom Repository and generates the code necessary to invoke the selected API in the proxy. The code produced depends on the programming language. For example, while for Java, it creates a *class* with an invoke *method* that internally calls the proxy API, for BPEL, it generates an invoke *statement*. Further, Code Generator also analyzes

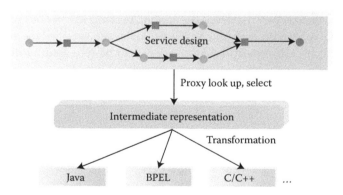

Figure 1.4 Code Generation.

the method signature of each API to understand the exceptions being thrown and organizes appropriate exception-handling blocks around it. Finally, Unit Test codes available with each proxy are included in the code generated.

SewNet has been implemented as a framework, including various Telecom proxies, a composition studio to develop services using these proxies, and generation of code in two different programming styles—Java and BPEL. The details of the implementation are available in [8]. Next, we provide an overview of related art in this space.

1.2.5 Related Work

Work related to service composition using Telecom functionality can be broadly classified into (1) Composition Using Web Services and (2) Tools for Mashup Applications.

■ *Composition Using Web Services:* The literature on Web service composition is extensive. Here we describe a few that we believe are broadly tied in with service composition using Telecom Functionalities. Triana [9] aims to facilitate Web service composition by providing a higher level of abstraction and guiding developers in creating composed services. The paper presents a case study that investigates how this environment can be used in a Telecom setting. Synthy [10] demonstrates the composition of a workforce management application enabled over a Telecom network, where core functionalities like Location are exposed as semantically annotated Web services and orchestrated using AI planning techniques. The Meteor-S [11] framework looks at a number of aspects related to the composition of Web services, including capture of semantic requirements of the process and choosing components under given constraints. Zeng et al. [12] propose a method for choosing component services during Web service composition based on a generic QoS model (based on price, duration, reliability, etc.) and established linear programming techniques.

As noted earlier, we focus on SewNet in this chapter as it provides a broader model than pure Web services to cover different protocols. SewNet's T-Rec proxy model is rich enough to apply many of these techniques. For example, each proxy can be annotated with QoS guarantees regarding its reliability, response time, and so on, using which developers can estimate the QoS parameters of their composed services according to [13].

■ *Tools for Mashup Applications:* One of the most commonly used terms for Web 2.0 applications is a *mashup*—an application that combines content from more than one source into an integrated experience. Content is picked up from multiple servers (data sources) using technologies like Ajax and REST, and composed (or rendered) at the same UI, typically a browser. There are several tools that aid in the creation of such mashup UIs. Examples are QEDWiki [14], Yahoo Pipes [15], Aqualogic [16], and PrestoStudio [17]. These tools cater to a class of application developers who prefer "drag-and-drop" operations to create their own mashups. It is worth noting that most of these tools are not Telecom specific and can benefit by incorporating SewNet's proxies.

Web21C [5] from British Telecom allows developers to integrate core Telecom functionality with other Web services into a single application, while allowing application-specific logic to bind the component services. Similar to our work, Web21C hides the complexity of Parlay-X or SIP by exposing these Telecom functions as higher-level APIs. Similarly, Connected Services Framework Sandbox [18] from Microsoft provides high-level Web service interfaces via its SDP that hide low-level Telecom protocol-specific constraints and allow the creation of Managed network mashups. However, both of these can benefit by incorporating a structured and rich format such as the T-Rec proxy model.

1.3 Service Control Layer

The ease and flexibility of service creation and orchestration (composition) offered by the SDPs have resulted in a plethora of advanced value-added services developed and deployed in Telecom networks. The management of these services is a very complex but very critical operation for the success of any service provider. The key management functionality in converged networks is *IP-based service control* that ensures many different services running over a single universal IP network, while controlling the customer access by dynamically configuring and provisioning them in the runtime. Thus, service control requires knowledge of what occurs in the network and the ability to effectively manage it in a flexible and customizable way. Considering the fast-changing Telecom services and application space, some desirable service control features are

1. Swift adoption to market changes and customer *preferences*.
2. Support of many different types of business models (e.g., changing the charging model of a service for a category of users).
3. Effective participation in a value network.

Today's approach to IP-based service control is typically inflexible and costly to maintain. That is, service control requirements are hard-coded into service logic (IN, SIP Servlet, etc.) during service development with limited configuration options. Where configuration is possible, it often requires highly specialized skills. The primitive service control in IMS based on SIP and SIP Application Servers (ASs) also provides a fair degree of control to applications. While this method is widely used, we believe it is insufficient for the service delivery models that emerge out of a next-generation application. First, it often leads to a "one-size-fits-all" model for service control which does not lend itself well to rapidly changing business and market requirements—especially in the case of mashup-based ecosystems. Further, current approaches tightly couple Telecom services with back-end IT applications such as Customer Relationship Management (CRM) and billing, by means of hard coding of service management policies and business logic into the application logic itself. In a mashup-driven ecosystem, this introduces downsides in third-party service delivery models for both the operator and the third party. From the operator's perspective, it might have to entrust service control—such as charging—to third parties for off-network services which may not always be desirable. From the third-party's perspective, it may have to adapt service logic to an operator's environment, which introduces complexity when seeking to partner with more than one operator.

This is further illustrated in Figure 1.5, which shows that today's service logic is tightly integrated with subsystems such as billing (1), adding complexity and redundancy to service logic. Further, some of the subsystems are tightly integrated with other subsystems (2) leading to *ad hoc* and inefficient management of these services. In either case, this often leads to a static and costly one-size-fits-all model. It is difficult, for example, to quickly change the way that services behave in response to an ever-changing business environment, for example, introduce variable charging schemes per user/user group, change business policies such as Service-Level Agreements (SLAs)—without resorting to costly and time-consuming code

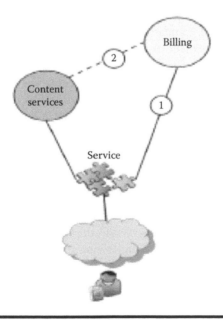

Figure 1.5 Traditional IMS service delivery.

changes. In a Web 2.0-oriented ecosystem, it not only becomes a nightmare for operators to effectively partner with third-party services, but also makes it infeasible to manage self-owned applications.

To address this shortcoming, we propose IMS-based service control architecture for offering next-generation Telecom services. This service control architecture can indeed be incorporated into the larger scheme of things of an advanced next-generation SDP, but here, we keep that out of the scope of this chapter.

IMS defines an overlay service architecture that merges Internet services into Telecom service development. The IMS architecture supports a wide range of services including traditional legacy services (Intelligent Network (IN) services), next-generation IMS services with SIP [7] as the underlying protocol, as well as SOA-based Web Services. Introduced by 3GPP [19], IMS uses the SIP in the signaling plane to provide assistance and control for multimedia sessions, established between two communicating peers.

Service control in IMS architecture is actually delegated to an entity called the Service Capability Interaction Manager (SCIM) [6]. The SCIM is envisioned to coordinate the diverse capabilities from different service domains, and orchestrate the interactions between them, using SIP as the control protocol in the signaling layer.* The SCIM can loosely be likened to an SIP-based AS that performs the functionality of runtime orchestration of interactions between capabilities in the converged network. Such interactions can occur between SIP features and legacy signaling features (e.g., SS7), between components that are wholly implemented as SIP proxies or user agents [7], or even with external business capabilities that are exposed using Web services (WSDL, Simple Object Access Protocol (SOAP)).

Surprisingly, the 3GPP standard only includes a loose definition of SCIM on its role of interaction management. According to this definition, an AS may contain SCIM functionality and other ASs. The SCIM functionality is an application which performs the role of interaction management. The implementation of the SCIM has been left up to the vendors to suite their individual application requirements. As a result, the literature has very little on the design principles of an SCIM that one can refer to. Thus, the SCIM with the limited scope of its definition falls short in realizing the full potential of service control in IMS networks.

1.3.1 Limitations in IMS Service Control

The primitive service control provided by the Initial Filter Criteria (IFC) in the Call Session Control Function (CSCF) server and the application chaining with SIP AS is fairly a rigid vertical because of the following reasons. The IFC can only match

---

* The reason behind the choice of SIP: First, SIP is based on a simple request–response interaction model that allows developers to interact with individual protocol messages. Second, SIP can start/manage/tear-down sessions for any media type, be it voice, video, or application sharing.

initial requests in a session such as INVITE, REGISTER, SUBSCRIBE, MESSAGE, and so forth. Subsequent messages (e.g., BYE, NOTIFY, re-INVITE) are not checked and filtered by the IFC. This means that IFC alone is not adequate to model and monitor the entire life cycle of a service thus making it incapable of providing greater control to the services. Besides, changing the IFCs is not trivial, hence not flexible enough. The SIP AS chaining is also quite inflexible as the control logic developed in such a chaining application is not easy to change and the interaction with the external capabilities are often hard wired rendering them inflexible in the face of any change in the interaction with the external world. These inflexibilities make the primitive service control in IMS unfit for the next-generation Telecom service and application development.

1.3.2 Service Control Layer Architecture

In contrast to the traditional one-size-fits-all model of service control, Service Control Layer (SCL) is a Telecom middleware technology that employs an innovative decoupled approach to service delivery. SCL enriches traditional service control on two fronts, namely (1) flexible, dynamic SIP service modeling and (2) transparent, flexible interface with external capabilities or business services. To begin with, SCL models a Telecom service as a state machine (Figure 1.6). During

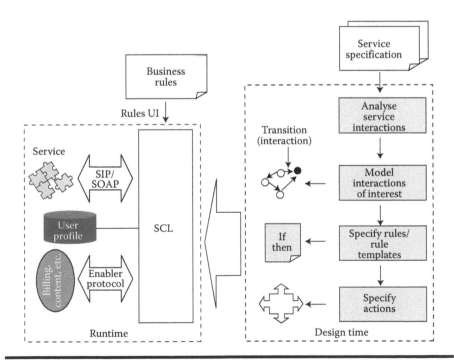

Figure 1.6   Flexible rule-driven service delivery in SCL.

the execution of an SIP service, SCL listens for the exchange of service interactions (e.g., SIP INVITE) of interest between parties (applications and users) engaged in a service). Rules are then attached to state machines, typically for actions on state transition, such that they are triggered for evaluation on processing of a given protocol message. Each action is, in turn, mapped to a service-oriented capability that needs to be blended during service delivery. These actions invoke service-oriented external business capabilities at runtime in order to control, enhance, or personalize the service. The external capabilities are invoked through a configurable interface or Control Enabling Proxy (CEP). The CEP abstracts out the functionality of the external services and manages the interaction between SCL and the services transparently.

1.3.2.1 System Architecture

Figure 1.7 illustrates the system architecture of the SCL. The core components and their functionalities are described as follows.

The *Protocol Adaptation* module handles different incoming protocol messages (e.g., SIP), and translates them into an internal format (known as Common Format Business Objects) for processing within the SCL. Additionally, it includes an *SIP*

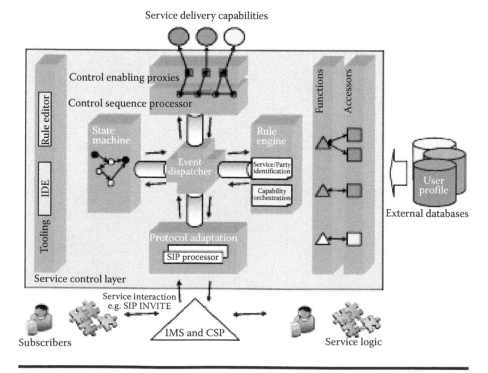

Figure 1.7   SCL architecture.

processor that inspects the SIP Initial Request (typically a SIP INVITE) and triggers one of the following modes for SCL to control a session:

- B2BUA Mode—Activates B2BUA mode for the SIP dialogue being processed, enabling the SCL to act as the mediator between two SIP endpoints.
- Terminating User Agent (TUA) Mode—Activates the TUA mode for the SIP dialogue being processed, enabling the SCL to respond directly to the originator of the service session—e.g., the SCL determines that a caller is not authorized to establish a service and sends SIP 4xx error in response to a SIP INVITE.
- Proxy Mode—Activates Proxy mode for the SIP dialogue being processed, causing all messages received to be proxied toward their intended destination following processing by the SCL.

Objects are then routed to the *Event Dispatcher*. The Event Dispatcher is responsible for identifying services on receipt of events belonging to new service dialogues and for dispatching these events between SCL modules. *State machines* are used to monitor the execution lifecycle of a service, or at least those stages in the lifecycle that are significant to the control or enrichment of the service by the SCL. Generally speaking, a state machine is used to model the protocol messages of interest in a given service. Rules can be attached to state machines such that they are triggered for evaluation by a *Rules Engine* on receipt of a given protocol message. External data such as user profile information can be queried during rules evaluation. An orchestration mechanism that composes one or more "to do" items expressed in an execution sequence list is generated by the Rules Engine. The *Control Sequence Processor* (CSP) manages the invocation of external business services according to this execution sequence list. Business services (e.g., online charging) are exposed to the CSP through *Control Enabling Proxy* (CEP). The CEP is responsible for resolving the execution sequence list to the appropriate business capabilities, providing the required input parameters as well as processing the responses received back from the operation.

1.3.2.2 How Does It Work?

SIP protocol messages are routed to the SCL as they traverse the IMS control plane from the source (e.g., a user's mobile device) to the intended destination (e.g., another user or an application).

On receipt of any message, SCL first determines whether a message belongs to an already active service session, or signifies the start of a new session. When a new service session is established (e.g., SCL receives an SIP initial Request message such as INVITE), SCL identifies the type of service by applying preconfigured service identification rules to the message received. Based on this information, SCL creates an active instance of a preconfigured state machine to direct subsequent SCL processing activity throughout the life cycle of the service session.*

After service and party identification, or on receipt of a subsequent message belonging to an already active service session (e.g., a SIP ACK that acknowledges a session establishment), SCL forwards the message to the appropriate state machine. This leads to a state transition in the active state machine. Preconfigured business rules may be executed during this transition to determine the service control required at this point in the service. Business rules are typically evaluated against data contained within the protocol message—for example, examination of SIP Request Uniform Resource Identifier (URI)—but may also utilize information from external data stores—for example, service and/or user profile information.

Service control or enhancement at this stage can be achieved by

1. The Rules Engine instructing an orchestration mechanism to call external business services; and/or
2. The Rules Engine instructing the SIP protocol adaptation module to change the service session from Proxy mode (default) [6] to either B2BUA [6] or TUA mode [6].

In scenario (1), the orchestration mechanism may call one or more external business services based on a "to do" list received from the Rules Engine—for example, the authorized user then initiates session charge. The "to do" list may contain directives to allow list processing to be dynamically changed based on the responses received by the orchestration mechanism from a given business service—for example, only initiate session charge if the user is authorized to access the service. Responses from calls to business services may be redirected to the Rules Engine for further processing—for example, instruct the SIP protocol adapter to terminate the service session if the user is unauthorized. Scenario (2) provides the SCL with the capability to control a service session either during session establishment (TUA mode) or during its life cycle (B2BUA mode). In the TUA mode, the SCL is able to tear down a service session by responding directly to the originating party—for example, the SCL determines that a user has insufficient credit to establish a media session with another user. In the B2BUA mode, the SCL is able to "transform" a service session into a back-to-back session, effectively acting as a bridging point between service endpoints. This enables the SCL to exercise control over a service session at any stage during its life cycle. For example, the SCL determines that a user who is trying to set up a VoIP session is on a prepaid plan, and so instructs the SIP protocol adapter to change the service session to B2BUA mode in case the user runs out of credit mid-session. If the user does exhaust her available credit and opts

---

* If the message is not relevant for service control requirement, it is simply forwarded to the destination.

not to top-up, the SCL is able to generate a SIP BYE message toward all endpoints, effectively tearing down the session.

The call to external business capabilities could be complex and a stateful one. Hence, every time the behavior of the external service changes the interaction between the SCL orchestration mechanism and the external service needs to be remodeled and implemented all over again. This is an expensive proposition, given the rapid changes expected in the external services due to enhancements and new features. In order to mitigate this problem, we introduce a proxy functionality provided by CEP, between the SCL orchestration mechanism and the external services. A key function of CEP is to map actions (and parameters, where specified) supplied by the Rules Engine to an appropriate interface (operation and arguments) supported by the target business service, including processing responses from the latter. A CEP can generate autonomous messages toward other SCL components, for example, a CEP communicating with a charging business service can inform the SCL after a stipulated time period that there is no credit left for a given user and that the SCL must therefore terminate the ongoing session. The introduction of CEP enables the decoupling of the external service interface with the SCL orchestration mechanism. The CEP can be configured for each individual external service, thus rendering flexibility in the interaction between SCL orchestration mechanism and the external services. The CEP consists of four major components as follows:

1. Interface to the SCL orchestration mechanism in CSP: This interface enables the CSP to send the directives to CEP required for leveraging the functionalities of external business services.
2. A state machine: It models all possible interactions with an external business service. There are as many state machines as there are external business services to interface with.
3. A timer module: The timer is internally managed with the CEP to trigger asynchronous events for both the SCL orchestration mechanism and the external services.
4. The proprietary interface with external business service: This interface hides all the details of the external service and exposes an abstract interface that accepts events from the CEP to trigger certain actions in the external service and send the response back to the CEP for further processing.

1.3.2.3 How Is It Configured?

The SCL has been designed to be configurable by users who undertake one of two distinct roles. These roles are loosely referred to that of a *technical expert* and that of a *business expert*. The former will typically be someone who is familiar with the protocol-level semantics of a service, while the latter will be concerned with implementing business policy and how a given service should be controlled or enhanced, for example, for a particular user or group of users. The two roles provide an important distinction in the modeling of rules within the SCL. Some configurations are deemed to be *design time*, that is, pertains to tailoring a service behavior and would require code to accommodate the same (e.g., integration of a new business capability like Internet Protocol Television (IPTV)). This change would need to be facilitated by the technical expert, and would typically involve the use of design-time tooling. Other configurations could be deemed as being *runtime*. These changes could be achieved through simple configuration of business rules (which are consulted by the Rules Engine), and would typically be undertaken through an intuitive UI. The design of the SCL has been mindful of the distinction of the two roles, and wherever possible, has provided a layer of abstraction so that (multiple) business policies can be implemented flexibly and transparently from protocol-level details of a service implementation.

The CEP is configured ideally by the external business service vendor who has knowledge about all the possible state transitions that the service can go through. This is captured in the state machine in the CEP, which accepts input from the SCL facing interface and sends directives to the external service through the service-specific interface. The timer module is also instantiated so that any asynchronous event can be fired or captured for possible state transition in the state machine.

1.3.3 Related Literature

IP service control is an active area of research in network service management. Some of the early works [20,21] in IP service control were targeted for legacy networks, mostly in an *ad hoc* manner. Later, IP-based service control was somewhat formalized [22] for next-generation networks by standard bodies such as 3GPP, resulting in overlay service control frameworks (namely IMS) for delivering advanced services. Service control in IMS, however, is very rudimentary and much of the flexibility in service control has been delegated to a fuzzy architectural component, namely SCIM, which has been left to the vendors for design and implementation. Stray attempts have been made to realize the SCIM functionalities and beyond. For example, JSR 289 [23], which is aimed at enhancing the capabilities of JSR 116 [24] SIP Servlet specification, is a step in the direction of standardizing the orchestration capabilities of SCIM in a converged Web-Telecom service layer. A key agenda of JSR 289 is to make HTTP (SOAP)-based services available on the same platform with SIP-based services. An early SCIM implementation has been reported in [25] but fails to address the flexible control requirements imposed by blended services. With the emergence of Web 2.0 and a large number of short-lived services in this paradigm, the requirement of flexible service control has become even more critical. A number of vendors have started offering service development and deployment platforms in this space. Examples are BEAs Aqualogic [16], British Telecoms Web21C [5], and Microsoft's Connected Services Framework Sandbox [18]. The key feature of all these offerings is the exposure of Telecom services as lightweight widgets, which are easy to incorporate in a mashup environment. However, none of these consider runtime control and mediation of the mashup services. To bridge this gap, the SCL architecture was proposed in [3] for executing Telecom mashups in a Web 2.0-oriented domain.

1.4 Illustrative Example

We now demonstrate how the twin components of Service Composition and Service Control can enable development, flexible control, and enrichment of a Telecom mashup service, using a combination of service modeling and rule-driven orchestration of SOA capabilities.

1.4.1 Call-a-Cab Scenario

The Call-a-Cab scenario is based on a mashup service called Business Finder [26], a next-generation, presence-enabled technology that leverages upon the underlying cellular infrastructure to provide efficient, on-demand, context-aware matching of customer requests to mobile vendors. Business Finder mashes up Location and Presence information, along with call control features, exposed by a Telecom operator with third-party Google Maps service. Using Business Finder, one can search for a *nearby and available* vendor for a particular service. In the Call-a-Cab scenario, we consider Linda who is visiting Barcelona and wants to take a cab back to the airport. Using her GPRS-enabled phone, she logs on to the Business Finder portal and issues a search for a cab. Business Finder uses the location of Linda's mobile phone to render a Google map showing all cabs which are available and within a few miles of her location, along with a star-rating for the cabs (Figure 1.8a). Linda selects one of the cabbies, Chris, uses the Click-to-Call widget available as part of the mashup. This widget is enabled by IBM Telecom Web Services Server. Model diagram for the Call-a-Cab scenario is depicted in Figure 1.9.

Figure 1.10 demonstrates the key functional elements of this setting, namely (a) Business Finder—a Telecom mashup service, (b) Telecom Web Services Server—an IBM middleware solution for delivering value-added Telco services such as 3PCC, Location, and so on, (c) Service Composition plane to bind various offerings together, (d) SCL—for runtime control, mediation, and interaction management of 3PCC service that is invoked as part of the mashup, and (e) a charging platform integrated with SCL and used to apply different charging policies to Business Finder users. IBM WebSphere Telecom Web Services Server (TWSS) [27] enables Telecom operators to provide third parties with controlled, reliable access to Telecom network capabilities such as location, messaging, presence, and call control. In particular, the 3PCC (third-party call control) functionality of TWSS is used by Business Finder to initiate a SIP call between the devices of Linda and Chris.

Telecom proxies can be used to compose mashups such as Business Finder involving Telecom functionality and third-party services. A code sample is demonstrated in Figure 1.11 that combines the Location proxy in JavaScript format along with the

(a)   Business Finder 2.0—your search results on Google maps

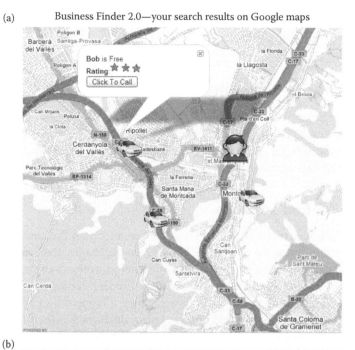

(b)

Figure 1.8 (a) Business Finder mashup, (b) visitor pass prompt to activate a visitor account at runtime.

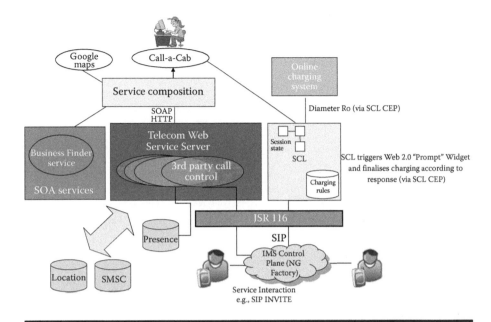

Figure 1.9   Call-a-Cab: device model.

Google Maps service. This simple code receives updates on the location information and displays it on Google Maps; note that this currently has no application logic component. However, the code highlights the steps needed to extend an existing mashup editor to enable seamless incorporation of JavaScript T-Rec proxies in the mashups designed. More specifically, the editor needs to add statements to import the implementation (`locationProxy.js` file that contains `getLocationUpdates()` function), include JavaScript code to invoke the proxy (call `getLocationUpdates()` that internally subscribes `locationCallback()` to receive location updates), define variables to capture the invocation output, and so forth. All this shall be automatically done through interpretation of the T-Rec proxy structure.

Now, let us see how the real-time charging operation can be done in a flexible way with SCL for the above scenario. Before the call is set up, the SIP INVITE is

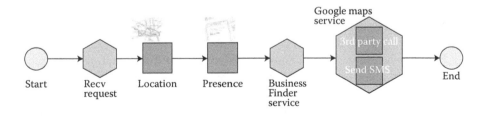

Figure 1.10   Call-a-Cab: integration architecture.

```
<html>
<head>
. . . . .
<script type="text/javascript" language="Javascript1.2"
src="scripts/locationProxy.js"></script>
. . . . .
<script type="text/javascript">
var latitude;
var longitude;
function callback(){
  if(req.readyState==4)      {
          if (req.status == 200) {
                    latitude=req.responseXML.getElementById
('lat');
                    longitude=req.responseXML.getElementById
('long');
                    }
      }
}
</script>
</head>
<body>
. . . . .
<!-- Application specific HTML and Javascript Logic -->
<!-- Take CellNumber as Input -->
<!-- javascript locationProxy returns latitude and longitude
respectively -->
<script type="text/javascript">
getLocation(CellNumber);
</script>
<script type="text/javascript">
if (GBrowserIsCompatible()) {
      var icon = new GIcon(baseIcon);
      icon.image = "http://www.google.com/mapfiles/marker.png";
      var point = new GLatLng(latitude, longitude);
      map = new GMap2(document.getElementById(" "));
      map.setCenter(point, 13);
      map.addOverlay(new GMarker(point, icon));
      }
</script>
. . . . .
</body>
</html>
```

Figure 1.11   Call-a-Cab: code fragment.

routed through SCL, which consults the service model programmed into it by the technical expert and uses the service identifier (i.e., Business Finder) and the subscriber mobile number (i.e., of Linda) to issue real-time charging for this call. The charging model, in turn, would depend on the subscription plan of Linda (prepaid, postpaid, premium, etc.) and hence it is typically a business decision taken and programmed into the SCL by business experts. Of particular interest, for example, is the scenario where Linda does not have an account with the operator. In this case, SCL pushes a SIP message to Linda's device—the message includes an HTTP URI which Linda can visit to buy a "visitor pass" with some credits in it (Figure 1.8b). Subsequently, SCL sets up a visitor account for Linda at runtime using the capabilities of the charging platform. Linda can now proceed to make the call and talk to the cabbie—to decide a time and location for the pickup. Once again, SCL (in a B2BUA mode) controls the entire call session and applies real-time charging to the call. Finally, when the call terminates (indicated by a SIP BYE), SCL pushes a second message to Linda's device. The message includes an HTTP URI which Linda can visit to confirm her appointment with the cab driver. SCL charges Linda and the cab driver based on the charging policies of Business Finder. Finally, Business Finder updates Chris's presence status to show that he is now busy picking up a passenger.

1.5 Discussion

A generic model of Telecom services is described in [2], where apart from core Telecom network functionality and third-party offerings, services also make use of device functionality and information such as calendar, user profile, and location (e.g., cell site information). Currently, there are a plethora of end-user devices in a Telecom environment, ranging from basic (offering only the capabilities of messaging and voice) to sophisticated, state-of-the-art devices such as the iPhone.* There are efforts to promote an open mobile phone software stack that abstracts hardware differences and offer a uniform set of APIs for accessing many of the functions of a mobile phone (calendaring, date, etc.). JSR 248 specification, contributed significantly by Sun Microsystems, is a step toward that direction. To compose richer Telecom applications, one can wrap the device functionality using the SewNet proxy model and make them available to a developer during the composition process. In this case, such "device-side" proxies will not only ease the access to device functionality, but also shield the developer from the heterogeneity of the underlying device itself.

SCL on the other hand aims to fill a technology gap in the architecture of current SDPs in terms of applying transparent and fine-grained service control requirements to next-generation IMS-based service delivery. The focus of SCL has been on processing SIP messages to apply service control. However, in a Web 2.0 domain, SCL might additionally need to process non-SIP (e.g., SOAP) messages generated from a user-driven session. Accordingly, the functionality of SCL can be extended to model such interactions and apply rule-driven blending of Web/Telecom capabilities. Further to this, the role of SCL could be extended in enriching a range of service delivery scenarios, for example, rich media sessions of a peer-to-peer nature that might otherwise be beyond the control of the service provider.

References

1. *Telco Web 2.0 Mashups: A New Blueprint for Service Creation.* http://www.network mashups.com/docs/ssi_0507.pdf.
2. S. M. D. Chakraborty, S. Goyal, and S. Mukherjea. On the changing face of service composition in Telecom. In *Proceedings of MNCNA: Middleware for Next-Generation Converged Networks and Applications*, California, 2007.
3. K. Dasgupta, N. Banerjee, and S. Mukherjea. Providing middleware support for the control and co-ordination of Telecom mashups. In *Proceedings of MNCNA: Middleware for Next-Generation Converged Networks and Applications*, Newport Beach, USA, 2007.
4. Open service access (OSA); Parlay X web services; part 1: Common. 3GPP TS 29.199–01.
5. Web 21c sdk. http://web21c.bt.com/

---

* http://www.apple.com/iphone/

6. IP multimedia subsystem (IMS); Stage 2. 3GPP Specification TS 23.228.
7. J. Rosenberg et al. SIP: Session Initiation Protocol. *RFC 3261*, June 2002. http://www. ietf.org/rfc/rfc3261.txt.
8. S. Mittal, D. Chakraborty, S. Goyal, and S. Mukherjea. SewNet—A framework for creating services utilizing Telecom functionality. In *Proceedings of the 17th International World Wide Conference*, Beijing, China, April 2008.
9. S. Majithia, M. Shields, I. Taylor, and I. Wang. Triana: A graphical web service composition and execution toolkit. In *Proceedings of IEEE International Conference on Web Services*, San Diego, USA, 2004.
10. V. Agarwal, K. Dasgupta, N. Karnik, A. Kumar, A. Kundu, S. Mittal, and B. Srivastava. A service creation environment based on end-to-end composition of web services. *Proceedings of the 14th International World Wide Conference*, Chiba, Japan, May 2005.
11. A. Sheth, K. Sivashanmugam, J. Miller, and K. Verma. Framework for semantic web process composition. In *Special Issue of the International Journal of Electronic Commerce (IJEC)*, 9(2), 2, 2004.
12. M. Dumas, J. Kalagnanam, L. Zeng, B. Benatallah, and Q. Z. Sheng. Quality-driven Web Services Composition. In *Proceedings of the World Wide Web (WWW) Conference*, Budapest, Hungary, 2003.
13. J. Miller, J. Arnold, J. Cardoso, A. Sheth, and K. Kochut. Quality of service for workflows and web service processes. *Journal of Web Semantics*, 1(3), 281–308, April 2004.
14. IBM QEDWiki. http://services.alphaworks.ibm.com/qedwiki/
15. Yahoo Pipes. http://pipes.yahoo.com/pipes/
16. BEA AquaLogic Family of Tools. http://www.bea.com/framework.jsp?CNT=index.htm&FP=/content/products/aqualogic/
17. JackBe PrestoStudio. http://www.jackbe.com/Papers/JackBe-StudioOverview.pdf
18. Microsoft Connected Services Framework Sandbox. http://www.microsoft.com/serviceproviders/solutions/connectedservicesframework.mspx
19. Third-generation partnership project. http://www.3gpp.org
20. A. Orita, M. Kakemizu, and M. Wakamoto. Unified IP service control architecture based on mobile communication scheme. *FUJITSU Scientific and Technical Journal*, 37(1), 81–86.
21. P. Trimintzios et al. A management and control architecture for providing IP differentiated services in MPLS-based networks. *IEEE Communication Magazine*, 37(5), 80–88.
22. M. R. Unmehopa, M. L. F. Grech, and M. Torabi. Service control architecture in the UMTS IP multimedia core network subsystem. In *Third International Conference on 3G Mobile Communication Technologies*, 22–26, London, 2002.
23. Jsr 289: Sip servlet v1.1. http://jcp.org/en/jsr/detail?id=289
24. Jsr 116: Sip servlet api. http://jcp.org/en/jsr/detail?id=116
25. I. Fikouras, K. Gronowski, R. Levenshteyn, P. Pettersson, T. Dinsing, G. Eriksson, and P. Wiss. Service composition in IMS using Java EE SIP servlet containers. *Ericsson Review*, Issue no. 03/2007.
26. S. Mittal, A. Misra, A. Gupta, E. Newmark, C. L. Oberle, D. Chakraborty, and K. Dasgupta. BusinessFinder: Harnessing presence to enable Live Yellow Pages for small, medium and micro mobile businesses. In *IEEE Communications Magazine, Special Issue on New Directions in Networking Technologies in Emerging Economies*, 45, 144–151, January 2007.
27. Ibm websphere Telecom web services server. http://www-306.ibm.com/software/pervasive/serviceserver

Chapter 2

Service Delivery Platform for the Next-Generation Network

Rolan Christian

Contents

2.1 Introduction

Traditional telecommunication networks (telco) provide a variety of voice and data services over legacy infrastructure. This infrastructure is limited due to the underlying network technologies, such as Time Division Multiplex (TDM) access, circuit-mode transports, service-specific platforms, and rigid Operations Support Systems and Business Support Systems (OSS/BSS). With the evolution of the access, transport, service platforms, and support systems the telco is able to provide improved services that combine multiple media formats including voice, video, and data. Specific examples of telco evolution include

- Access and transports support packet-based delivery of services, but still interwork with legacy circuit-mode mechanisms.
- Service platforms are more flexible by using new IT-based tools and software components to develop, deploy, and deliver any type of service.
- OSS/BSS migrate to loosely integrated IT systems to manage end-to-end business processes, network capabilities, diverse data sources, and service platforms.

The current phase of telco evolution is the convergence of Information Technology (IT)-based enterprise networks and Internet networks with the telco. This convergence fuels changes in telco infrastructure, but equally influences changes in the fundamental telco business. For example, new telco business models incorporate external partners, who provide applications or content to enhance telco service offerings to its customers. Also, telco transports are migrating toward packet-based mechanisms that operate across the Internet and both fixed and mobile legacy networks. As a result, the telco network is transforming into a Next-Generation Network (NGN) [1].

The NGN is a telco concept that aims to manage and benefit from the integration of telco, IT, and Internet technologies. The NGN is also service centric and aims at providing a rich set of services to customers anywhere, anytime using any capable device. These services not only include, as a subset, legacy voice and data services but also advanced services such as video streaming, instant messaging, and presence-related services. To support the development of these services, the NGN must provide a wealth of service enablers. Service enablers abstract access to and usage of the converged network resources and capabilities. The Service Delivery Platform (SDP) can be used to provide rich service enablers for the NGN.

The SDP [2] represents a telco service platform concept that allows IT applications to interwork with telco network services. In addition, the SDP enables IT applications to use diverse telco data that resides across decentralized heterogenous sources. Examples of IT application include traditional data-centric application, business processing applications, and messaging applications. However, by using telco network services, new IT applications can be created, such as communication applications, account management applications, and content delivery applications.

To satisfy Telco–IT interworking, the SDP provides a rich set of software-based service enablers that are used by applications to abstract the underlying telco network functions. These functions are complex in nature and technology specific. The SDP service enablers offer interfaces to applications. The interfaces hide their service enabler implementation details as well as the technology complexities of the lower network functions they abstract.

Current SDP implementations are proprietary. As a result, these SDPs provide proprietary service enablers that hinder interoperability between SDP implementations. Also, the enablers hinder the portability of applications across SDP implementations. To overcome these problems, industry requires a standardizable SDP architecture that provides standard-based service enabler interfaces. These formal interfaces are exposed to applications and provide a consistent abstraction of underlying telco network functions and data.

To aid the definition of a standards-based SDP architecture, a generalized Service-Oriented Architecture (SOA) [3] is used. The SOA is an IT concept that aims to create distributed system architectures using Web service technologies. In this chapter, we reuse technology-neutral concepts that are extracted from the SOA and constitute the Generic Service-Oriented Architecture (GSOA) [4]. By using the GSOA, generic service platform architecture is defined and named the SDP framework. The SDP framework represents a meta-architecture that can be manipulated in a consistent manner to create specific SDP architectures. The framework also promotes standards-based service enablers with interfaces that structure the NGN and formalize relationships between NGN architectural components.

In this chapter, we first provide a background on the NGN concept and its standardization. Second, we define the current SDP interpretation, its architecture and limitations. Third, we define the GSOA and demonstrate its usefulness in building service platform architectures, independent of technology, implementation, and distribution. Fourth, we present our SDP framework using layers, domains, and planes that map to multiple GSOAs. Fifth, we discuss an implementation of the SDP framework using standards-based technologies. Finally, we evaluate the SDP implementation and determine the appropriateness of the framework for structuring the NGN.

2.2 Next-Generation Network

The NGN provides a packet-based transport that delivers diverse services to fixed and mobile customers, reliably, securely, and with associated guarantees. These services include legacy telco services, such as telephony, free phone, voice-mail, and abbreviated dialing. In addition to legacy services, the NGN aims to deliver a range of advanced services to customer. These advanced services include Internet Protocol Television (IPTV), Video on Demand (VoD), information on demand, and Internet-like services. To fully support service delivery, the NGN enables interworking between telco networks, IT-based enterprise networks, and the Internet.

To enable development, delivery, and management of services, the NGN demands separation of telco transport functions from service implementations. This loose coupling ensures services operate independently across all network infrastructures. The NGN defines a technology-neutral reference model to visualize service and transport independence. The reference model is shown in Figure 2.1 and is modified from [5] to include reference points, discussed in Section 2.2.1.

The reference model depicts the separation between services and transports using two strata. The *network stratum* contains the complex physical network resources. Also, the stratum contains network-oriented functions that abstract access and usage of both legacy and new transports and systems. These functions abstract capabilities offered by network databases, signaling gateways, and OSS/BSS. The network stratum intersects with the customer domain, thereby abstracting customer terminal capabilities. Also, the stratum intersects with other network transports, thereby illustrating transport network convergence between the telco, the enterprise, and Internet networks.

The *service stratum* contains the NGN service platform that uses network stratum functions. The service platform manages the service-oriented capabilities of the NGN. It enables various types of services to be developed, deployed, and managed independently of the underlying networking infrastructure. As a result, services are portable across SDP implementations. The separation between services and network is achieved by abstracting access to technology dependent network functions using service enablers. The reference model also shows the service stratum intersecting with the customer domain. This enables customers to access and use services offered by the service platform. In addition, the service stratum intersects with other network operators' service platforms. This allows the telco to expose its service enablers to external sources, such as service developers. This exposure also supports standardized interworking between SDP implementations when different telcos want to interwork or federate across services and service enablers. As a result, the service stratum also includes OSS/BSS capabilities to securely manage and administrate exposure and interworking of services and service enablers.

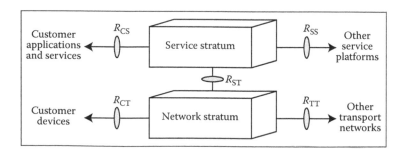

Figure 2.1   Modified NGN Reference Model.

2.2.1 Standardization

We add reference points to the NGN reference model in Figure 2.1, to formalize relationships between and across the strata. These reference points promote standardization of the NGN. The R_{CS} reference point specifies the relationship between customers and the NGN service platform. The R_{SS} reference point ensures external access to the NGN service platform and its enablers are standardized. The R_{ST} reference point provides the service platform with unified standard-based access to underlying network functions. The R_{CT} and R_{TT} reference points enable standardized end-to-end delivery of services across all customer terminals and converged networks.

The NGN reference model does not prescribe specific technologies to implement strata and reference points. However, it is essential that service enabler technologies have open standards-based interfaces to ensure diverse NGN implementations remain consistent and interoperability. A current example of standardizing the network strata reference points, R_{CT} and R_{TT}, is the Internet Protocol Multimedia Subsystem (IMS) [6].

The IMS defines necessary network functions that satisfy most network stratum requirements. These functions provide various capabilities, such as managing access control, mobility, data management, and service triggering. The IMS functions communicate using various standards-based telco and Internet protocols, such as adapting to CAMEL Application Part (CAP) [7] and a profile of the Session Initiation Protocol (SIP) [8]. The collection of these protocols implements the R_{CT} and R_{TT} reference points. The IMS also attempts to implement the R_{ST} reference point by using the same protocols to interface with various service platforms.

Open standard-based service platforms for the NGN service stratum are limited. Most service platform standards are now considered legacy. For example, the Intelligent Network (IN) [9] focuses on voice services operating over circuit-mode transports. The IN does not take advantage of the underlying converged network resources and capabilities, since it implements the R_{ST} reference point using its legacy INAP [10] protocol. In addition, legacy service platforms do not satisfy service stratum requirements, such as exposing service enablers to external entities. As a result, they do not implement the R_{CS} and R_{SS} reference points. Other service platforms use proprietary mechanisms and therefore hinder NGN standardization. Hence, a standards-based service platform is required for the NGN service stratum. We motivate the use of the SDP for the NGN service platform. The SDP architecture will also provide the structure needed for the NGN service platform architecture.

2.3 Service Delivery Platform

The SDP represents a concept with no agreed definition. However, a popular interpretation specifies the SDP as an IT-based platform used by telcos to provide services to a variety of customers. [2]. Both fixed and mobile telcos may use an SDP to provide services. These services include traditional telco services, such as telephony, and advanced data services, such as those provided by IT applications.

Expanding this definition, [2] defines a minimal set of SDP requirements:

1. Is service-oriented because it manages service creation, provisioning, execution, and billing.
2. Supports the delivery of services in a network and device-independent manner, by abstracting the underlying network infrastructure.
3. Provides a single standardized point for developers to find and use diverse service enablers and content.
4. Provides external developers and IT using enterprises with open and secure access to telco capabilities.

These requirements align with the NGN service stratum requirements. So, the SDP can be used for the NGN service platform. The SDP would require some architecture to satisfy these requirements, and also promote its structure for the NGN service platform architecture.

Currently, the SDP has no standardized architecture. However, diverse architectural interpretations exist. These interpretations are influenced by standards and technologies defined for the telco and IT-based enterprise networks. Also, many of these interpretations are based on vendor-specific solutions. Some vendor-driven architectures focus on satisfying specific SDP requirements, such as converged billing, content delivery, or service creation.

A general SDP architecture is proposed in [2]. We use this general SDP architecture but add detail that includes converged networks, interfaces, and a distribution plane. The more detailed SDP architecture is shown in Figure 2.2. The architecture also shows standard-based technology mappings to its interfaces. For example, existing standards implement vertical interfaces but standards for horizontal interfaces are lacking.

The SDP architecture identifies various distributed service platforms that constitute a SDP. These platforms abstract and simplify telco resources and capabilities as service enablers. These platforms and their enablers are

- *Network Abstraction Platform* houses network service enablers that provide a common point of access to heterogenous network resources and capabilities.
- *Content Delivery Platform* houses content service enablers that provide and deliver content to customers. We regard content as a network resource, which is either provided by the telco or external providers.
- *Management Platforms* contains management service enablers that abstract telco OSS/BSS functionality.
- *Service Execution Platform* contains telco applications that provide telco services to customers. These services include voice, data, or multimedia-based

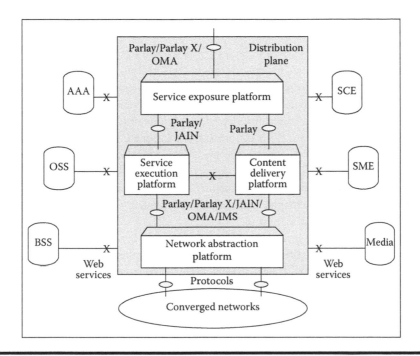

Figure 2.2   Popular SDP Architecture.

services. Telco services make use of network, content, and management services. We treat these telco applications as a set of service enablers since they expose interfaces to other platforms.

■ *Service Exposure Platform* houses service enablers that provide external entities with simplified access to all platforms' services enablers.

In the figure, we show service enabler interfaces being offered by the respective platforms. These interfaces are mapped to Application Programming Interfaces (API) standards. For example, Parlay, Parlay X [11], and OMA [12]. APIs implement the interfaces offered by the service exposure platform. However, Parlay APIs also implement the interfaces offered by the network abstraction platform, service execution platform, and content delivery platform. JAIN [13] APIs also implement the service execution platform interfaces. For the interfaces offered by the network abstraction platform, API standards such as Parlay, Parlay X, JAIN, and OMA are used. Apart from standard-based APIs, IMS and its related protocols can be used but to implement the capabilities in the network abstraction platform that interact with the lower converged networks and higher platforms.

In Figure 2.2, nonstandard-based APIs are depicted as an X. This illustrates the lack of standards-based APIs used to integrate capabilities, such as OSS/BSS functions, into an SDP.

2.3.1 Limitations

The SDP architecture in [2] generalizes various products, such as [14–16], and is influenced by their specific requirements and technologies. These solutions focus on varying SDP requirements, such as OSS/BSS integration, content delivery, and external partner management. As a result, the current SDP architecture is a simplification of common functions offered by various products.

The SDP architecture in [2] is found to be incomplete when mapped onto the NGN reference model and reference points. No appropriate customer platform is defined to satisfy the NGN service stratum requirements and implement the R_{CS} reference point. Therefore, communication between the device, SDP, and converged networks is not managed. Also, platforms require further *decomposition* to uncover their architectural structure.

The SDP architecture in Figure 2.2 maps standards-based APIs onto the architecture in [2]. However, different APIs are used to implement a single platform. In addition, proprietary Web-based APIs are used to expose OSS/BSS and data store functionality to platforms. Thus, when implementing the general SDP architecture, multiple API standards and resulting technologies span multiple platforms, service enablers, and their interfaces. As a result, SDP implementations become inconsistent and provide limited interworking with each other and various networks. This limits the implementation of the NGN reference model, especially the R_{SS} and R_{ST} reference points.

The SDP architecture in Figure 2.2 also incorporates the SDP within a distributed environment that is supported by a distribution *plane*. This plane supports communication between diverse implementations of platforms, service enablers, and interfaces. However, [2] does not provide detail on the distribution of the SDP, the distribution plane, and the technologies required. This further reduces the flexibility of the current SDP architecture.

To overcome these limitations, we structure a SDP framework that allows standard-based SDP architectures to be structured and consistently implement NGN reference points. We define this "meta-architecture" by using a generic architectural concept, that is, the GSOA.

2.4 Generic Service-Oriented Architecture

The GSOA represents a collection of technology, implementation, and distribution neutral concepts that are at a greater level of abstraction than the Web services-based SOA [3]. The GSOA is a distributed system architecture that provides client entities with technology-independent interfaces to services. These clients represent applications that require use of one or more service to provide some form of functionality to customers. The GSOA, with clients and services, is illustrated in Figure 2.3.

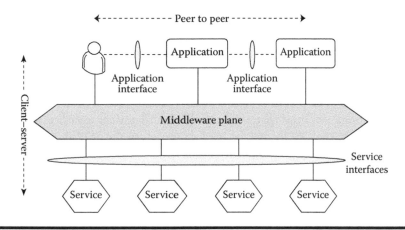

Figure 2.3 Representation of the GSOA.

The GSOA contains diverse services that abstract various complexities of all underlying support facilities. For instance, services abstract the access and usage of other software entities, databases, network resources, and network capabilities. As a result, services hide the technology and distribution of underlying resources they abstract. Services also abstract complexities from applications accessing their functions. For instance, services offer interfaces to hide their implementation and distribution from applications. Hence, GSOA services are similar to service enablers required by an SDP or NGN service platform.

Like the Web services SOA, the GSOA models a distributed system. So, the GSOA includes abstractions to manage the complexities associated with distribution. A key abstraction is the middleware plane. The middleware plane enables applications across the GSOA to communicate in a "peer-to-peer" manner via their own interfaces. In addition, communication between applications and services are perceived as direct client–server communication via service interfaces. As a result, the middleware plane hides communication details between applications and services. These details include the diverse transports, computing platforms, and devices that connect applications and services together. Besides hiding distribution, the middleware plane provides its own services with interfaces that are used by applications and GOSA services. These middleware services may implement administrative functions, such as monitoring or managing the usage of GSOA services. Hence, the middleware services are similar to service enablers.

Architectures that can be generalized by the GSOA occur in various networks, including the telco, Internet, and IT-based enterprise. For example, IT-using enterprises implement the GSOA as a Web services SOA with an Internet-based transport. In the telco domain, the GSOA is also implemented using the Web services SOA. This is evident in the Parlay X standard that uses Web services to expose simple telco capabilities to external third parties. However, the telco also implements the GSOA closer to the network. For instance, a GSOA may be implemented using the Parlay standard with CORBA [17] technology. Parlay provides rich logic required to support Parlay X Web services and abstract underlying network functions.

An advantage of using the GSOA is that it represents a generic design pattern used in structuring other types of architectures [4]. For example, GSOA may be used to model service platform architectures that abstract network complexities, expose functions to third parties, and deliver abundant services to various customers. Thus, the GSOA may be used to structure the SDP in a technology, implementation, and distribution neutral manner.

2.5 Service Delivery Platform Framework

We show a general representation of our SDP framework in Figure 2.4. The figure expresses the layers and domains that provide the foundation of our SDP framework. Layers and domains represent design patterns used to structure the framework independently of technologies. Layers group hierarchies of service enablers, while domains distribute layers and their service enablers across various infrastructure

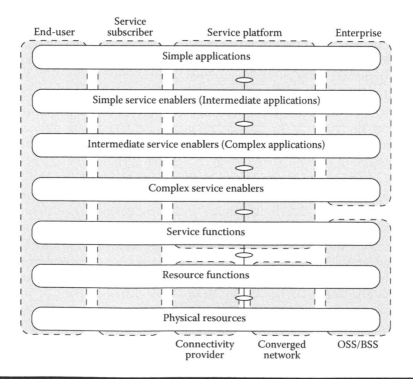

Figure 2.4   SDP Framework Expressing Layers and Domains.

boundaries. The SDP framework's upper layers decompose the NGN reference model's service stratum, while lower layers for the network stratum are defined.

Current attempts at standardizing telco service enablers provide varying levels of network abstraction. Classification of these levels of service enabler abstraction defines the framework layers. For instance, Parlay X defines simple Web services that expose highly abstract functionality to external sources. We classify service enablers providing simple functionality to external sources into a *simple service layer*. Standards such as Parlay and OMA, define rich service enablers that support Parlay X Web services by simplifying access to network resources. We classify service enablers providing rich functionality to simple service enablers into an *intermediate service layer*. The IMS defines network-oriented service enablers that support Parlay and OMA by providing an interface to network functions. We classify service enablers providing richer functionality to intermediate service enablers into a *complex service layer*.

The IMS also provides functional entities that abstract the service and resource-oriented capabilities of the underlying network. We classify these network-related abstractions into a service function layer and resource function layer. The lowest layer of the framework contains the physical network resources.

All service layer abstractions (service enablers) offer access to their capabilities using standards-based interfaces. Hence, higher-layer service enablers invoke lower-layer service enablers via their interfaces. Another abstraction that invokes service enablers interfaces are applications. Applications do not offer interfaces to higher layers, but invoke lower-layer service enabler interfaces to provide specific functionality. Varying levels of applications may be defined based on the service enabler interfaces they invoke. For instance, applications invoking simple service enabler interfaces constitute the simple application layer. Similar to simple service enablers, some applications may invoke intermediate service enabler interfaces. Hence, these applications form part of the simple service layer. In addition, applications invoking complex service enabler interfaces are similar to intermediate service enablers. Hence, they form part of the intermediate service layer.

In the SDP framework we divide layers across various areas of interest. These areas are modeled as distributed domains. We use an end-user domain to illustrate the consumer of applications. The domain includes the end-user terminal hardware and software that provides access to a communications network and applications. The service subscriber domain represents an IT-using enterprise that provides the end-user with access to a variety of applications. The connectivity provider domain is a telco or enterprise that provides the physical connectivity required for end-users to access applications. The converged network domain is the integrated collection of transports including telco, enterprises, and Internet. The service platform domain represents the heart of the SDP framework. This domain contains the collection of simple, intermediate, and complex services. The enterprise domain represents one or more IT-using enterprises that access SDP service enabler interfaces. The OSS/BSS domain represents the integrated collection of telco management systems.

2.5.1 Applying the Generic Service-Oriented Architecture

To finalize the SDP framework, we use multiple GSOAs. For each of the service layers we provide a GSOA to structure applications, service enablers, interfaces, and distribution. For the simple service layer we use a GSOA to house simple service enablers. The intermediate service layer is modeled as a GSOA containing intermediate services enablers. Also, for the complex service layer we use a GSOA to house complex service enablers. The GSOA design pattern is also applied to the service and network function layers. As a result, we provide GSOAs to structure service functions and network functions. The integrated collection of GSOAs constituting the SDP framework is shown in Figure 2.5.

The layered GSOAs provide containers for the various applications. The GSOAs contain distributed applications that communicate with each other horizontally (peer to peer) via their interfaces. These applications provide specific functionality to satisfy end-users' requirements. For example, the GSOA used for the simpler service layer also contains client, server, and third-party applications. These distributed applications may be used to provide telco, IT, or integrated services to end-users. The SDP framework also shows lower-layer intermediate and complex applications. These applications provide richer functionality by using intermediate or complex services.

The layered GSOAs house SDP service enablers and manage access to their interfaces. Service enablers contained in a GSOA are accessed and used by multiple distributed applications via their interfaces. For example, simple service enablers are accessed via their interfaces by applications contained in the same GSOA. In Figure 2.5, we show service enablers that are self contained or invoke lower GSOA service enablers. For instance, some simple service enablers invoke intermediate service enabler interfaces to fulfill application requests, while others provide functions local to their GSOA applications. In the figure, local service enablers are those that have one interface attached to a single middleware plane. We expand on these service enabler interactions in Section 2.6 by using standard-based APIs and protocols.

The SDP framework GSOAs manage access to application and service interfaces using their middleware plane. In Figure 2.5, the GSOAs middleware planes are distributed across the various domains. Communication across these domains and between the middleware parts are performed by standards-based mechanisms, such as software interfaces or protocols. However, these details are hidden from the service enablers and applications that use the middleware planes.

All layered GSOAs, with the help of their middleware planes, hide lower technology complexities. These complexities include the communication and computing infrastructure that implement applications, service enablers and the middleware itself. In the SDP framework GSOAs also abstract each other. For instance, the GSOA with simple service enablers', abstract access to the GSOA containing intermediate service enablers. The GSOA with intermediate service enablers abstracts

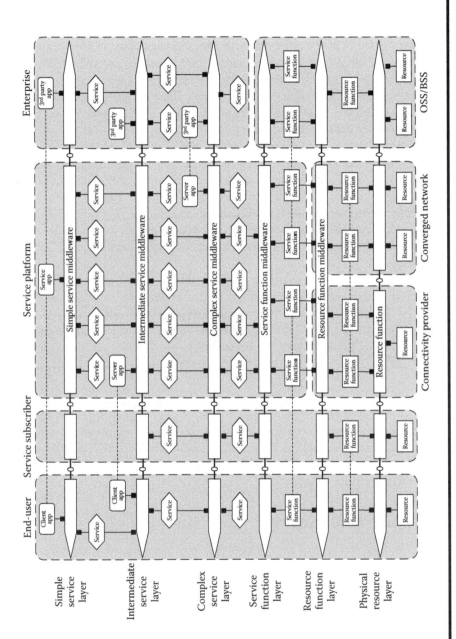

Figure 2.5   Complete SDP Framework Using GSOAs.

access to the GSOA with complex service enablers. The GSOA containing complex service enablers abstracts all lower function-oriented GSOAs.

Each of the SDP framework's GSOAs contain service enablers with interfaces that implement NGN reference points. The R_{CS} reference point is implemented across all layered GSOAs. This is evident in the end-user domain that groups various GSOA applications and service enablers used to implement the reference point. The R_{SS} reference point is implemented across the GSOAs modeling the simple, intermediate, and complex service layers. These GSOAs expose their service enabler interfaces to external IT-using enterprises. These enterprises are represented as service subscriber and enterprise domains in the SDP framework. The R_{ST} reference point is implemented across the GSOA containing service functions. The interfaces offered by this GSOA hide the complexity of physical network resources used within the end-user, service subscriber, connectivity provider, converged network, and OSS/BSS domains. The R_{CT} and R_{TT} reference points are implemented across the GSOA housing the resource functions and the physical resource layer. Both physical resources and their functions span across end-user, service subscriber, connectivity provider, converged network, and OSS/BSS domains.

Therefore, in the SDP framework, the service-related GSOAs house varying levels of service enablers with technology neutral interfaces that adequately implement NGN reference points. For completeness, we also evaluate the SDP framework against the general SDP architecture in Figure 2.2 to show the framework's benefits. This evaluation is shown in Table 2.1.

2.6 Implementing a Service Delivery Platform

To realize the SDP framework, we implement all its GSOA layers across all domains, using standards-based technologies. These technologies include the IMS functional entities and the Parlay and Parlay X set of API standards. The resulting SDP architecture is shown in Figure 2.6.

2.6.1 Mapping Technologies

Using the full SDP framework as a foundation, we map the existing network resources to the physical resource layer. For instance, physical resources are implemented using the access network Gateway GPRS Support Node (GGSN), media gateway, signaling gateway, media stores, and billing systems. The IMS functional entities detailed in [18] are used for the lower resource and service function GSOAs. For the resource functions the Policy Decision Function (PDF), Media Gateway Controller Function (MGCF), Breakout Gateway Control Function (BGCF), and Media Resource Function Processor (MRFP) are used. In the OSS/BSS domain, the IMS provides the Charging Gateway Function (CGF).

Table 2.1   Comparing the Generalized SDP Architecture and Our Proposed SDP Framework

Criteria	Generalized SDP Architecture (Figure 2.2)	SDP Framework (Figure 2.4)
Customer platform	Not catered for. Assumes customers access services via multiple layers and platforms	Explicitly caters for customer access and management since a customer domain is defined. Also all GSOAs extend enablers into this domain
Mapping of APIs (implementing reference points)	Multiple APIs can be used across multiple layers and platforms. This leads to diverse and inconsistent implementations	GSOAs can be implemented using standard-based APIs consistently across specific layers and domains. GSOAs remove ambiguity when mapping APIs
Distributed environment	Not catered for, but by its nature the SDP is a distributed system	Each GSOA design pattern provides a middleware plane to manage distribution transparency
Technology neutral	Aims to be agnostic by generalizing vendor products which can make implementation difficult when using standards-based technologies	GSOAs are technology and implementation neutral. These properties are inherited by the SDP framework
Abstraction and ease of decomposition	Medium level of abstraction since the architecture abstracts only vendor products. The model does not decompose smoothly to ease implementation	High level of abstraction achieved using GSOAs. Each GSOA is easily decomposed when a standard technology is chosen to implement the service enablers, their APIs and the middleware plane

The service functions provided by the IMS include the Call Session Control Functions (CSCFs) and Home Subscriber Server (HSS). The IMS also provides end-user, media processing, and OSS/BSS service functions, such as the SIP user equipment (UE), Media Resource Function Controller (MRFC) and Charging/Event Collection Function (CCF/ECF), respectively. The IMS service functions

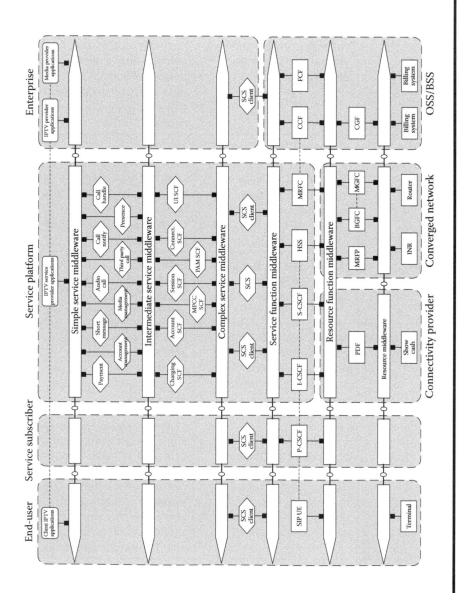

Figure 2.6 Implementing the SDP Framework.

invoke resource functions using protocols, such as SIP and Diameter [19]. The protocols hide limited distribution complexities. Hence, these protocols provide limited implementations for the GSOA middleware planes supporting the resource and service functions.

The IMS also contributes SIP application servers to the realization of complex services. However, we use the Parlay Service Capability Server (SCS) to encapsulate these SIP application servers and expose an API. The Parlay SCS implements a complex service enabler offering a complex interface. The intermediate service enablers are implemented using the Parlay Service Capability Functions (SCFs). These SCFs expose their capabilities using their APIs. Both SCS and SCFs are implemented using CORBA-based technologies. These technologies abstract numerous distribution complexities and provide rich standard-based implementations for the GSOA middleware planes supporting intermediate and complex service enablers.

Simple service enablers and interfaces are implemented using the Parlay X Web service APIs. By using Web services the GSOA middleware plane, supporting simple services, is implemented as an Enterprise Service Bus (ESB) [20]. The bus provides functionality to abstract distribution complexities. Using the Web services across the ESB are various Web-based applications.

2.6.2 *Alternative Mappings*

In Parlay X, Web services may be used to directly access network resources and capabilities [21], at the cost of increasing their implementation complexity. The SDP framework may use Parlay X Web service APIs, rather than SCF and SCS APIs, to offer access to service functions. As a result, we may collapse lower-layer GSOAs (intermediate and complex) into a single higher-layer GSOA (simple). Alternatively, we may completely remove Parlay X Web services and use Parlay SCFs to expose network abstractions to external enterprises via their APIs. Hence, we may extend a lower-layer GSOA (intermediate) to replace higher-layer GSOAs (simple).

Similar to manipulating GSOA layers, we may alter SDP framework domains. For instance, the service subscriber domain may be removed if end-users are within his/her home IMS network. So, the P-CSCF and resources are removed and not accessed from the customer domain. Also, a telco may assume the access provider role for end-users, provide OSS/BSS capabilities to charge end-users and provide converged network functions and resources. As a result, the connectivity provider, OSS/BSS, and converged network domains may merge into a single telco domain.

Therefore, mapping standardized technologies to the SDP framework may require manipulation of layered GSOAs and distributed domains. As a result, specific SDP architectures are created. Though incorporating technologies, these derived architectures inherit the generic concepts, principles, structure, and abstractions of the SDP framework.

2.6.3 *Simulating an Application*

To determine the appropriateness of the implemented SDP framework, we create, deploy, and simulate the execution of an application. The chosen application is Internet Protocol Television (IPTV). With this application, end-users are able to view real-time and on-demand television content. In addition, end-users may use other telco services while viewing their content. These services include voice calling and messaging with presence. The proposed SDP architecture that is derived from the SDP framework to implement the IPTV application and required service enablers is shown in Figure 2.6.

To ease the development of the IPTV application we reused, rather than rebuild, a collection of standard-based service enablers that provide functions to make voice calls, send messages, and manage presence. These enablers include the Parlay X Web services and Parlay SCFs. The Web services included the third-party call [22], presence [23], and messaging [24] Web services. These Web services abstract the Parlay call control [25], presence and availability [26], and user interaction [27] SCFs, respectively. By using Parlay X Web services, the SDP offers a set of service enablers to external IT-using enterprises. For the IPTV application these enterprises included IPTV application providers and television media providers.

We could not identify a standardized Parlay X Web service to manage the delivery of television content to end-users. However, Parlay provides a data session control SCF [28] that can be reused to deliver content within a data session. Currently, the SCF API is limited and provides basic functions to manage network data sessions that have already been set up by network functions. Hence, we modify the data session SCF API to initiate, control, and manipulate data sessions within the network for content delivery purposes. Using Parlay X principles we abstract the modified SCF to define our own content delivery Web service. Table 2.2 provides details on service enablers, their APIs, and our modifications to implement the IPTV application across the SDP framework.

Figure 2.6 also shows gaps where no service enablers are defined. These gaps are evident in layer and domain intersections such as where the intermediate service layer and complex service layer meet the end-user domain, service subscription domain and enterprise domain. This represents areas of concern since customized service enablers and interfaces were created to support the IPTV application, and also to consistently support the abstraction hierarchy between layers and across domains. These gaps can be interpreted as opportunities for standardization.

Middleware used for Parlay X includes an ESB. ESB solutions are not fully standardized as yet. Hence, we only incorporated standards-based technologies to hide distribution complexities. These technologies included basic protocols, such as SOAP [29] and HTTP [30]. As a result, the ESB did not provide additional functionality that benefited the Parlay X Web services.

Table 2.2 Summary of IPTV Implementation Results

Parlay X Web Services	Parlay SCF	IMS Functions	Notes
Payment	Charging	CCF	Provides a service billing and charging interface into the OSS/BSS
Account management	Account management	HSS	Allows customer/provider account management by interfacing with account databases
Short message	User interaction	CSCF	Delivers messages between customers
Audio call			Setup audio call between customers
Third-party call			Setup conference calls between multiple customers
Presence	Presence and availability	CSCF/HSS	Enables customers to share, publish and notify others of their online presence
Subscription management[a]	Framework	HSS	Allows customer service subscription management
Service management[a]			Provides an interface to customer service profiles
Policy management[a]			Manages and enforces policies on customers, providers, services, and network
Content management[a]	Content management[a]		Allows content providers to register content
Content delivery[a]	Data session control[b]	CSCF	Delivers content via a data stream to customers with negotiated quality of service
	SCS[b]	CSCF/ HSS/CCF	Manages the conversion between IMS-related protocols and Parlay API invocations

[a] Newly defined enabler and API.
[b] Modified enabler API to provide additional functions.

Due to the lack of SCS specifications, we provide our own API and implementation. Also, the interworking between the SCS and IMS functions are not fully standardized. For instance, no standard is defined to translate data session API invocations to SIP messages. We provided our own mapping that uses existing SIP messages, such as SIP invites, to initiate data sessions. However, it is complex to work between API invocations and SIP messages. As a result, we create software components called SCS clients that facilitate communication between the IMS functions and SCS on its API. These clients reduce complexity by managing state and data needed to move between APIs and SIP messages. Also, the SCS clients distributed, manage lower network events, simulate network events and communicate with the SCS and IMS functions across the framework domains.

2.6.4 Results

We successfully implemented an SDP using Parlay X, Parlay, and IMS. As shown in Table 2.1 the SDP framework and its GSOAs have overcome the limitations of the generalized SDP framework. Via an IPTV application we also evaluated Parlay X, Parlay, and IMS service enablers as well as there open interfaces for standardizing the SDP. We illustrated gaps with these service enabler standards since we defined and modified specific service enablers and their APIs. The modifications are summarized in Table 2.2. The newly defined or modified service enabler APIs suggest that additional specification is required to ensure a complete standards-based SDP when using Parlay X, Parlay, and IMS to implement the GSOAs.

We can evaluate other standards-based service enablers and their APIs, by mapping them onto the SDP framework and implementing the various GSOAs. The mapping and implementation will aid in identifying gaps within these standards and their integration to form a SDP. This process will help telcos when defining and procuring a SDP solution that must integrate into their network and IT systems. Also, vendors can map their SDP architectures against the framework GSOAs to determine their offerings compliance to standards, unique properties and identify any gaps not exploited within their solutions. Therefore, the SDP framework provides an architectural foundation used in building standards-based SDP implementations and evaluating the appropriateness of service enablers and their interfaces.

2.7 Conclusion

The SDP framework provides a foundation of concepts and abstractions that contribute to SDP standardization. This is achieved using multiple-layered GSOAs to structure the SDP framework and consistently implement NGN reference points. This contributes to the structuring of the NGN service stratum as service platform architecture. The framework's GSOAs represent containers for simple to complex service enablers that abstract network capabilities and expose technology neutral interfaces to distributed applications. The SDP framework inherits the GSOAs technology, distribution, and implementation-independent properties. The SDP framework is generic and easily extended to produce specific SDP architectures. The implementation of the framework, using standard-based service enablers and interfaces, to support an IPTV application has illustrated the flexibility of the framework as a means to generate SDP architectures. The implementation also aided in the evaluation of the standards-based service enablers and APIs for use with the SDP framework.

Acronyms

AAA	Authentication, Authorization, and Accounting
API	Application Programming Interface
BGCF	Breakout Gateway Control Function
BSS	Business Support Systems
CCF	Charging Collection Function
CGF	Charging Gateway Function
CSCF	Call Session Control Function
ECF	Event Collection Function
ESB	Enterprise Service Bus
GGSN	Gateway GPRS Support Node
GSOA	Generic Service-Oriented Architecture
HSS	Home Subscriber Server
IMS	Internet Protocol Multimedia Subsystem
IN	Intelligent Network
IPTV	Internet Protocol Television
IT	Information Technology
MGCF	Media Gateway Controller Function
MRFC	Media Resource Function Controller
MRFP	Media Resource Function Processor
NGN	Next-Generation Network
OSS	Operations support Systems
PDF	Policy Decision Function
SCE	Service Creation Environment
SCF	Service Capability Function
SCS	Service Capability Server
SDP	Service Delivery Platform
SIP	Session Initiation Protocol
SME	Service Management Environment
SOA	Service-Oriented Architecture
UE	User Equipment

References

1. ITU-T, General overview of next generation network, Recommendation Y.2001, December 2004.
2. The Moriana Group, *Service Delivery Platforms and Telecom Web Services*, Thought Leader Report, last accessed on 01/11/2009, http://www.morianagroup.com, June 2004.
3. D. Booth et al., *Web Services Architecture*, Working Group Note, W3C Web Services Architecture Working Group, February 2004.
4. R. Christian and H. Hanrahan, Towards a standards-based service delivery platform using service oriented reference points, in *Convergent Service Delivery. Tenth International Conference on Intelligence in Service Delivery Networks (ICIN)*, Bordeaux, France, May 2006, pp. 96–101.
5. ITU-T, General principles and general references for the next generation network, Recommendation Y.2011, October 2004.
6. 3GPP, *Third Generation Partnership Project; Technical Specification Group Services and System Aspects Service Aspects; Service Principles (Release 8)*, Technical Specification TS 22.101 V8.0.0, March 2006.
7. ETSI, *Digital Cellular Telecommunications System (Phase 2+); Customised Applications for Mobile Network Enhanced Logic (CAMEL); Service definition—Stage 1 (GSM 02.78 Version 8.0.0 Release 1999)*, Technical Specification GSM 02.78 v8.0.0, July 1997.
8. J. Rosenberg et al., *SIP: Session Initiation Protocol*, RFC 3261, IETF Network Working Group, June 2001.
9. ITU-T, *Principles of Intelligent Networks*, Recommendation Q.1201, October 1992.
10. ITU-T, *General Aspects of the Intelligent Network Application Protocol*, Recommendation Q.1208, September 1997.
11. Parlay Group, Parlay Group Home Page, last accessed on 01/11/2007, http://www.parlay.org.
12. OMA, OMA Home Page, last accessed on 01/11/2009, http://www.openmobile alliance.org.
13. Sun Microsystems, *JSLEE and the JAIN Initiative*, last accessed on 01/11/2009, http://java.sun.com/products/jain/, 2007.
14. Appium, Appium Home Page, last accessed on 01/11/2009, http://www.appium.com.
15. Erricsson, *Service Delivery Platforms*, last accessed on 01/11/2009, http://www.erics son.com/, 2006.
16. Hewlett-Packard, *Service Delivery Platform*, last accessed on 01/11/2009, http://www.hp.com, 2007.
17. O.M.G, *Common Object Request Broker Architecture: Core Specification*, Formal Version 3.0.3, March 2004.
18. 3GPP, *Technical Specification Group Services and Systems Aspects; Network Architecture (Release 8)*, Technical Specification TS 23.002 V8.2.0, December 2007.
19. P. Calhoun et al., *Diameter Base Protocol*, RFC 3588, IETF Network Working Group, September 2003.
20. S. Craggs, *Best-of-Breed ESBs*, White Paper, Enterprise Application Integration (EAI) Industry Orginization, June 2003.
21. The Parlay Group, *Parlay X 4.0: Parlay X Web Services White Paper*, White Paper V1.0, http://www.parlay.org, December 2002.

22. ETSI, *Open Service Access (OSA); Parlay X Web Services; Part 2: Third Party Call (Parlay X 2)*, ETSI Standard ETSI ES 202 391-2 V1.2.1, December 2006.

23. ETSI, *Open Service Access (OSA); Parlay X Web Services; Part 14: Presence (Parlay X 2)*, ETSI Standard ETSI ES 202 391-14 V1.2.1, December 2006.

24. ETSI, *Open Service Access (OSA); Parlay X Web Services; Part 4: Short Messaging (Parlay X 2)*, ETSI Standard ETSI ES 202 391-4 V1.2.1, December 2006.

25. ETSI, Open Service Access (OSA); Application Programming Interface (API); Part 4: Call Control; Sub-part 3: Multi-Party Call Control SCF (Parlay 5), ETSI Standard ETSI ES 203 915-4-3 V1.2.1, January 2007.

26. ETSI, Open Service Access (OSA); Application Programming Interface (API); Part 14: Presence and Availability Management SCF (Parlay 5), ETSI Standard ETSI ES 203 915-14 V1.2.1, January 2007.

27. ETSI, Open Service Access (OSA); Application Programming Interface (API); Part 5: User Interaction SCF (Parlay 5), ETSI Standard ETSI ES 203 915-5 V1.2.1, January 2007.

28. ETSI, Open Service Access (OSA); Application Programming Interface (API); Part 8: Data Session Control SCF (Parlay 5), ETSI Standard ETSI ES 203 915-8 V1.2.1, January 2007.

29. M. Gudgin et al., *SOAP Version 1.2 Part 1: Messaging Framework*, Recommendation, W3C, June 2003.

30. R. Fielding et al., *Hypertext Transfer Protocol—HTTP/1.1*, RFC 2068, IETF Network Working Group, January 1997.

Chapter 3

Moving the SDP to the Cloud

Luis Angel Galindo Sánchez and
Joaquín Salvachúa Rodríguez

Contents

3.1 Introduction

A Service Delivery Platform (SDP) usually refers to a set of components within telecom operator service architecture that manages easing service creation, deployment, and execution utilizing enablers (network capabilities) and service capabilities. The keystone objective of a SDP would be to reduce the time taken to market of a service and the key characteristic for an SDP is thus, the ability to ease the integration task of application logic with the enablers and capabilities in the telecom service architecture for the delivery of a service to users. The SDP thus enables the rapid transformation of a service concept into a full product including all the related service life cycle management and commercial interactions.

There is no global standard for SDP implementation, although there are some industrial *de facto* conventions. Four interfaces can be considered for describing SDP following a compass-inspired interpretation, as it is depicted in Figure 3.1.

- *North interface:* An SDP provides clear service interfaces for exposing telecom capabilities, both for in-house development as well as enabling external agents to develop services and applications using the operator as the platform. In this case, operators open APIs of internal features/capabilities.
- *South interface:* It is where the SDP simplifies the process of integrating an application container with the network infrastructure and enablers. Thus, the SDP provides abstractions for invoking telephony, media, and other service features.
- *West and East interfaces:* It is the way where the SDP simplifies the integration of a new application toward the commercial and operation systems of the company. The SDP includes connectors to the OSS and BSS systems of the operator, as well as for other back-office applications. By means of this connection to OSS/BSS, the services deployed over the SDP can be integrated into the company service and product life cycle management, including operations like provision and activation of services; supervision and failure management, administrative validations, billing, and QoE measurement.

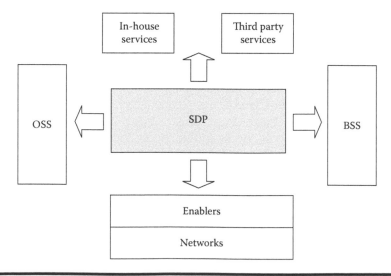

Figure 3.1　SDP interfaces.

Nevertheless, the SDP typically provides a service creation environment for service design, and a service orchestration and execution environment, which represents the core of the SDP. Together with this core functions, there is a set of components in the SDP to enable policy control, security, user directories, and other functions for business/operation process adherence and integration.

One of the most common situations in the SDP deployments is the adoption of Service Oriented Architecture (SOA) {SOA} in the SDP. SOA simplifies enabler deployment and facilitates other SDP requirements for composition, reuse, delegation, orchestration, policy enforcement, and business/operation process adherence and integration. SDP and the election of SOA for its implementation have been deemed necessary in order to cope with a series of business drivers; in particular, the SDP focuses on the reduction of time-to-market for new services, as already mentioned. It is also expected that the SOA-based SDP allows the creation of a wider range of value-added services in a cost-efficient way, the implementation of an open-service marketplace for third-party interactions, and the efficient integration with OSS/BSS.

3.2 WIMS 2.0 Initiative

3.2.1 What Is WIMS 2.0?

WIMS 2.0 (Web 2.0 and IMS) {WIMS 2.0} is an open initiative founded by Luis Angel Galindo and David Moro. WIMS 2.0 seeks convergence between the Web 2.0 and the telecom new-generation services based on IMS (IP Multimedia Subsystem), to create innovative, appealing, and user-centric services and applications. These services will combine features from both Web and telecom worlds. On the one hand, relevant features such as interactivity, ubiquity, social-orientation, user participation, and content generation will be adopted from Web 2.0 Services; on the other, IMS will enhance WIMS 2.0 services features with a collection of capabilities such as multimedia telephony, media sharing, push-to-talk, presence and context, online address book, and so on, all of them being applicable to mobile, fixed, or convergent telecom networks. In addition, in order to provide more versatile and enriched services and a short-term implementation, WIMS 2.0 also considers the utilization of pre-IMS Telco capabilities such as SMS/MMS messaging, circuit-switched video and voice calls, networked address books, and so on.

WIMS 2.0 comprises the strategies, technologies, and service platforms that will allow telecom operators to achieve this convergence with innovative Web 2.0 services in All-IP networks. Following the Web 2.0 revolution and the Telco 2.0 initiative {Telco 2.0}, a change in Telco traditional business model is recommended following the adoption of a user-centric and open-garden philosophy consisting in the enhancement of the openness, flexibility, and freedom in the fashion services are provided. IMS, along with its associated set of service enablers, has been identified as the base platform to support this convergence and business model change.

The two main goals of the WIMS 2.0 initiative {WIMS 2.0} are to create innovative services mixing the Web 2.0 and Telco World, and to create a new fair ecosystem: the WIMS 2.0 ecosystem {WIMS 2.0 ecosystem}: a fair ecosystem, where all of us actors can play a sustainable role. Currently, we are competing not only, as usual, between operators but also with those new Internet players. In this new ecosystem, we are speaking more about co-opetition than competition.

3.2.2 WIMS 2.0 Strategic Lines

WIMS 2.0 has defined two strategic lines:

■ Offering Telco capabilities to the Web 2.0 community
■ Exploiting Web 2.0 for telecom services

Offering Telco Capabilities to the Web 2.0 Community

The main objective of this convergence approach is to offer Telco capabilities from the telecom network to the Web 2.0 community. The final service is actually provided by a Third-Party located in the Web 2.0 world with the added value of the operator. Two general guidelines are considered for this approach:

1. *Incorporation of Telco capabilities into Web 2.0 services through open Web APIs*, allowing the integration of Telco into Web 2.0 mash-ups. This, potentially applies to any Telco capability, which would therefore be usable by/from any Web 2.0 service. Two different, but interrelated, strategies are proposed here:

   a. *Mash-ups based on widgets or PSEs (Portable Service Elements).* Telco applications can be incrusted into Web 2.0 services in the form of Web-widgets, which would be the technology-specific implementation of a PSE. The PSEs provided by the operator (or a Third Party) can be easily integrated by end users into Web 2.0 sites and, once incrusted, they are able to handle the interaction with the operator's open Web APIs, thus, enabling the use of Telco communication capabilities from Web 2.0 sites. Specific application examples are a Call-Me button for Blogger {Blogger} or a more sophisticated Facebook {Facebook} application providing voice and presence services for the members of a group.

   a2. *Mash-ups based on APIs.* Once Telco capabilities are exposed through open Web APIs, they can be directly incorporated into any Web 2.0 service in order to provide complementary functionalities, just like any other regular mash-up. The final outcome is a complete integration of telecom capabilities in the resulting service. A specific service example could be "Real-time sharing of YouTube {YouTube} videos to mobile users," while maintaining an Instant Messaging session among the involved users.

b. *Telco-enabled user-generated content publication and generation.* This convergence guideline aims to obtain new ways for content publication through the usage of Telco multimedia transmission capabilities. Operators can provide new solutions to receive the multimedia content from the user and automatically upload it to Web 2.0 sites. An example application could be "YouTube realtime video generation," where users videocall YouTube to create a spontaneous clip.

2. *Exploiting Web 2.0 for telecom services.* In this approach, the operator keeps the role of final service provider. Accordingly, two different guidelines are considered for the enrichment of telecom services:

a. *Inclusion of Web 2.0 contents and events in the operator services.* The introduction, integration, and distribution of videos, advertisements, podcasts, news, and so on from the Web 2.0 within Telco services is of great interest for the end-user. As example services, we may consider special feed readers for mobile handset, the inclusion of social networks events into Telco Presence information, or news, video, and music distribution.

b. *IMS online applications (virtual terminal).* Through the use of AJAX {AJAX} and other technologies, Web 2.0 applications have achieved major advances in the field of user interfaces. WIMS 2.0 favors for the usage of these technologies in order to build IMS online applications. The benefits of such kind of IMS applications are very important: ubiquity and a great simplification of service development and deployment, since service logic resides on the network and not on the terminal.

3.2.3 WIMS 2.0 SDP

The WIMS 2.0 SDP is an intermediary for the adaptation between both worlds, enables the convergence in the desired terms. The proposed reference model is reflected in Figure 3.2.

The above reference model defines a framework that identifies the required logical entities, as well as the relationships existing among them.

Two main types of entities are identified in the reference model: adaptation entities and entities for the provision of IMS-online applications.

Within the *adaptation entities*, we also find two further sub-types: *entities for the exposure of IMS capabilities* and entities *for the exchange of multimedia content and events*. As a relevant difference, for the interaction with the Web 2.0 world, the entities of the first group use APIs defined by the operator, while the entities of the second group use the APIs defined by the Web 2.0 services to be contacted. In fact, since the entities of the latter group are actually using functionalities of Web 2.0 services along with Telco capabilities, they can be considered as platforms for the development and execution of service mash-ups by the operator itself.

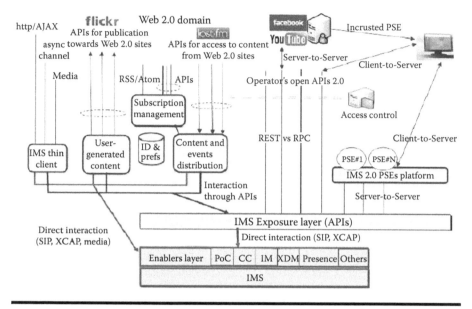

Figure 3.2   The WIMS 2.0 SDP reference model.

The entities for the exposure and exploitation of IMS capabilities, and their functionalities, are the following:

- *IMS Exposure Layer:* This entity exposes IMS capabilities to the outer world through open Web APIs. On the outer side, it manages HTTP messages associated with the execution of a specific procedure, while, on the inner side, it interacts with IMS capabilities using the appropriate protocol (SIP or XCAP). Obviously, the interactions on each side must be properly inter-related. It can be clearly seen that it is actually a gateway between HTTP and IMS protocols. The WIMS 2.0 holds the IMS Exposure Layer as a major axis for the architecture.
- *IMS 2.0 PSEs Platform:* This entity holds and serves the operator's PSEs to be incrusted into the Web 2.0 sites. PSEs implement the required logic to inter-act with a specific group of open IMS APIs (the ones considered above).
- *Access control entity:* To secure the use of the operator's APIs, this entity performs the required access control mechanisms.

The entities for the exchange of multimedia content and events, and their functionalities, are the following:

- *Subscription Management enabler:* It subscribes to content and events of any kind generated in Web 2.0 services. Subscriptions may be made on behalf of the operator or directly on behalf of the final user. This entity talks with the

Content and Events Distribution enabler to inform about the existence of new content/events to be obtained.

■ *Content and Events Distribution enabler:* It downloads, adapts, and transmits content and events from Web 2.0 services to Telco users.
■ *User-Generated Content enabler:* It receives content from terminals and, after adapting its format, it uploads this content to Web 2.0 services.
■ *IDs and Preferences*: Database that stores the relationships between Telco identities and the identities used by Web 2.0 services. The other entities of this group consult this database to perform the conversion of identities when using the Web 2.0 services APIs.

Finally, we find an individual *entity for the provision of Telco online applications.* On one side, this entity interacts with IMS&Telco capabilities, acting as the final Telco client. On the other side, it shows a rich multimedia Web interface to the user, according to the service being provided. This entity must also handle a continuous media interface toward the user, as well as an asynchronous interface to signal the need to refresh the Web interface.

3.2.4 Web 2.0 Technologies in the WIMS 2.0 SDP

The WIMS 2.0 platform eases the technology transition from the Telco technology domain toward the pure Internet technology domain aiming at facilitating the dialogue between a telecom world and the Web 2.0, from the operator perspective.

Figure 3.3 depicts a summarized view of the entire Web technologies considered representing the domain of scope of each one of them.

Before describing each of the different groups identified in Figure 3.3 and the relationship with the implementation of the WIMS 2.0 service platform, it is important to stress the principles that organize this technology map:

1. Technologies have been located following the client–server paradigm that governs the Web world. Within the server and the client sides, the different groups comprise technologies for the implementation of similar functionalities which might be complementary, for example, the AJAX set of technologies, or represent an alternate choice to provide the same functionality, for example, JavaScript and ActionScript.
2. HTTP protocol is the base for all interactions between remote elements, independently of the nature of those elements. These interactions are retrieval and manipulation of Web page contents and the usage of Web-ready APIs.
3. As a featured concept in Web 2.0, remote APIs are of paramount importance to support the mash-up of functionality within a given service. In the map, there are two variants of the usage of Web-ready APIs:
   a. Client-server scenario, where a browser directly invokes the API by sending HTTP request to a remote server.

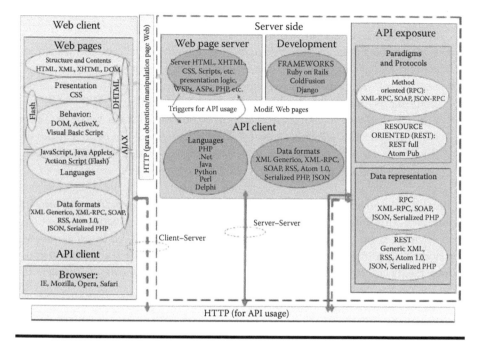

Figure 3.3   The WIMS 2.0 SDP technology map.

b. Server–server scenario, where a server is the actor invoking the APIs provided by another server. The invoking server is acting as a Web client as far as the usage of API is concerned. Normally this scenario is used to prevent the Web browser to realize this kind of actions, and thus, all the service logic to handle the remote API is managed by the server.

Once principles are explained, technologies are grouped as follows:

Client side: The client side comprises all technologies that are typically used by Web browsers. The following groups and subgroups are considered.

■ *Web pages technologies* related to presentation and processing of Web pages. There are three technological levels:
  – Structure, comprising all technologies to support information, both contents and semantics, and the Web page structure. HTML, XHTML, XML, and DOM are technologies within this level.
  – Presentation, comprising technologies to control how the information is presented. CSS is the most representative one in this level.
  – Behavior, which gathers all technologies used to implement the logic that control local actions and dynamic behavior of the Web page. This dynamic activity can result in modifications over the presentation and structure

levels. VisualBasic Script, Javascript, ActionScript, ActiveX are some of the ones considered.

- *Technologies for Web API clients in the Web browser* include the technologies in charge of the process, form and send HTTP request toward remote APIs and to receive and interpret responses according to specific data formats. Since these actions can be considered as dynamic behavior of the Web page, languages which rule this level are the same than those utilized in the Web page behavior level.
- *Browsers* are those alternative options as client platforms, for example, Internet Explorer, Mozilla Firefox, Opera, Safari, and Google Chrome. Even if functionalities presented by these platforms are similar, some browses can present specific features that may affect the usage of certain technologies considered for the client side.

Server side: This comprises technologies typically utilized in network infrastructures. Even though they are mostly associated to a Web server, the role of a Web API client shall also be considered. The following categories apply:

- *Web page server technologies*, which groups technologies to serve Web pages such as software containers, servlets, JSPs, ASPs, and so on. These are mature technologies which does not represent a key differential aspect of the Web–telecom convergence.
- *API technologies:* The exposure of telecom capabilities in the form of Web APIs enables the offering of distributed functionalities within the Web world. In this group, technologies for implementing the serving API logic are considered, and then, they are valid for implementing both the client-server and server–server scenarios. There are different options when considering how the HTTP protocol is employed and how the functionality behind the API is modeled. In the WIMS 2.0 technology map, the RPC {RPC} and REST {REST} approaches are considered. The map also considers options to encode the data exchanged between the API client and the API server, for example, XML, JSON, ATOM, and so on.
- *API client technologies for the server side:* As previously mentioned, in the server–server scenario, Web servers act are API clients, and in this particular case of the Web loom, they act as Web clients in order to invoke remote Web APIs. Technologies are those languages at the server side to rule API invocation such as PHP, Ruby, Java, Python, and so on as well as the data formats that the specific API is utilising to encode the data exchange. As mentioned before, this is a feature of the API server and there are a variety of possibilities, most of them based on different variants of XML.
- *Development frameworks*, which includes environments geared at easing the Web service creation. These frameworks feature a high level view on the

development process and a rapid deployment over different server platforms like Java or .NET. A popular framework for Web 2.0 application is Ruby on Rails.

3.2.5 Key Entity in the WIMS 2.0 SDP

The axial concept for convergence within the WIMS 2.0 SDP is the exposure of capabilities through Web-friendly APIs. The entity in charge is the IMS Exposure Layer. There are two aspects for the technical realization of this entity: on one side, the Web-friendly APIs that exposes IMS service capabilities toward Web applications and, on the other side, the internal logic to interwork and interact with the underlying IMS infrastructure in order to execute the functionalities exposed via the API.

The IMS Exposure Layer shall implement a gateway adapting the HTTP-based API primitives into SIP and XCAP protocol operations to enable communication between the Web world and the IMS. This mission affects two levels of communication in the telecom side: the signaling plane and the media plane. Wherever possible, both planes need to be adapted from a pure technical perspective, although continuous real-time media is not supposed to be adapted by means of API approaches. This means, that audiovisual media exchange between a Telco endpoint and a Web component is not processed through an API and thus, not handled by the IMS Exposure Layer.

The following IMS service capabilities constitute a nonexhaustive list of capabilities that can be exposed via the IMS Exposure Layer, and thus, the underlying gateway shall be able to communicate with the enabling servers implementing these service capabilities:

- Presence: An API client might be able to at least subscribe, consult, or modify Presence information for a given user.
- XDM: to manage lists and contact items in the IMS domain.
- Instant Messaging both in pager mode as well as session mode.
- Multimedia Telephony, to provide third-party call ability to Web applications and call control features of sessions established between two endpoints.
- Multimedia conferencing to provide a Web application with the possibility to schedule and create multimedia multiparticipant calls and manage them.
- Rich CS calls, to provide features related with media sharing combined with circuit switched calls.
- SIP *Push*, to provide Web applications the ability to push events and contents to IMS terminals.
- *Customized Alerting Tone* enables a Web applicaton the control over ringback tones and personalized alerting tones (ringforward tones).
- Multimedia inbox control.

To expose capabilities, the APIs of the IMS Exposure Layer will implement a resource-oriented model as a basis. That is, the REST philosophy will be utilized to model the APIs for IMS capabilities. Since most IMS services inherit a session orientation, some RPC techniques might be added to the basic REST representation in order to obtain a similar result as when modeling telecom services as WebServices. However, REST provides a simpler approach which reduces the entry barrier for Web developers to integrate and utilize telecom service features in Web applications. Most Internet players in the Web 2.0 ecosystem provide APIs utilizing this resource-orientation philosophy.

Besides, since REST is naturally stateless, scalability increases in comparison with pure RPC alternatives like SOAP and WebServices. Beyond that, the processing for a lightweight option like REST is not comparable to the SOAP computational cost.

For the technical realization of the REST APIs of the IMS Exposure Layer, there are several protocol alternatives. Among all of them, the *Atom Publishing Protocol* or AtomPub is one of the most suitable ones, besides it is a standardized mechanism by the IETF.

Figure 3.4 presents an example of the IMS Exposure Layer exposing Presence by means of REST API. In this example, the Presence APIs rely on some of the

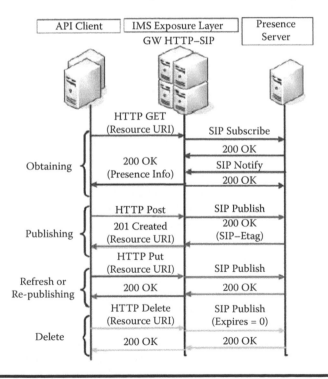

Figure 3.4   Example REST API for Presence.

intrinsic features of REST, that is, it works with resources, for example, presentity. The resources can be freely defined and can keep states (e.g., presence status). The client acts on the resources: get, modify (e.g., state), delete, and so on and these resources are always univocally identified by URIs. The actions on the resources are performed through standard HTTP verbs: GET, POST, PUT, and DELETE, which has a specific meaning in the context of each API. For instance, in this presence example HTTP GET is utilized to retrieve a presentity's status; HTTP POST can be used for publishing presence information; HTTP PUT for refreshing soft-state presence information; and HTTP DELETE to erase this softstate information. The application of REST principles to IMS Presence is thus immediate.

Aside from the protocol aspects, the API exchange information between the API client and the API server. To accommodate the expectations of a bigger number of developers, APIs shall provide different data formats so the user can select the most suitable one according to the programming language and development tools the user utilizes. The IMS Exposure Layer at least shall provide the following data formats:

■ Client–server scenario: XML, Atom 1.0, RSS, and JSON.
■ Server–server scenario: XML, Atom 1.0, RSS, and serialized PHP.

This allows utilizing the remote APIs and the exchanged data from most of the languages being utilized in the Internet.

Furthermore, the Telecom Exposure Layer might be simpler to utilize if the developers are provided with language libraries that handle the API client behavior. That means that the developer does not need to deal with HTTP request and response construction and processing. Most Internet players providing functionalities via API are adopting this best practice, offering client libraries in a limited number of programming languages and also the bare REST interface with different data formats available for those programmers utilizing other programming languages besides those considered by the client libraries.

For instance, the Telecom Exposure Layer should offer client libraries in JavaScript and ActionScript languages for client-server scenarios, and Java, PHP, Python and Ruby for server–server scenarios. For developers utilizing other languages, they still can access the functionality by handling the REST APIs directly.

3.2.6 WIMS 2.0 SDP Deployment

Previous sections have shown the reference model of the WIMS 2.0 SDP. In order to deploy onto the service delivery domain of an operator, it is required to analyze how the different entities of the WIMS 2.0 reference model can be either implemented by existing entities in the telecom Service Architecture or how new ones—specific for WIMS 2.0—fits into the service delivery domain. Figure 3.5 shows the functional entities of the WIMS 2.0 reference model that are new in the service delivery domain of an operator (IMS thin client, UGC 2.0 platform, IMS PSE 2.0

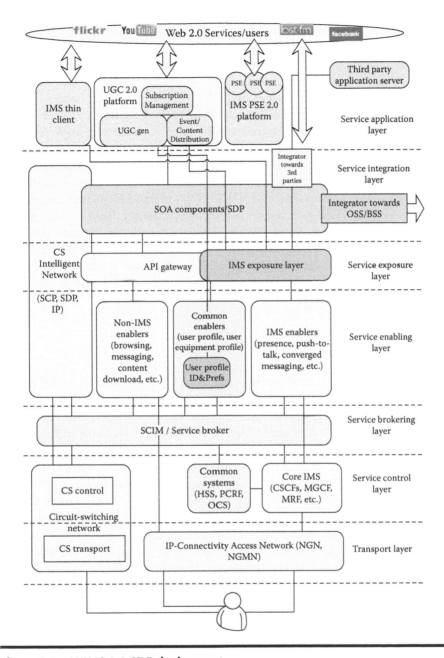

Figure 3.5 WIMS 2.0 SDP deployment.

platform, IMS Exposure Layer, and User Profile) and also the existing ones (IMS Enablers and Core IMS). For instance, a case for the latter would be the "access control entity" in the WIMS 2.0 reference model; here it is mapped onto a general gateway for securing the access of third parties to the service delivery domain, that is, the integrator toward third parties of the SDP of the layered architecture.

As for WIMS 2.0 specific entities, the IMS Exposure Layer is located at the Service Exposure Layer of the layered architecture, while the rest of WIMS 2.0 functional entities are considered service-platform and thus, they are represented in the Service Application Layer. That would be the case for the IMS Thin Client, containing service logic for providing end-user application online representations of an IMS phone; the functional entities of WIMS 2.0 related with content and media delivery or generation are gathered under a service platform called UGC 2.0 at the Service Application Layer, except for the database containing the user IDs and preferences, which is stored together with other user data at the User Profile directory (a common enabler) at the Service Enabling Layer; and finally, the IMS 2.0 PSE platform is regarded as a container for end-user applications, in fact, the Portable Service Elements.

3.3 SDP in the Cloud

One of the most important transitions from the Grid Infrastructure to the cloud infrastructure has been the change of the W3C Web Services (WS-*) for the REST {REST} way in designing services. This makes a small philosophical change from the plain SOA to a ROA (Resource Oriented Architecture) where the different services emerge as a set of Resources to be reused and composed by the different players; usually as a mash-up of resources. This approach has been proven quite reasonable and sound over the different Web 2.0 start-ups and all the collaborative services that have shown up all along Internet. Cloud Computing is based on the virtualization of the resources, mainly computing power and storage. It allows using the resources in a very flexible way and being dynamically able to add new resource with a fine grain measure. It provides an elastic way of paying as you grow for the resources needed. This has been used into private clouds, where it virtualized all the resources available in the computer department in order to allow users to gain access exactly with the resources they need. Also, there have appeared public clouds where users may be able to buy, quite cheap, the resources needed on demand. This solution is useful for end-users, helping them in the implementation of new services in an important issue: how to dimension your computing resources.

Most of the technologies in cloud computing are based on using virtual operating system images into the cloud server and instantiate, with the proper parameters, an image of the operating system and services needed.

This image is quite closer to the one provided by the SDP layer into IMS architecture. In the SDP, there are different services available, usually by an API or enablers, without need to care where are deployed or the real details for the configuration.

Cloud computing has been deployed into three layers:

■ Infrastructure as a Service (IaaS)
■ Platform as a Service (PaaS)
■ Software as a Service (SaaS)

Most of the Internet Web 2.0 applications fall into the Software as a Service layer. Applying this model to the IMS architecture, it is possible to find similarities and, in the case of SDP, it can be considered as a PaaS or a SaaS, depending if the SDP services are taking into account like a final service. This approach may be quite convenient from the architectural perspective and provides a convergence of the Telco and Internet world. This proposal currently does not consider that any operator may take into account the outsourcing of their capabilities to a public cloud; nevertheless, this may be used to organize Telco resources in a virtualized way. In this way, it is possible to have a fine grain tuning of the resources in use and the exact cost of them, clarifying if the revenues from the services are in the guidance of the forecast business plan. From the point of view of OPEX (Operational Expenditures), it also may be a great cost reduction, since it unifies procedures and maintenance interfaces.

One of the most powerful skills from the Telcos is the ability to manage nonstop services. In that way, the five 9s paradigm is a Telco characteristic. This knowhow could be a great contribution from Telcos in order to operate Cloud services. So, it can foresee new business lines with a hybrid approach, where Telcos can offer cloud infrastructure to their end users also deploying the different services.

One of the current missing capabilities for the cloud is "Communications as a Service" (CaaS). This is an Achilles-heel for its usage into the Telco world, although it cannot be ensured any Quality of Service on the communication. Most of the current providers just bypass the situation declaring that their communication infrastructure is just enough for any application. Also in this case, the IMS services have special needs in the areas of the communication and interconnection with the operator infrastructure. This means that the interconnection of actual infrastructure may not be as easy as it is in other Internet services; most of them are based on asynchronous interactions over HTTP. One remarkable aspect is that into Internet a wide development of the Quality of Services technologies has failed till now, mainly due that it has not been required by end-users. Perhaps with a convergence of Internet with IMS, Internet networks may gain capabilities that guarantee some of the communication skills without losing the freshness of Internet.

The convergence of both worlds implies to consider the characteristics of the SIP protocol: synchronous interactions and a protocol that is not just question–response like HTTP, which implies that most of the cache infrastructure will not be reusable. Under this consideration it may not be able to deploy real SIP services integrated into an operator infrastructure into a public or hybrid (mixing public and private infrastructure) clouds. This does not mean that there is no a place for the cloud computing paradigm and technology into the IMS architecture. The Service Brokering Layer may be the right interface where to interconnect a cloud service with the in-company SIP communication infrastructure.

Under this approach, it is possible to offer SDP capability to other operators using the Cloud computing paradigm and business model. For example, for a virtual operator, it can be offered a "pay as you grow" model or even not paying for the "Infrastructure as a Service," but for "SDP as a Service" where virtual operators or those which have still not deployed the SDP may pay for the capability of implementing the services provided in the SDP.

In this line of work, it is quite important to enable the creation of mash-ups by third parties with the virtual services offered. This is why now the WIMS 2.0 proposal is even more interesting. IT will provide a standard way of joining both worlds with the right set of APIs coming from the Telco world, also allowing implementing the Web 2.0 needs.

3.4 WIMS 2.0 SDP in the Cloud

Previous sections have shown how a traditional SDP can be migrated to a cloud computing architecture. This section provides the strategy to follow in the way of moving the real WIMS 2.0 SDP to the cloud.

As it has been exposed in Section 3.2.6, the different entities that compound the WIMS 2.0 SDP have been placed in existing or new entities under the traditional SDP architecture. Analyzing the proposal for migrating, a traditional SDP to a cloud computing architecture done in Section 3.3, it is understandable that the migration for the WIMS 2.0 SDP can be easily performed. This migration can follow two paths:

■ Considering that both SDPs, the traditional and WIMS 2.0, are moved to a cloud computing architecture. In this path, the point of contact with the in-company infrastructure of the operator is the Service Broker. Under this consideration, some access control functionalities should be included in this entity that ensures that SLA agreement is fulfilled.

■ Considering that the traditional SDP remains under the infrastructure of the operator and the WIMS 2.0 SDP is moved to the cloud. This way requires a multilevel integration and it is more complex. Entities like IMS thin client, the UGC 2.0 platform, and the IMS PSE 2.0 platform interact with the in-company SDP in the same way, but other like the IMS exposure layer or the User profile entity requires an adaptation. The *IMS Exposure Layer* is a platform where the IMS APIs are exposed to third parties. In this case, this platform requires a connection with the IMS and common enablers. The *User Profile* entity is a database that stores the relationships between Telco identities and the identities used by Web 2.0 services and it requires or connecting both databases or to have a replica of the Telco identities in the WIMS 2.0 SDP. Under this consideration, the access control functionalities should be included in the SDP framework.

Any of the two options mentioned is possible and will provide a collection of mixed Telco–Internet services in a cloud computing strategy, enabling a scalable solution for those operators which are not in the position to tackle huge costs in the deployment of a SDP allowing the convergence of both worlds, the Telco and Internet.

3.5 Conclusions

SDP is a key component in order to deploy commercial Telco services especially when the operator is planning to launch services using advanced enablers like Presence, XDM, PoC, and so on.

A deployment of the SDP architecture is expensive in CAPEX and requires expertise during all the phases of the project. After deployment, OPEX could represent a huge amount, overall in a Telco–Internet convergent scenario where operators are demanding the same quality of service levels that normally they offer in a pure Telco service. Thus, moving this infrastructure to a cloud computing architecture mitigates the expenses in both terms, capital and operational.

Other important issue to remark, as it is reflected by Win Win Consultores {Win Win Consultores}, is the reality for most of the operators in the world: they spend more than nine months in time to market (TTM) for a new service. Having a deployed "Cloud SDP," scalable, with a predefined O&M SLA, connected with the operator's provisioning systems, ensures that the TTM decreases due to a part of the operator's processes are prevalidated or removed, allowing a quicker commercialization of services without forgetting that it enables to create a richer service portfolio.

With users demanding convergent Telco and Web 2.0 services, it seems that capabilities of both worlds should merge in a common platform. Thus, the Cloud WIMS 2.0 SDP is a proposal following this line of convergence. With this approach, a quick launch of new innovative and convergence services is ensured, but moreover, with the Cloud computing functionalities, the WIMS 2.0 SDP becomes a multioperator platform, where new operators using this SDP could offer most of the services running in it. Users are demanding more customization and remixability in order to satisfy their necessities {Win Win Consultores}. As it has been shown, the technical solution is available; now it is needed to be introduced in the operator's strategy.

References

{SOA} http://en.wikipedia.org/wiki/Service-oriented_architecture

{WIMS 2.0} WIMS 2.0 initiative. Luis Angel Galindo, David Moro

{http://www.wims20.org}, 2008

{WIMS 2.0 ecosystem} WIMS 2.0 ecosystem. Luis Angel Galindo, David Moro {URL
http://www.iirevents.com/IIR-conf/Telecoms/DocumentView.aspx?EventID=1800&
DocumentID=2691}, 2008

{Telco 2.0} Telco 2.0 {http://www.telco2.net}, 2009

{Blogger} {1999} Pyra Labs {http://www.blogger.com}

{Facebook} {2004} Facebook {http://www.facebook.com}

{YouTube} {2005} YouTube Broadcast Yourself {http://www.youtube.com}

{AJAX} Arthur Gittleman: Review of "Foundations of Ajax by Ryan Asleson" athaniel
Schutta, APress, 2005, ISBN 1590595823. ACM Queue 4(1): 62 (2006)

{RPC} R. Srinivasan, "RPC: Remote Procedure Call Protocol Specification Version 2", IETF
RFC 1831, August 1995.

{REST} C. Pautasso, O. Zimmermann and F. Leymann, "RESTful Web Services vs. "Big"
Web Services: Making the Right Architectural Decision". WWW 2008, April 21–25,
2008, Beijing, China. ACM 978-1-60558-085-2/08/04.

{Win Win Consultores} Win Win Consultores {http://www.winwinconsultores.com}

Chapter 4

Enabling Service Delivery in Next-Generation Networks toward Service Clouds

Anett Schülke and Toshiyuki Misu

Contents

4.1 Introduction

Creating a service market place characterized by fast time to market, innovation for new services, and a one-stop-shop for service providers and customers has been the desire in the Telecom and Internet market over a decade, exploiting new technologies, business models, and market opportunities. Telecommunication companies struggle to move from large system providers to new business approach to align to the more flexible Internet in order to adapt from a single-service provider to a multiple service provider.

This big challenge has been responded by the industry with a programmable all-IP (Internet protocol) architecture delivering voice and multimedia functionalities—the IP multimedia subsystem (IMS). This technology is competing with Web 2.0 and peer-to-peer technology representing the major competitors in today's technology landscape. Converging of networks drives the evolution of new business models further, like to well-known on-demand model into the recently hyped cloud computing. While each approach has its advantages and disadvantages, those different service delivery paradigms compete on the level of cost efficiency for development time and operation as well as their capacity to create and deliver new services in a fast and automatically manageable way.

This article discusses the evolution of the IMS as service delivery layer of the Next Generation Networks (NGN), it challenges under the demand to align the service provisioning between legacy systems and the new Web 2.0 area, up to the evolution of Telecom Service Application Programming Interfaces (APIs) for faster service creation and the placement of those technologies toward the new cloud-enabled business models.

4.2 Service Delivery in Telecommunication Networks

4.2.1 Overview of NGNs

In recent years, reconstruction of existing Telecom operator networks into new IP-based networks, that is, migration to NGN is gaining momentum. In the background of this movement, there is the need for both the fixed and mobile operators to lower the network operation cost by turning the legacy network into IP network, and to provide new types of communication service to make differences from competitors. Specially, the Fixed Mobile Convergence (FMC) is expected to be the next revenue source for Telecom operators, because by merging fixed and mobile telephone services, it will be possible to create new value-added services and convenient solutions.

In NGN, various services are provided by IMS, which is the service control function located in the service stratum. Introducing IMS enables providing IP-based multimedia services without depending on the access network to mobile phones, Wireless LAN (WLAN) terminals, and so on. Therefore, in NGN, IMS is expected to be used as the service control architecture common to a fixed and mobile network.

4.2.1.1 NGN Service Stratum

The standardization of IMS had been carried on by 3rd Generation Partnership Project (3GPP/3GPP2) taking the leadership, on assumption that the IMS should be used mainly on the mobile network. Then later, ESTI TISPAN and ITU-T reviewed and extended the specification of service platform common to fixed and mobile network, to be used as the service control function for NGN. Today, 3GPP reviews the specification of functions common to fixed and mobile networks, as common IMS.

IMS consists of components such as Call Session Control Function (CSCF) controlling session and services, Multimedia Resource Function (MRF) controlling multimedia resources, Home Subscriber Server/Subscription Locator Function (HSS/SLF) executing user profile management, Media Gateway Function/Media Gateway Control Function (MGF/MGCF) performing interwork with existing networks, and Application Server (AS) providing applications used in IMS. Out of these components, the CSCF is the server controlling sessions and services via Session Initiation Protocol (SIP). The CSCF performs access control of user terminals, roaming control, and activation of services provided by AS, playing the central role in the IMS architecture. Figure 4.1 shows the NGN architecture overview. Detailed architectural descriptions can be found, for example, at the International Telecommunication Union (ITU) [1] as well as 3GPP [2].

4.2.1.1.1 S-CSCF

Using the user profile information and user location information managed by HSS, Serving-CSCF (S-CSCF) performs routing based on terminating address specified by originating user (SIP-URI, telephone number, etc.) to establish, manage, and release the sessions between originating and terminating users. Also, via IMS Service Control (ISC), the SIP-based standard interface, the S-CSCF is connected to common enablers shared by different services (general-purpose functions used to realize each service such as presence and messaging), and the AS performing control for each service. Therefore, S-CSCF interworking with multiple ASs enables the Telecom operators to provide various services in a flexible

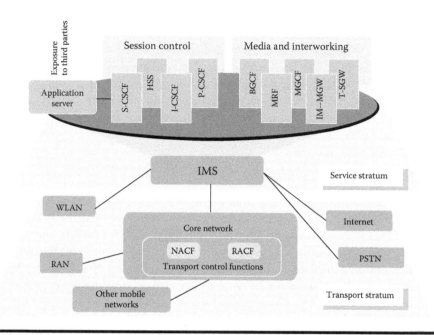

Figure 4.1 System view of the NGN and IMS architecture.

manner. Moreover, as the solution to easily realize cooperation among various ASs, the ISC function provided by the S-CSCF can be converted via a gateway into logical API, which is more accessible from applications. Through such API, various ASs can use the functions provided by the IMS such as setting and releasing of sessions, and acquisition of user location information, development, and deployment of services are simplified and accelerated. In NGN, the application service support function located in the service stratum realizes this function, enabling connection to external applications provided by a third party via Application-Network Interface (ANI).

4.2.1.1.2 P-CSCF

Proxy-CSCF (P-CSCF) is connected to a user terminal via the access network, and it is allocated upon the connection setup with the user terminal. In W-CDMA networks, the connection-to-user terminal is established via a packet switch, which is called Gateway GPRS Support Node (GGSN). In fixed broadband access and WLAN access, the connection is established via a relay router in the access IP network. The user terminal connected to the access network is authenticated at connection setup. An IPSec tunnel is set up with the user terminal and the SIP messages from the user is transferred over the IPSec, and the validity of these messages is verified at P-CSCF.

4.2.1.1.3 I-CSCF

The Interrogating-CSCF (I-CSCF) is located in the home network. It accesses the SLF if required at user registration to identify the HSS in which the subscription information of that user is stored. Then, following the instruction from the HSS, registration processing for that user is taken over by the S-CSCF. I-CSCF facilitates the load balancing when there are multiple HSSs, and also plays the role to hide the network structure from outside.

4.2.1.2 NGN Transport Stratum

In addition to the session and service control, there are other NGN features, which are Quality of Service (QoS) control and authentication functions. These functions are provided by Resource and Admission Control Functions (RACF) and Network Attachment Control Functions (NACF), respectively. RACF and NACF perform QoS control and authentication, in cooperation with the service control function in service stratum, and are located in the NGN Transport Stratum.

4.2.1.2.1 RACF

While the quality was guaranteed for each service in conventional networks, in NGN all-IP is realized and various services and media are handled in an integrated manner. Even though all services are using IP network, there is a need to provide these services surely, according to Service Level Agreement (SLA). In particular, voice services without delay and loss must be provided, and emergency calls must be prioritized, and so end-to-end QoS control is required. In NGN, when the service traffic passes through the network, packets are transferred according to their priority order. In this case, the priority can be controlled per call session with the authentication information of that session, and so different service levels can be provided, such as emergency and ordinary calls within voice calls, or depending on the media type or the user type.

RACF also performs admission control to ensure QoS. It manages network bandwidth to secure bandwidth used by prioritized traffic which was admitted by admission control, and restricts malicious overflow traffic. RACF controls the edge router to open and close the Real-Time Protocol (RTP) port to coordinate with call control, so that it can restrict the calls in congestion and spoofing traffic which did not go through regular SIP negotiation.

4.2.1.2.2 NACF

In NGN, there are two levels of authentication defined. They are access level authentication that allocates the IP address, and service application level authentication to use SIP. Out of these, NACF performs the network access level authentication based on user profile. It also has functions such as managing the IP addresses and connection information of currently connected terminals, and providing QoS profile information to RACF.

An example of services provided by NGN based on such architectures is IPTV. In ITU-T, FG-IPTV has been discussing three types of architectures to provide IPTV, which are non-NGN, non-IMS-based NGN, and IMS-based NGN. In September 2008, the standardization of IPTV requirement condition Y.1910 was approved, which completed the basic specifications to provide IPTV by IMS-based NGN. Examples of the advantages of providing Internet Protocol based Television (IPTV) service using NGN as service control platform are

- QoS (bandwidth/ priority control) is guaranteed.
- User authentication, charging, and profile management functions can be used.
- Mobility, FMC are realized.
- Services can be provided to roaming users on different types of network.

Also, by cooperating with other services, realization of "blended services" such as receiving voice call and displaying messages while watching IPTV, and seamless broadcast control in which broadcast is optimized for the functions of the terminals used to watch IPTV are being discussed.

4.2.2 NGN IMS Service Provisioning and Third-Party Service Access

The objective for the NGN service creation and provisioning is a flexible service framework for implementation of value-added services using network capabilities. These capabilities are accessed via standard application interfaces. The NGN Release 1 supports the following classes of value-added service environments:

- IN-based service environment (INAP, CAMEL, WIN, . . .)
- IMS-based service environment
- Open service environments (OSA/Parlay, Parlay X, OSE)

These new services either reside in the user's home network or in a third-party location. The ASs of the IMS-based service environment may influence and impact the SIP session controlled at the S-CSCF on behalf of the services supported by the operator's network, or may host and execute services. Hereby, the ASs can be

- SIP AS, which may host and execute services.
- OSA AS, which does not directly interact with the IMS network entities but through the OSA Service Capability Servers (SCSs). The OSA framework provides a standardized way for third-party access to the IM Subsystem.

■ CAMEL IM-SSF whereby the CAMEL network feature enables the use of operator specific services by a subscriber even when roaming outside the home network.

For details at the NGN service provisioning architecture governed via the IMS, please refer to the respective technical specification of the 3GPP TS23.002 [2].

4.2.2.1 Third-Party Access via OSA/Parlay

Value-added service creation via standardized APIs to network services is the commonly understood way to develop and deploy new services over network service architectures. The Parlay Group [3] is a consortium formed to develop open, technology-independent APIs that enable the development of applications that operate across converged networks. Parlay integrates Internet multimedia networks and intelligent networks (IN) with IT applications via a secure, measured, and billable interface and has been widely deployed in Telecoms networks globally. The goal of Parlay was to enable the creation of vast numbers of new applications in and around the network for both enterprise and consumer markets by utilizing network features and capabilities, while it asked to enable the worldwide software development community to exploit the features of current and future networks and to bring the IT developer community into the Telecoms space. The Parlay framework was intended to make it easier for network operators and service providers to deploy many new applications and services, and to enable new business models for service providers. During the course of time, the standards bodies 3GPP, Parlay Group, and finally also ETSI joint forces to develop consistent API specifications jointly. After completion of the NGN Release 8 in 3GPP, the responsibility for the evolution of Parlay specification had been transferred to the Open Mobile Alliance (OMA). Parlay/OSA is up to now *the* open standard for bringing Telecoms and IT together into one value chain. Parlay-X Telecom Web Services is the application interface that is published to developers and partners outside the Telecom network. The Parlay-X Web Services interface is attractive to many application developers, because of the strong support for Web Services in IDEs.

The Parlay APIs are understood as

■ A mediator API between Telecom networks and third-party applications
■ A secure interface between Network Operators and ASs
■ Open standards (specified by the Parlay Group and 3GPP, OSA)

They raise the programming abstraction level and allow multinetwork applications. The OSA/Parlay APIs are made up by defining:

Framework interface set: Common functions required to enable services to work together in a coherent fashion, for example, authentication, discovery, manageability.

Resource interfaces: Interfaces used between a Parlay Gateway and network elements (not specified in Parlay).

Service interface set(s): Common functions that deliver whole complex services or subcomponents of services, for example, Call Control, User Interaction, Content-based Charging, Location, Presence and Availability, Messaging, Policy Management, Quality of Service.

Transport interface: CORBA, WSDL (Web Services).

The Parlay/OSA Gateway consists of several SCS: functional entities that provide Parlay/OSA interfaces toward applications. Each SCS is seen by applications as one or more Service Capability Features (SCF): abstractions of the functionality offered by the network, accessible via the Parlay/OSA API. Sometimes they are also called services. The Parlay/OSA SCFs are specified in terms of interface classes and their methods. One of the Parlay/OSA SCSs is called the Parlay/OSA Framework, and is always present, one per network.

The functionality of the OSA Framework is grouped in two main sets of interfaces:

- Interfaces to the third-party applications, which provide mechanisms for controlling their access to the network (authentication and authorization), for ensuring the integrity of the network (e.g., load and fault management), and for the applications to discover the functionality that the operator's network exports via the OSA API.
- Interfaces to the other SCSs, which provide mechanisms for registration in the framework of their SCFs so that the applications may discover them; they also provide the necessary access control mechanisms to allow the framework functionality to be provided by an independent business entity (multidomain).

To enable the availability and usage of the Parlay APIs, there was the requirement to publish the capabilities of the network for software developers, and on the other hand the demand for enterprise applications to exploit the capabilities of the Telecom network. Hereby, Web services are a key software development technology which triggered the evolution of Parlay X.

Parlay X is a subset of the Parlay APIs, and provides a higher-level Web services abstraction of the more complex Parlay/OSA APIs. While Parlay/OSA provide rich function in an asynchronous programming model, are CORBA-based implementations, and are primarily used for Service provider-hosted applications, the Parlay X Web services provide high-level function in a largely synchronous programming model, are given as Web service-based implementations, and are intended primarily for third-party access. Parlay X defines a set of easy-to-use Web services, which provide simple and high-level access to widely used telecommunication functions. In general, both are used together in deployments. More detailed information to the Parlay and Parlay X specifications can be found under [2–4].

In recent times, the API standardization in the Telecom world experienced an advent of popularity, given the need for new services and the converging of Telecom and IT world. OMA picked up the Parlay development from 3GPP, ETSI, and the Parlay Group in 2008. The specifications are about to evolve with new interface technology for Representational State Transfer (REST) bindings in addition to Parlay X Web services, as well as with extension with new functionalities developed for the new Next-Generation Services Interfaces (NGSI) in OMA [4], as well as with respective profiles for the interface sets as triggered by the GSMA ACCESS group for the GSMA One API [5].

4.2.3 Service Broker, Service Capability Interactions, and Service Orchestration

Accelerating the time to market for new services is consistently one of the drivers for the acquisition of new service creation and delivery architectures. In order to bring new services more quickly, the industry realized the need of reuse of components as part of the new service mix. The ability to support new product assembly from reusable service components has also been one of the rationales to invest in IMS and Service Delivery Platforms (SDP). Both support the concept of reusable components into new products and services in a quick manner, and it is expected to provide this simpler, and in a more cost-efficient way. This expectation is feed as both platforms support the separation of the application from the core network through an abstraction layer in the SDP and through standardized interfaces in the IMS, which ease the developers' responsibility in assembling and deployment of new services. Besides this advantage, the matter complicates through the possible vendor-selection for reusable components as well as through the demand to bring IT, Internet, and so-called Web 2.0 services that evolve from the "cloud" beyond an operator's boundaries into the new product and service assembly mix. This demand brought the need for new service broker technology to the light.

4.2.3.1 Network Service Brokering and Service Capability Interaction Management

Service brokering as well as terms like Service Capability Interaction Management (SCIM), service chaining, and so on, are aiming to describe the technology to create "blended services," "combinational services," and product assemblies up to "Telco 2.0 mash-ups," which are all targeting the mixing of multiple services together into new product offerings. All these terms are used slightly different in different consortia. The 3GPP [2] distinguishes the terms such that service brokering belongs to legacy networks while the new term of SCIM belongs rather to IMS. An IN service broker is a function in the legacy IN network which sits between the service/feature/application and the switch to overcome the SS7 signaling problem that only one service can control the call. This way, the service broker seems to be a single functional service control point but it actually coordinates the interaction between multiple services and features. Through the way to a centralized service trigger management, the evolutionary path to IMS was given where the service invocation takes place at the signaling core. The functional architecture for the management of the user's service-related information via the SCIM as been proposed by 3GPP for the IMS; details are given in TS 23.002 [2]. Hereby, an SIP AS may contain the SCIM functionality and other ASs. The SCIM functionality is an application which performs the role of interaction management. The internal structure of the SCIM is not standardized. For the IN service brokering as well as the IMS-based SCIM, the main characteristic is that they trigger the service invocation from each other's independent services according to an interaction script/order/ rule. As the messaging protocol is standardized in either case, no changes to the service internal logic are needed in order to interact with the service broker. The service itself only interacts with the service broker instead with many different services. So far it appears simple, but for the separation of the network for the service layer, the understanding of the service invocation of the SCIM for IMS networks are not as straightforward as in the IN service brokering understanding. As for now, the details of the SCIM have therefore, not been defined and agreed upon, and left for each vendor's understanding.

4.2.3.2 Service Orchestration

From the functional perspective, there is no much difference between service brokering and service orchestration, except for the applicability of the domain. The term Service Orchestration describes the IT-level service brokering and is closely linked to the services and concepts of the service-oriented architecture (SOA). It is directly associated with service composition and service component reuse. Service orchestration is a way to describe how loosely coupled are service fragments—the so-called SOA services—offered by open, standard interfaces cooperate within a new single large composite service. SOA services are invoked remotely through standard protocols, have well-defined interfaces, and are self-contained. SOA service orchestration uses standard Web service protocols like Simple Object Access Protocol (SOAP) for invoking services, Web Service Description Language (WSDL) for service definitions, and the Universal Definition, Discovery and Integration (UDDI) for service discovery. In the IT domain, the service broker is called the Enterprise Service Bus (ESB), which refers to a software architecture construct, implemented by technologies found in a category of middleware infrastructure products usually based on Web services standards that provides foundational services for more complex service-oriented architectures via an event-driven and XML-based messaging engine [6]. In this way, an ESB is a flexible connectivity infrastructure for integrating applications and services.

Given the trend toward converging of network and enterprise/Web2.0 service domains, operators have begun to change their SDP landscape in order to build SOA-bases service component sets that will empower their next-generation service and product assembly. There is a need to define these components' functionality and requirements such that they are suitable for the Web 2.0 style of exposure.

4.3 Network/IT Convergence: Changing the Service Delivery Methods

Over the past few years, new technologies like IP networks, Web service technologies, and virtualization have emerged and reasonably matured in order to drive a new computing platform aiming to be highly distributed but with a central management, accessible remotely as well as enabling new business models for digital services and infrastructure support.

This new computing platform—discussed under the term *cloud computing*—is enabled to stretch over all service markets, and supports the consumption of service and infrastructure on demand, and is aiming to reduce the cost of IT infrastructure and a faster time to market for new services. A possible definition provided by NIST [7] for cloud computing can be given as following: "Cloud computing is a model for enabling convenient, on-demand network access to a shared pool of configurable computing resources (e.g., networks, servers, storage, applications, and services) that can be rapidly provisioned and released with minimal management effort or service provider interaction."

The definition and concepts of cloud computing are still evolving and not yet well defined to all aspects; however, it has already gained a substantial momentum for service-delivery technologies toward a global service market place serving the large and small–medium enterprise market as well as the Telecom consumer market. This section shall provide a short view on selected topics of the still emerging cloud computing technologies.

4.3.1 Evolving of New Service Delivery Technology for the XaaS Trend

Currently, the most known form of cloud computing is the on-demand support of the usage of IT and network infrastructure (such as processing power, bandwidth, and storage facilities) called Infrastructure as a Service (IaaS). Examples are AT&T's Synaptic Hosting or Amazon EC. The other popular type is the Software as a Service (SaaS) model. Hereby, a service provider offers applications on an "on-demand" basis (like for rental) over the IP network in real time, representing the offer of these applications *from the cloud*. This model is typically associated with the on-demand delivery of software for small–medium business with limited IT resources and includes originally applications like ERP, CRM, sales, backup, security, and messaging tools. SaaS can here be understood as evolution from the application service provider (ASP) delivery concept, offering a similar service delivery business model. This model has been redusted, and mixed up with the Web mash-up emergence. It is all about providing any piece of software into a service for exposure to the consumption using mash-up technologies. As smashups live on the Web, the interaction with the cloud service delivery concept is a natural fit. The third prime type is the Platform as a Service (PaaS) concept. A PaaS is a collection of software development and testing tools in order to allow the creation of new applications using the providers' APIs. Currently, existing PaaS offering nonstandardized APIs resulting in problem that applications developed at one PaaS cannot be transferred into another PaaS. For standardized Telecom service APIs, refer to Section 2.2.

In addition to those above-mentioned concepts, various others are emerging in the light of the cloud business opportunities, like Communication as a Service, Database as a Service, Desktop as a Service, and alike, transferring the trend into *Everything* as a Service.

4.3.2 Cloud Computing: IT-Based Service Delivery Business

Considering the various business opportunities for large enterprises, small–medium business, for customer support or internal usage, for large-scale consumer market—three types of cloud for IT-based service delivery can be identified:

Private cloud is a corporate system emulating a cloud environment in private networks to deliver applications and services to users (e.g., employees or business partner). It can be provided by a third-party provider, or network operators. The important aspect is that all resources within a private cloud are not shared with any external customer. It uses technologies to reduce the customer's opex, for example, by virtualization, and dynamic provisioning with special appliance to the private setting requirements. The effort to host and operate private clouds fits well in large enterprises aiming to optimize their own IT systems (for cost, centralized service management, etc.)

Public cloud is a system that provides dynamic resource provisioning on a self-service on-demand basis over the Internet. The resources (hardware, software, services) are accessed over Web portals. Applications can be offered for free or a minimal fee when provisioned via the providers Web site. Those resources are normally addressed to the consumer or Small–Medium Enterprise Business (SMB) users, and in nature accessed in an *ad hoc* manner. On the other hand, resources can also be accessed on a contract basis via a defined sign-in mode. This is normally applied in SaaS business models, where parameters like service quality, provisioning, or guaranteed availability is defined by policies in the form of Service Level Agreements (SLA).

Hybrid cloud is a model which combines aspects of private and public clouds. It is a merging model where some resources are offered via public clouds and combined with private cloud services. An example can be a business-critical application which is integrating public services with privately owned and hosted user data. Another example is where privately provided cloud services make use of publicly offered storage or computing resources for nonsensitive data storage and computing. In this way, the cost reduction promise of public clouds can be obtained while security and privacy measures for business data are fulfilled in the private cloud.

4.4 Cloud-Enabled Business Models: New Chances for the Telecommunication Market

Besides the strong evolvement of cloud computing from the IT domain, the Telecom network and service environment provide a natural environment to serve consumer as well as corporate domains with trusted and secure cloud environments. This spans from an evolution of managed Telecom services usually offered in a shared mode toward Network as a Service adapting to the IaaS model, up to the challenges to address the so-called long-tail market [8] in order to utilize the service creation potential of reusable service components in the horizontal NGN service layer toward a new service market place offered via service creation PaaS solutions and the SaaS aggregation business opportunity for Telecom providers.

In today's economically demanding times, flexibility in business, lowering costs, and increasing search for new revenues, Telecom providers start to invest more in the opportunities for cloud computing aiming to address all market segments. As they are well placed to deliver software and services on demand and on a pay-per-use basis to their enterprise and SMB customers, Telecom providers have a unique opportunity to exploit the various cloud types integrating their Telecom assets in network infrastructure, service portfolio for integration in new services, and their trusted environment given by their owned subscriber base.

One of the most promising XaaS models exploited so far by Telecom providers is the investigation into the value chain for the SaaS model. Given their experience in service delivery platforms, brining those platforms into a carrier cloud promises an acceleration of new service creation market via PaaS/SaaS business models. The Telecom provider appears not only as a host of SaaS services and applications, but mainly opens up as SaaS aggregator via a PaaS-enabled SDP with open standardized Telecom service APIs for open service creation. Business customers can be provided with a one-stop-shop for their applications needs combined with single-sign-on services which offers to support any business model combining public, private, or hybrid cloud models.

A SaaS aggregation platform needs to enable support for the full service life cycle including customer life cycle, for example, for provisioning, activation, deactivation of services and respective customer support, security, and policy management, charging and billing support, performance and SLA management up to the service catalog services, and customer self-service portals for service ordering, provisioning, and administration. The TeleManagement Forum (TMF) has defined the blue-print of a service delivery framework (SDF) defining the whole service management ecosystem for service delivery and life-cycle management [9]. The SDF service model is complemented by data representation to support SDF Service Lifecycle Management which aims at covering also contextual information of a service in relation to the business and operation environment, as well as the unified life cycle management of services that they or their dependencies may be managed in different frameworks, regardless of

- How the service has been created, for example, SDP, IMS, Internet access, Web 2.0, cloud computing, in broadband network, in service provider domain, or in another domain.
- How the service shall be operated (end-to-end or partially outsourced/managed, service components operated in other service provider domains, SOA style OSS or not, etc.).
- How the service should be commercialized (individual or part of a product/service bundle, syndicated with service providers' offerings, charged by usage, etc.).

The TMF SDF reference architecture is a generic architecture integrating SDP technology experiences from various industry groups and represents an applicable service delivery model for traditional as well as cloud-based service delivery. The main differentiators for SaaS aggregation platforms are identified by the multi-tenancy support to reduce cost, ensure customer separation, and openness for different reselling models, trustful operational procedures, open APIs, support for various business models (e.g., from flat-rate and subscription up to different levels of personalization), and a certain degree of flexibility to allow for customization for user segments for a customer. Today, all those new evolving cloud types and models are on the stage of promise to and investigating by the industry. The proof of the market success is yet to come.

Acronyms

3GPP   3rd Generation Partnership Project
ANI   Application-Network Interface
API   Application Programming Interface
AS   Application Server
CSCF   Call Session Control Function
FMC   Fixed Mobile Convergence
HSS   Home Subscriber Server
IMS   IP Multimedia Subsystem
ISC   IMS Service Control
ITU   International Telecommunication Union

MGCF   Media Gateway Control Function
MGF   Media Gateway Function
MRF   Multimedia Resource Function
NACF   Network Attachment Control Functions
NGN   Next Generation Network
OMA   Open Mobile Alliance
OSA   Open Service Access
QoS   Quality of Service
RACF   Resource and Admission Control Functions
SDF   Service Delivery Framework (defined at TMF)
SIP   Session Initiation Protocol
SLF   Subscription Locator Function
SMB   Small-Medium Enterprise business
SOA   Service-Oriented Architecture
TMF   TeleManagement Forum

References

1. ITU-T Recommendation Y.2012, Functional requirements and architecture of the NGN, International Telecommunication Union (ITU), http://www.itu.int/home/
2. 3GPP TS 23.002, Network Architecture http://www.3gpp.org/; http://www.3gpp.org/ftp/specs/html-info/23002.htm
3. Parlay Group Consortium, http://www.parlay.org/en/index.asp
4. Open Mobile Alliance, http://www.openmobilealliance.org/
5. GSMA ACCESS project. http://www.gsma.org
6. See Wikipedia, available at http://en.wikipedia.org/wiki/Enterprise_service_bus
7. NIST Cloud Computing Definition, http://www.thecloudtutorial.com/nistcloudcomputing definition.html
8. See Wikipedia, available at http://en.wikipedia.org/wiki/The_Long_Tail
9. TM Forum Service Delivery Framework Reference Architecture, TMF 061, http://www.tmforum.org/TMF061ServiceDelivery/39341/article.html

Chapter 5

How to Model Dynamic Service Composition Using UML 2.x and Composition Policies*

Judith E. Y. Rossebø and Ragnhild K. Runde

Contents

* From J. E. Y. Rossebø and R. Bræk. Using composition patterns to manage authentication and authorization patterns and services. Avantel Technical Report 3/2008, NTNU, 2008 and J. E. Y. Rossebø and R. K. Runde. Specifying service composition using UML 2.x and composition policies. Technical Report 378, Department of Informatics, University of Oslo, 2009. With permission.

5.1 Introduction

During the past two decades, the telecommunications environment has evolved from centralized dedicated telephony networks to a distributed environment with a vast number of service providers, third-party vendors, and a range of network technologies. In addition to the traditional telephony services, there is now a range of different services, Internet-based services and Web Services, with a strong drive for convergence, to deliver all services over a single IP-based core network.

In this evolving, convergent environment, there is a need for rapid development and deployment of services. As the service environment changes, the services are also required to be adaptable. This is because, of what makes sense at the time of deployment, many would not be valid after some years of use. For example, for a multimedia over IP (MMoIP) service, in order for the provider to be able to ascertain the identity of and the number of users accessing a service at a specific time, some level of authentication may be required. Initially, the operator may for instance choose a username password-based solution, as this seems to make sense at the time of deployment. However, a period of usage may demonstrate that this authentication is too weak, and with a significant number of unauthorized accesses where a large number of calls are made with the end user not known or incorrectly identified. As a result, there may be widespread fraud and misuse of the service, with a need to upgrade the service with stronger authentication to counter these threats. In order to avoid disrupting the service, ideally this upgrade should be deployed dynamically. Indeed it may be desirable for the service to be monitored by a management system, which automatically determines the need to change or upgrade the service.

As can be seen by this example, it is not enough to deploy services rapidly, as the dynamic service environment is changing continually, requiring services to be adaptable at runtime also. By this we mean, that in addition to being able to rapidly deploy new services, it must be possible to dynamically change existing services at runtime to adapt to conditions in the service environment that were not foreseen when the service was originally deployed.

This chapter addresses the problem of rapid development and dynamic deployment of services in the converged service environment. We present a policy-driven approach to service modeling, design, deployment, and implementation. In the approach, policies are used both for monitoring, and for describing the composition of services, where a policy is a set of rules defining choices in the course of the behavior of a system.

A service is an identified functionality (i.e., behavior) involving a collaboration among entities in order to achieve some desired goals/effects for end users or other entities. One important aspect of services is that they normally are partial functionalities that cross-cut the component structure so that a service may involve several collaborating components while each component may participate in several services. Service composition addresses the means of composing the partial behaviors of collaborating components in order to achieve the overall service. We model a service as collaboration among roles with associated behavior, using the Unified Modeling Language (UML) version 2.x [1]. The basic building blocks for composed services are elementary service collaborations between two entities collaborating on an interface to provide a service or a service feature [2]. The visible interface behavior associated with an elementary collaboration is modeled using UML 2.x interactions.

The policy-driven approach to model, design, and deploy the dynamic composition of services, that we present in this chapter, was first introduced in Rossebø and Bræk [3]. For modeling the choreography of service collaborations, the approach uses the concept of a policy enforcement state machine (PESM) and composition policies, which we elaborated on in Rossebø and Runde [4], and in Rossebø [5]. Essentially, the global PESM is similar to a UML 2.x activity diagram with collaboration uses as nodes.

The approach is presented in this chapter along with a methodology explaining how to use UML 2.x, PESMs, and composition policies to model, design, and deploy the dynamic composition of services. We exemplify the approach and methodology using a tele-consultation service.

The rest of the chapter is organized as follows: In Section 5.2, we provide an overview of the policy-based approach to dynamic composition, and in Section 5.3, we introduce the tele-consultation service used in the remainder of this chapter. In Sections 5.4 and 5.5, we explain how we model service composition, introducing some basic terminology and presenting first the global PESM diagrams at the service model layer, and then their transformation into local PESM diagrams, one for each composite role, at the design model layer. The methodology is presented in Section 5.6 and is explained in Section 5.7 as to how to use the methodology to change and adapt existing services. Related work is presented in Section 5.8, before concluding in Section 5.9.

5.2 Overview of the Policy-Based Approach

A graphical overview of the model-driven, policy-based approach to dynamic service composition is provided in Figure 5.1. At the service model layer, services are

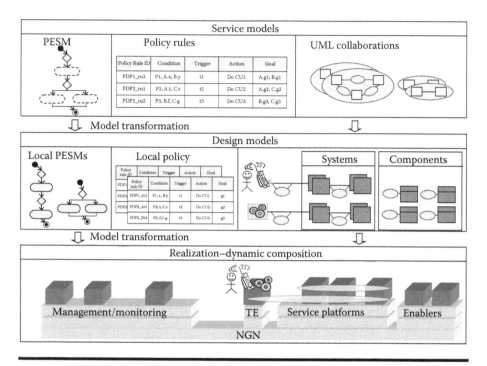

Figure 5.1 Overview of the policy-enabled approach to service composition.

modeled structurally using UML 2.x collaboration diagrams [1], and policy rules stored in a table are used for governing dynamic choices among collaborations, that is, dynamic composition. The PESM diagram graphically depicts the choreography of the collaboration uses involved in the composed service. While the service model layer provides a description of the service as a whole, the design model layer describes the service behavior as it is distributed to each separate component. The individual role behaviors are given by UML 2.x state machines, and the coordination of these is specified by local policy rules and local PESM diagrams, derived by transformations from the global models and rules at the service model layer. From the design models, implementations can be generated, which in combination with a policy-management system in the runtime system, enable the actual runtime realization of the dynamic composition. In Figure 5.1, we have illustrated deployment of the policy-based approach delivered over next-generation networks (NGN) which consist of common IP-based networks and services in a distributed environment [6]. The policy-management system for enabling dynamic composition in the runtime system is illustrated on the left, to emphasize that the policy management is a separate management layer in the deployed system. Services are dynamically composed across user terminal equipment (TE) and service platforms which make use of enablers in the NGN. As shown in Figure 5.1, the policy management is a separate management layer in the deployed system.

A policy-based dynamic composition approach requires a policy-specification model to express the composition strategies. We have chosen event-condition-action declarative policy rules [4,7]. There are two main reasons for this choice. First, it is natural to use signaling in order to set up a service session involving a set of components. As signaling is an event-oriented activity, an event-driven model is a natural candidate for notifying distributed components of changes that occur. Second, specifying the declarative policies separately from the behavioral specifications of the various service parts results in more freedom for the dynamic composition with respect to which separately specified behaviors to compose.

In our approach, events specified as policy triggers are monitored, and these events are used to trigger the evaluation of one or more policy rules. We allow for policy rules with the same trigger to have overlapping conditions, so that more than one of them may be true at a given point in time. If this is the case, then the choice between which one of the possible actions to be performed is made nondeterministically. This means, that if the conditions of a policy are matched, then the action specified by the policy rule may be executed.

For deploying the approach in the runtime system, an architectural model for the policy-enabled dynamic composition system, and a process for runtime dynamic composition using the policy-based approach are required.

For the architecture, we have been inspired by the Poema [8] approach, although in Montanari et al. [8], the policy-managed adaptation is used for dynamic re-configuration of Mobilcode applications specifically, and not to enable the dynamic composition of services in general. The Chisel adaptive process described in Keeney [9] provides a guide for defining a process for runtime dynamic service composition. However, the actual way that the adaptation is performed in Chisel involves manipulation of the existing behavior in the system, using metatypes to adapt existing objects. In our approach, we do not change the existing building blocks (elementary roles), but change which are combined together and in which order they are executed. Further details of this layer may be found in Rossebø et al. [10].

A family of services will run dynamically in the system, according to a global PESM description and policy rules. But additionally, the policy rules and PESM diagrams may be changed by a central administrator. Changes made centrally are then distributed to the different parts of the system, and in this way services may be dynamically adapted and changed. Additionally, in the system, a system security policy, such as an availability policy may be violated causing a need to dynamically adapt the service. In such a case, the system would look up in a repository, the appropriate PESM and policies to apply. For example, if the management system registers $x \geq maxlimit$, the number of unauthorized uses of the service exceeds the limit of such unauthorized behavior allowed by the availability management policy; this may trigger the requirement for stronger authentication to gain access to the service. In this case, the service must be dynamically adapted and composed with a sufficiently strong authentication pattern. Depending on which policy rules are satisfied, different authentication patterns will be called.

5.3 Description of the Tele-Consultation Service

The tele-consultation system is designed to allow chronically ill patients to have intermediate consultations from their homes, so as to reduce repeated travel to the doctor's office/hospital for routine consultations when routine tests may be equally well performed remotely. Equipment is provided in the home for carrying out tests. This equipment collects data from, for example, medical sensors and this information is transmitted/communicated to the doctor. In order to request a consultation with a doctor, the patient needs to register to an electronic tele-consultation service desk, which then sets up a remote consultation with the patient's doctor, depending on availability. The patient may register by communicating directly with the electronic tele-consultation service desk, or if the patient is in need of special assistance to register, a human receptionist may be contacted. The receptionist then gets the registration information from the patient, and provides this information to the electronic tele-consultation service desk. We will use the tele-consultation service to exemplify the model-driven, policy-enabled approach. For the purpose of clarity in presentation, we will consider a simplified view of the service involving a doctor, a receptionist, a patient, and the electronic tele-consultation service desk. We will also use the tele-consultation service to illustrate the steps involved in the methodology, in particular, explaining how to change the existing tele-consultation service by adding authentication to the composed service and by adding a choice between three specializations of the call collaboration used in the existing service.

5.4 Service Models

In this section, we introduce some basic terminology, exemplified using the tele-consultation service. Then we introduce the global PESM diagrams and composition policies which are modeled at the service layer.

5.4.1 Collaborations and Roles

A composite service is described at the service model layer using a UML 2.x collaboration. Such an overview diagram describes the distributed parts involved in a service and which roles are to be played by the different service parts, providing a means of statically and structurally representing the binding of collaboration uses involved in a composed service to the actual service parts. Figure 5.2 shows the collaboration overview diagram for the basic tele-consultation service.

An *elementary collaboration* is a service collaboration with exactly two elementary roles that collaborate on an interface to provide a service or a service feature [2]. An *elementary role* is the smallest modeling element. These are predefined behavioral specifications (elementary role state machines) that are not changed under dynamic composition. Figure 5.3 shows an elementary collaboration for the registration

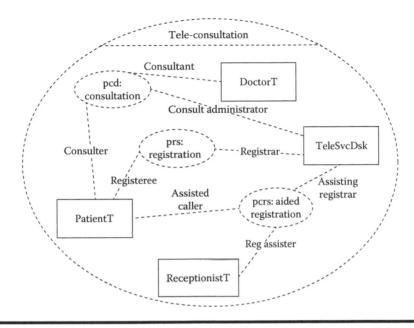

Figure 5.2 Collaboration overview diagram for the basic tele-consultation service.

service. The registration collaboration models the registration of a patient's request for consultation with a doctor. The two elementary roles (registeree and registrar) are bound to the PatientT and TeleSvcDsk composite roles, respectively. The visible interface behavior associated with an elementary collaboration may be modeled using, for example, UML 2.x interactions or semantic interfaces [2]. For ensuring correct dynamic composition in the runtime system, the visible interface behavior between the two elementary roles should be validated with respect to safety and liveness properties [11,12].

Composite roles are dynamic compositions of elementary roles and may also be compositions of composite roles. PatientT is a composite role playing the registeree elementary role, and the consulter and assisted caller

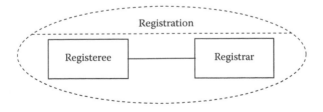

Figure 5.3 Elementary collaboration for a registration **service.**

composite roles. Composite roles play elementary roles and/or composite roles governed by local composition policies as described later in Section 5.5.

In Figure 5.2, there are three subcollaborations used in the composed service. The `registration` collaboration is an elementary collaboration for which the two elementary roles (`registeree` and `registrar`) are bound to the `PatientT` and `TeleSvcDsk` composite roles, respectively. There are two composite subcollaborations that are used in the `teleconsultation` collaboration. One of these, shown in Figure 5.4, is the `aided Registration` collaboration for which the composite roles `assisted caller`, `assisting registrar`, and `reg assister` are bound to the `PatientT`, `TeleSvcDsk`, and `ReceptionistT` composite roles, respectively. The other is the `consultation` collaboration for which the composite roles `consulter`, `consultant`, and `consult administrator` are bound to the `PatientT`, `Doctor T`, and `TeleSvcDsk` composite roles, respectively.

The collaboration overview diagram in Figure 5.2 describes the subcollaborations involved in the tele-consultation service, but not how their associated behavior should be composed in order to achieve the desired behavior for the composed service. We use global PESM diagrams and global composition policies to describe this, as explained in the next section. As will be specified later by the PESM diagram, only one of `registration` and `aided Registration` will be performed during a single run of the tele-consultation service.

In the next two sections, we describe how composition policies and PESM diagrams may be used to describe how the associated behavior should be composed in order to achieve the desired behavior for the composed service.

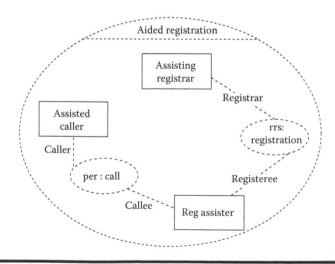

Figure 5.4 Collaboration for the `aided registration` service used in the `teleconsultation` service.

5.4.2 Global Composition Policies

A composition policy is a set of policy rules specifying how a given set of collaborations can be composed to obtain a given set of composite services. As such, a composition policy applies to a family of related services, where the number of possible services in the family depends on the choices governed by the policy rules. Choices are made to decide which set of collaboration uses will be composed in order to deliver the selected service.

In terms of the family of services defined by a composition policy, a UML 2.x collaboration overview diagram graphically represents a high-level view of the service possibilities, where all or some of the collaborations may have more concrete specializations as will be demonstrated in Section 5.7.

In Rossebø and Runde [4] we defined a Backus–Naur Form (BNF) grammar for policy rules, which is given in Figure 5.5. `Signal` represents the actual content of a message, `C-role` and `E-role` are names, representing composite and elementary roles (or in the case of a hierarchical PESM, subcomposite roles), respectively. `Pred` is a Boolean predicate, `C-name` is the name of a collaboration, and `Cu` is the name of a collaboration use.

As can be seen from the definition, a policy rule consists of four parts: trigger, condition, action, and goal. The definition of these rules is derived from the event–condition–action rule paradigm of active database systems [13]. In order to ensure correct dynamic composition of a service, the system is obliged to execute the given action when a policy rule applies. An example of a global policy rule is given in Figure 5.6.

In a policy rule, the trigger T and condition C together define when the policy rule is applicable. The trigger is the event of receiving or sending a message sent to/ from a composite role, which will then be the role discovering the trigger and

```
<Policy rule>  →   T : <Trigger>
               →   C : <Condition>
               →   A : <Action>
               →   G : <Goal>
<Trigger>      →   <Event>
<Condition>    →   Pred, C-role.Pred,{C-role.Pred}+,
<Action>       →   Cu ( E-role → C-role , { E-role → C-role }+, ) : C-name
<Goal>         →   C-role.Pred , { C-role.Pred }+,
<Event>        →   { ! | ? } <Message>
<Message>      →   ( Signal , C-role , C-role )
```

Figure 5.5 Extended BNF syntax for global policy rules, with +, denoting one or more separated by commas with trailing comma optional, | denoting alternatives, and { } as meta parentheses. (With kind permission from Springer Science+ Business Media: J. E. Y. Rossebø and R. K. Runde. Specifying service composition using UML 2.0 and composition policies. In *MoDELS 2008*, Vol. 5301, LNCS, pp. 520–536. Springer, Berlin, 2008.)

```
PDP4_Rule
T  : ! ( RegistrationOK, TeleSvcDsk, PatientT )
C  : PatientT is a Patient of Doctor, TeleSvcDsk.online,
     DoctorT.AvailableForConsultation, PatientT.registered
A  : pcd (consult administrator → TeleSvcDsk, consultant → DoctorT,
     consulter → PatientT ) : consultation
G  : TeleSvcDsk.svcAccessed, DoctorT.consultationComplete, PatientT.consulted
```

Figure 5.6 Example global policy rule with name PDP4_Rule1. The trigger T defines that the policy rule applies when the TeleSvcDsk composite role sends the RegistrationOK signal to the PatientT composite role. Furthermore, condition C states that the rule only applies when the instance playing the PatientT composite role is a Patient of the Doctor represented by the DoctorT composite role, DoctorT role is available, and that the Patient is registered. The action A then defines that the collaboration to be executed is consultation, with the TeleSvcDsk as consult administrator, the DoctorT as consultant, and the PatientT as consulter. Finally, the goal G states that the goal when executing the collaboration is for the TeleSvcDsk to access the services (e.g., patient journal), the DoctorT to complete the consultation of PatientT, and PatientT to have been consulted (by the Doctor).

responsible for enforcing the policy rule. For easier transformation from global to local policies, the condition is split into parts: one condition relating the roles, and one condition for each of the composite roles involved.

The action A defines the collaboration use to be performed when the trigger event has occurred, given that the constraints in the condition hold. The action also defines the necessary role bindings from elementary (or subcomposite) roles to composite roles.

The goal G defines the desired result of the action when the policy is applied and is a Boolean predicate expressing when the goal is achieved from the perspective of the elementary collaboration as a whole. To allow for deriving the role goals for each of the policies that will follow the roles in the local PESMs, the goal of a composition policy rule is split into parts: one for each composite role, similar to what is done for conditions. The goal provides flexibility of choice between different collaborations with the same goal to provide options for ensuring that reaching the goal can be achieved so that the service can continue to execute correctly. For example, if one authentication pattern is not achievable, for example, because constraints cannot be satisfied, another alternative with the same goal can be chosen so that the service can execute correctly.

5.4.3 Global PESM Diagrams

Graphically, a global PESM is used to depict the dynamic composition of collaborations according to the composition policy. In Figure 5.7, a PESM diagram for the tele-consultation service is given. The global PESM diagram and composition policies describe how the associated behavior of the subcollaborations should be

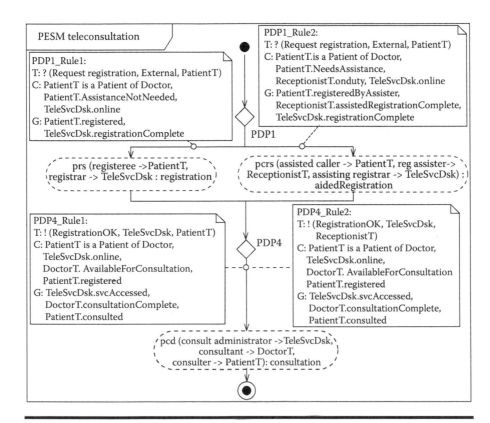

Figure 5.7   Global PESM diagram for the tele-consultation service.

composed. The PESM is modeled as a graph defining the flow of invocations of collaborations represented as nodes.

As nodes, the PESM contains collaboration uses, each consisting of a collaboration use name (e.g., `prs`), a collaboration name (e.g., `registration`), and the roles as bound to composite roles (e.g., `registeree` bound to `PatientT` and `registrar` bound to `TeleSvcDsk`).

The collaboration uses are linked by transitions and choice nodes. The initial state represents the start of execution of the composite services represented by the PESM. The nodes are collaboration uses linked by transitions (–>) and choice nodes (<>). The choice nodes are labeled as numbered policy decision points (PDPs) so that the policy rules can be stored with a pointer to the PDP they apply to. Each PDP refers to a set of policy rules (one or more). For the convenience of the reader, each policy rule is shown as a note annotated to the appropriate branch of the PESM diagram. At each PDP, policy-ruled choice is evaluated, and the outgoing transition chosen is policy ruled.

In Figure 5.7, the first policy decision point (PDP1) points to two global policy rules, governing a choice between execution of `registration` and `aided` `registration`. The choice is made in PDP1, when the `PatientT` composite role receives a request for registration. This request is the trigger event for both policy rules in PDP1, and which outgoing transition to be taken will then depend on the conditions in the policy rules.

In this example, the condition C of `PDP1_Rule1` states that the policy only applies when the instance playing the `PatientT` composite role is a Patient of the Doctor represented by the `DoctorT` composite role, `PatientT` does not require assistance, and that the `TeleSvcDsk` is on line. If the condition of a rule is evaluated as true after the trigger event has occurred, the outgoing transition corresponding to the action part of the rule (e.g., `prs` in the case of `PDP1_rule1`) is taken. In the general case, policy rules with the same trigger may have overlapping conditions, so that more than one of them may be true at a given point in time. In that case, the choice between the possible actions is nondeterministic.

For presentation purposes, the PESM diagram in Figure 5.7 describes successful execution only. Unsuccessful or exceptional executions (e.g., registration failure) maybe described using additional policy rules and possibly involving other collaboration uses.

The actual (successful) execution of the tele-consultation service will follow one of the choices depicted by the PESM, and the choice is made dynamically each time.

For each node of a PESM that is a composite subservice used in the service (e.g., the collaboration use `pcrs` of `aided Registration` in the `teleconsultation PESM`), a global PESM diagram is defined. For the tele-consultation service, there are two composite subcollaborations, the `aided registration` collaboration and the `consultation` collaboration. The PESM diagrams for these subcollaborations are given in Figures 5.8 and 5.9, respectively.

Figure 5.8 shows the global PESM diagram for the aided registration service used in the tele-consultation service. For the convenience of the reader, each policy rule is shown as a note annotated to the appropriate branch of the PESM diagram. In the diagram, we use a simplified notation for the collaboration uses, where, for example, `pcr(caller → assisted caller, callee → reg assister):call` stands for the collaboration use `pcr` of the collaboration `call`, where the elementary roles `caller` and `callee` are bound to the composite roles `assisted caller` and `reg assister`, respectively.

During `aided Registration`, the receptionist obtains the registration information from the patient during the call, and provides this information to the electronic tele-consultation service desk. As such, these collaborations are not synchronized, and horizontal coordination between the collaboration uses is not shown in this view, as the receptionist is responsible for translating information received from the patient to the appropriate form required for entering into the electronic service desk. This is why the two collaboration uses for `call` and `registration` are shown in parallel in Figure 5.8 (using the UML 2.x notation where a heavy bar represents a fork or a join). Evaluation of `PDP2_Rule1` is triggered

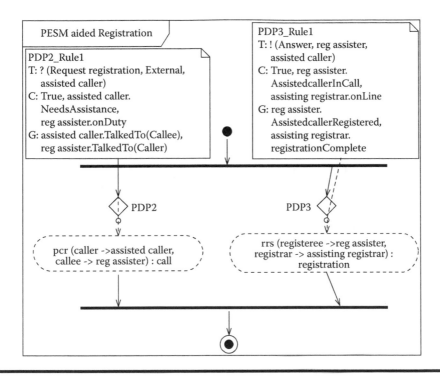

Figure 5.8 Global PESM diagram for the aided registration service.

when the `assisted caller`, bound to the `PatientT` composite role, receives a request for registration. If the given condition evaluates to true, the call collaboration should be performed. The goal is that afterward both the patient and the receptionist should have reached a status of having talked to each other. Similarly, the evaluation of `PDP3_Rule1` is triggered when the `Answer` message of the call collaboration is sent from the `reg assister` to the `assisted caller`, bound to the `ReceptionistT` and `PatientT` composite roles, respectively.

During consultation, the doctor is in a call with the patient. At the same time, the doctor is interacting with the electronic tele-consultation service desk in order to obtain patient data as well as record information in the patient journal.

As can be seen in Figure 5.9, the composition policy for consultation gives that the administration collaboration can be performed when either one of the two rules `PDP6_Rule1` and `PDP6_Rule2` applies. As part of the behavior of the administration collaboration, the consult administrator role, which is bound to the `TelesvcDsk` composite role, then triggers the evaluation of policy rule `PDP5_Rule1` by sending the `StartConsultation` message to the consultant role, which is bound to the `DoctorT` composite role. The `administration` collaboration runs in parallel to the `call` collaboration as the doctor will need to look up information in the patient journal and record patient status and test

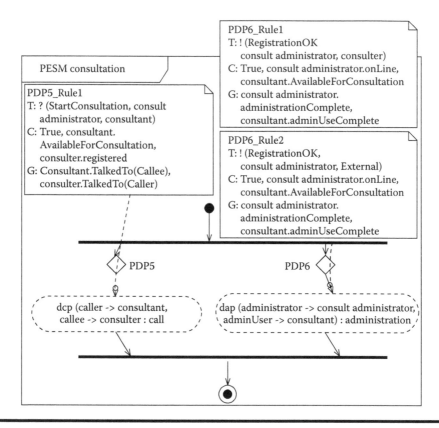

Figure 5.9 Global PESM diagram for the consultation collaboration.

results during the entire consultation by communicating with the electronic tele-consultation service desk. The call collaboration can be performed when the policy rule `PDP5_Rule1` applies, that is, when the trigger has been received and the condition part of the policy rule is true.

The semantics of UML diagrams are stated informally in the UML standard. As the PESM is an extension of the UML 2.x modeling language, there may be ambiguities that may lead to different interpretations of the meaning of the diagrams. In order to provide a precise understanding of the meaning of the PESM and composition policies, we have defined a formal semantics for global policy rules and for global PESM diagrams. See Rossebø and Runde [14] for details.

5.5 Design Models

At the design model layer, we describe the behavior as distributed to the components of the system. The behavior of each elementary role is given as a UML 2.x state machine. Their coordination into composite roles is described using local composition policies and local PESM diagrams, one for each composite role. In Rossebø and Runde [4], we have defined a formal semantics for local policy rules and for local PESM diagrams. For details, see Rossebø and Runde [14]. In this chapter, we explain how the local composition policies and PESM diagrams may be automatically derived from the global ones.

5.5.1 Local Composition Policies

Each global policy rule at the service model layer corresponds to two or more local rules, one for each of the composite roles involved in the collaboration specified in the action part. The role receiving the trigger is referred to as the initiating role, while the other one or more roles are called participating roles. In addition to performing its role in the collaboration, the initiating role is responsible for sending a role request demanding the participating role to play its part, for each participating role.

Note that the initiating party is not necessarily the first one to act within the collaboration. This is to allow for providing a choice between collaborations to execute regardless of the behavior specified within the collaboration. Separating the initiating role from the collaboration to be initiated thus allows for greater flexibility in dynamic composition.

The local policy rules are similar to the global rules in that they consist of a trigger, a condition, an action, and a goal. The main difference is that the actions to be performed are now given as state machines instead of collaborations. The global goal is distributed to the roles, and similarly for condition, where the initiating role should be responsible for checking both its local condition and the condition relating the roles.

The BNF-grammar for local policy rules is defined in Figure 5.10, where `Pred`, `E-role`, `C-role`, and `Signal` are as for global policy rules described in

```
<Policy rule>      →    T : <Trigger>
                   →    C : <Condition>
                   →    A : <Action>
                   →    G : <Goal>
<Trigger>          →    <Event>
<Condition>        →    Pred {,  Pred, C-role.Pred +,  }?
<Action>           →    {{ ! <Role-request> } +, ,   }? PlayRole(E-role)
<Goal>             →    Pred
<Event>            →    { ! | ? } <Message >
<Message>          →    ( Signal, C-role,  C-role ) | <Role-request>
<Role-request>     →    ( RoleReq(E-role), C-role, C-role )
```

Figure 5.10 Extended BNF syntax for local policy rules, with { }? denoting optional parts. (With kind permission from Springer Science+Business Media: J. E. Y. Rossebø and R. K. Runde. Specifying service composition using UML 2.0 and composition policies. In *MoDELS 2008*, Vol. 5301, LNCS, p. 531. Springer, Berlin, 2008.)

Global rule	Local rule, initiating rule a	Local rule, participating role b
T: t	T: t	T: ?(RoleReq(f), a, b)
C: c, a.c_a, b.c_b	C: c, c_a, b.c_b	C: c_b
A. Cu(e→a, f→b) : C	A:!(RoleReq(f),a,b, PlayRole(e))	A: PlayRole(f)
G: a.g_a, b.g_b	G: g_a	G: g_b

Figure 5.11   From global to local policy rules. (With kind permission from Springer Science+Business Media: J. E. Y. Rossebø and R. K. Runde. Specifying service composition using UML 2.0 and composition policies. In *MoDELS 2008*, Vol. 5301, LNCS, p. 531. Springer, Berlin, 2008.)

Section 5.4. `RoleReq(E-role)` should be understood as a special signal, namely a request for playing the elementary (or subcomposite) role given as parameter. This means that a role request is a special kind of message. For a given global policy rule, the corresponding local rules may be derived automatically following the pattern given in Figure 5.11 as illustrated in Figure 5.12.

5.5.2 Local PESM Diagrams

Local PESM diagrams are used to graphically depict local coordination and enforcement of the local composition policies. From the global PESM diagrams at the service layer, it is possible to derive a number of local PESM diagrams, one for each composite role governed by the policy.

The transition is made by traversing the global PESM diagram by focusing on one composite role at a time. The collaboration uses, which the composite role participate in, are replaced by the corresponding role state machine, while collaboration

PDP2_Rule1	PDP2_P_Rule1	PDP2_R_Rule1
T: ?(Request registration, External, assisted caller)	T: ?(Request registration, External, assisted caller)	T: ?(RoleReq(callee) Assisted caller, Reg assister)
C: True,assisted caller.NeedsAssistance, reg assister.onDuty	C: True, NeedsAssistance, reg assister.onDuty	C: onDuty
A. pcr(caller→assisted caller, callee→reg assister) : call	A:!(RoleReq(callee), assisted caller, Reg assister), PlayRole(caller)	A: PlayRole(callee)
G: assisted caller.TalkedTo(Callee) Reg assister.TalkedTo(Caller)	G: TalkedTo(Callee)	G: TalkedTo(Caller)

Figure 5.12   Example of deriving local policy rules.

uses that do not involve the composite role are removed from the diagram along with the preceding PDP.

For the tele-consultation service shown in Figure 5.2, four local PESM diagrams may be derived from the global PESM diagram for the `PatientT, Doctor T, ReceptionistT`, and `TeleSvcDsk` composite roles, respectively by applying the transformation rules as explained above. The local PESM diagram for the `PatientT` composite role is shown in Figure 5.13, annotated with policy rules for each of the PDPs. Policy rules for each of the PDP are shown alongside the diagram, for clarity. The hierarchical structure of the global PESM is preserved when deriving the local PESM diagrams. The nested local PESM and local policies for the `assisted caller` subcomposite role are derived from the `aided Registration` PESM and policies by applying the transformation rules recursively. Similarly, the nested local PESM and local policies for the `consulter` subcomposite role are derived from the `consultation` PESM and policies.

At `PDP1_P`, each of the two policy rules trigger `PatientT` as initiating role and therefore it is required that role request messages are sent to the `TeleSvcDsk` and `ReceptionistT` composite roles.

The local PESM diagram for the `DoctorT` composite role is shown in Figure 5.14, and the local PESM diagram for the `ReceptionistT` composite role is shown in Figure 5.15. The specifications for the `TeleSvcDsk` composite role, not shown here, may be derived similarly from the global PESM diagrams in Figures 5.7 through 5.9.

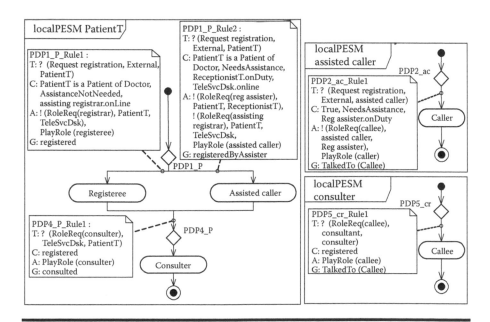

Figure 5.13   Local PESM diagram for the `PatientT` composite role.

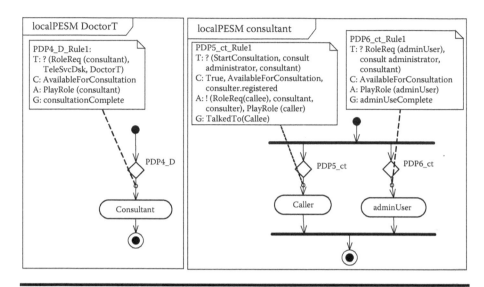

Figure 5.14   Local PESM diagram for the `DoctorT` composite role.

In the next sections, we present the general methodology for model-driven policy-enabled dynamic composition of services. The practical usage of the methodology is explained through its use in the tele-consultation service.

5.6 Methodology

The methodology follows the three-layered approach from Section 5.2.

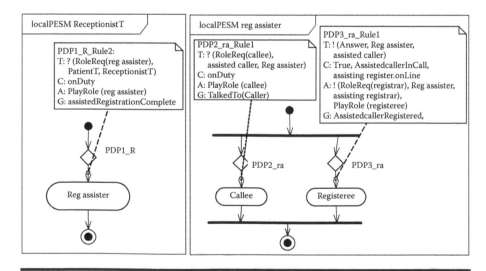

Figure 5.15   Local PESM diagram for the ReceptionistT composite role.

5.6.1 The Service Model Layer

In this section, we address the steps involved at the service model layer.

Step 1: Specify the service (or family of services)

Use a UML 2.x collaboration overview diagram to describe

1. The distributed parts involved in the composed service, given as one composite role for each part.
2. The set of collaboration uses involved in the composed service. Each collaboration use is specified in one of the following ways:
   a. Either, as an elementary collaboration, that is, a collaboration between exactly two elementary roles,
   b. Or, as a composed (sub-)service described recursively as a UML 2.x collaboration overview diagram in the same manner as described here.
3. The binding of the (elementary or composite) roles in the collaboration uses to the composite roles from Step 1.

The resulting diagram then provides a means of statically and structurally representing the binding of collaboration uses involved in the composed service to the actual service parts. Any of the collaboration uses involved in the composed service may be selected from a repository of reusable service specifications.

Step 1 applied to the tele-consultation service: The UML 2.x collaboration overview diagram produced for the tele-consultation service is shown above in Figure 5.2. In order to arrive at this collaboration overview, the following steps were taken:

1. The distributed parts involved in the composed service are `PatientT`, `ReceptionistT`, `DoctorT`, and `TeleSvcDsk`, given as composite roles in the collaboration overview diagram.
2. The tele-consultation service is composed of three collaboration uses: one for consultation, one for registration, and one for aided registration.
   a. Several elementary collaborations are used in the tele-consultation service. The `registration` collaboration is shown above in Figure 5.3. Here we also show the `call` collaboration in Figure 5.16. This elementary collaboration, which is used in the `aided Registration` and `consultation` collaborations, models a two-way multimedia communication between two parties such as a telephone call.
   b. The collaboration diagrams for the two composite subservices used in the tele-consultation service are `aided Registration`, shown above in Figure 5.4, and `consultation`, shown here in Figure 5.17. In addition to using the elementary `call` collaboration, `aided Registration` also uses the elementary `registration` collaboration and `consultation` uses the elementary `administrate` collaboration.

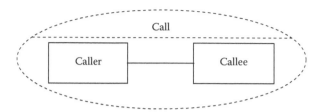

Figure 5.16   The `call` collaboration.

3. The binding of the (elementary or composite) roles in the collaboration uses to the composite roles, `PatientT`, `ReceptionistT`, `DoctorT`, and `TeleSvcDsk` has been explained above in Section 5.4.

Step 2: Specify the behavior of each elementary collaboration

For each of the elementary collaborations used in the UML 2.x service specification in Step 1, if collaboration is not already specified and stored in the repository of reusable service specifications, specify the visible interface behavior as a UML 2.x sequence diagram. For ensuring that the behavior associated with the elementary collaborations can execute correctly during dynamic service execution in the run-time system, the visible interface behavior between the two elementary roles should be validated with respect to safety and liveness properties. This may be done using static analysis, for example, by checking the type compatibility of signals exchanged, or using dynamic analysis, that is, taking into consideration the order of events occurring during dynamic execution [2,11,12].

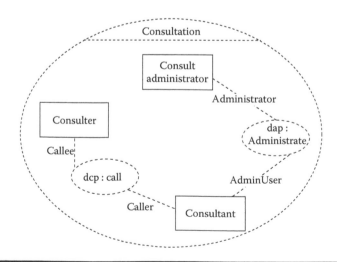

Figure 5.17   The `consultation` collaboration.

Step 2 applied to the tele-consultation service: We illustrate this step by specifying the UML 2.x sequence diagram for the `call` collaboration from Step 1. This sequence diagram is shown in Figure 5.18. Here, we use the STAIRS [15,16] operator `xalt` to express that each of the alternatives are mandatory in the final implementation. We also use UML state invariants declared on the lifelines to express role goal assertions for each of the roles, for example, `Caller.TalkedTo(Callee)` and `Callee.TalkedTo(Caller)`, and UML continuation labels, for example, `Connected`, to express collaboration goal assertions as in [2].

For examples demonstrating validation of the visible interface behavior between the two elementary roles in the call collaboration with respect to safety and liveness properties, we refer to [2] which uses a very similar call specification to demonstrate the validation techniques.

Step 3: Specify the global PESM diagram

Use a global PESM diagram together with a global composition policy to describe the choreography of the behavior associated with the UML 2.x collaboration overview diagram from Step 1 using

1. A global PESM diagram for describing the ordering of the collaboration uses.
2. A set of global policy rules for each PDP in the global PESM diagram, using the precise syntax given in Rossebø and Runde [4]. Together, these policy rules constitute the composition policy for the composed service.

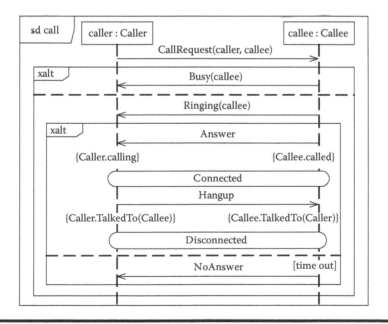

Figure 5.18   UML 2.x sequence diagram for the `call` collaboration.

Step 3 applied to the tele-consultation service: The global PESM diagram for the tele-consultation service is described in this step along with the global PESM diagrams for each of the composed subservices, `aided Registration` and `consultation`, used in the tele-consultation service. The PESM diagrams and policy rules have been explained in Figures 5.7 through 5.9 of Section 5.4.

5.6.2 Methodology: The Design Model Layer

At the design model layer, the overall service behavior described at the service model layer is broken down for each composite role separately. In this section, we address the methodological steps performed in order to achieve this.

Step 4: Create elementary role state machines

For each elementary collaboration not taken from the repository but specified in Step 2, create one elementary role state machine for each of the two roles involved in the collaboration. The transformation from sequence diagrams to elementary role state machines may be performed manually or automatically as described in Whittle and Schumann [17] and Seehusen and Stølen [18].

Step 4 applied to the tele-consultation service: As examples, Figure 5.19 shows the elementary role state machines for the `caller` and `callee` roles involved in the call collaboration. The elementary role state machine for `caller` is generated from the `caller` lifeline of the call sequence diagram in Figure 5.18, so that each message sent/received in the sequence diagram results in a send/receive signal in the state machine, and the three mandatory alternatives in the sequence diagram is reflected by three alternative paths in the state machine. Similarly, the elementary role state machine for `callee` is generated from the `callee` lifeline in the sequence diagram.

Step 5: Create the local PESM diagrams

Based on the transformation rules specified in Rossebø and Runde [4]:

1. From the global PESM diagram, create one local PESM diagram for each composite role involved in the service (as given in Step 1).
2. From the global policy rules, create the local policy rules to be associated with each PDP in the local PESM diagrams.

Again, this step should be applied recursively to each composite service specified as part of the composed service.

Step 5 applied to the tele-consultation service: We illustrate this step by using the transformation rules specified in Rossebø and Runde [4] to derive the local PESM diagram and local policy rules for the `TeleSvcDsk` composite role.

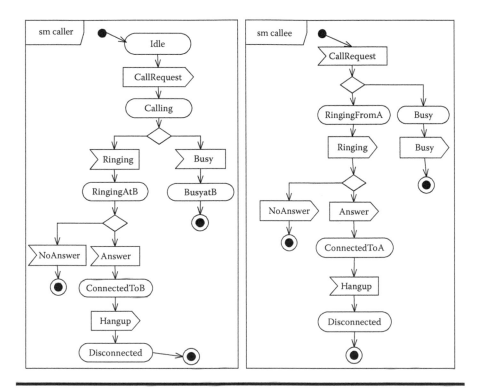

Figure 5.19 Role state machines for the `caller` and `callee` roles. (With kind permission from Springer Science+Business Media: J. E. Y. Rossebø and R. K. Runde. Specifying service composition using UML 2.0 and composition policies. In *MoDELS 2008*, Vol. 5301, LNCS, pp. 520–536. Springer, Berlin, 2008.)

As shown in Figure 5.20, the hierarchical structure of the tele-consultation global PESM diagram is preserved. The right side of Figure 5.20 shows the `assisting registrar` and `consult administrator` composite roles, derived from the `aided Registration` and `consultation` PESMs, respectively. As for the global PESM diagram, the policy rules to be associated with each PDP in the local PESM diagrams are illustrated as notes. The local PESM diagrams derived for the `PatientT`, `DoctorT`, and `ReceptionistT` composite roles are given in Figures 5.13 through 5.15, respectively, in Section 5.5.

5.6.3 Methodology: The Realization Layer

At the realization layer, implementations are generated from the design models at the design model layer. In this section, we address the methodological steps performed in order to achieve this.

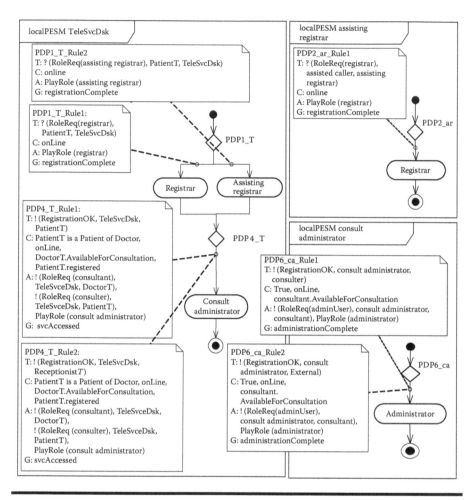

Figure 5.20   Local PESM diagram for the `TeleSvcDsk` composite role.

Step 6: Create and distribute service role implementations

For each elementary role state machine taken from the repository or created in Step 4:

1. From the elementary role state machine, generate the code if this is not already done. Such code generation may be performed automatically as described in Kraemer [19, 20].
2. Distribute the relevant code to each of the subscribing components of the system.

By subscribing components, we mean the system components interested in playing this role. Exactly which system components the roles should be distributed to is platform-specific information and not a part of the methodology described here.

Step 6 applied to the tele-consultation service: A schematic diagram showing the tele-consultation system deployment is given in Figure 5.21. The figure shows the distributed parts of the system: the patient TE located in the patient's home, the doctor TE located at the doctor's office, and the electronic tele-consultation service desk and the receptionist TE, each located in the hospital. Local policy management is deployed on each of the distributed parts of the system which are connected to the central policy management via the network. Although not described in this chapter, we have defined an architectural framework for a policy-based dynamic composition system; for information on this, we refer to [10].

For each elementary role state machine, the code is generated and distributed to each of the subscribing components. In this example, the code for the roles to be played by the `PatientT` composite role are distributed to the patient's terminal, the roles to be played by the `ReceptionistT` composite role are distributed to the receptionist's terminal, and the roles to be played by the `TeleSvcDsk` composite role are distributed to the tele-service desk terminal.

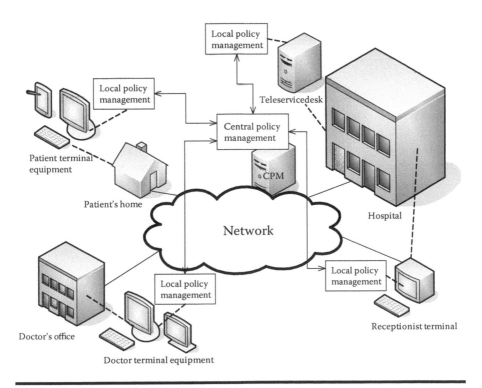

Figure 5.21   The policy managed tele-consultation deployment.

Step 7: Create and distribute PESM implementations

1. For each local PESM diagram from Step 6:
   a. Generate the local PESM description. This involves translating the control flow described by the PESM to a data format that can be used by the local policy management.
   b. Distribute the description to the subscribing components.
2. For each of the local policy rules from Step 6:
   a. Translate the policy rules to a data format that can be used by the local policy management. A policy parser is required to do this. The parser takes the policy rule file and converts the rule to the data format required.
   b. Distribute the rule to the subscribing components.
   c. Extract the trigger event and distribute it to the subscribing components.

Step 8: Perform dynamic composition of the service

Dynamic composition is now enabled in the runtime system. This involves

1. Dynamic service execution, composing the distributed roles as governed by policy and PESM descriptions
2. Dynamic composition, for example, involving the use of a user interface to change policy rules and include dynamic adaptation of services

How Steps 7 and 8 are actually done is platform specific and therefore we do not address this as part of this example. For a description of the policy-management system, and for results of investigations of the feasibility of the implementation of parts of the tele-consultation service and the some of the functionality of the policy management using the Arctis and Ramses tools suites [20], we refer to [10].

5.7 Using the Methodology to Change the Existing Service

In this section, we demonstrate that the methodology can be used to introduce changes to an existing service to allow for additions of new behaviors to the service dynamically.

5.7.1 Changing the Existing Service by Specialization

For any of the subcollaborations specified in the UML 2.x collaboration static structure specification of a service of Step 1, the subcollaboration may be specialized to allow for choice in dynamic composition. This is done during Step 1, or during a reiteration of Step 1. By introducing specializations as a reiteration of Step 1, the eight steps are then reiterated also.

For example, for the tele-consultation case, in the subcollaboration use `pcr` of the `call` collaboration, the designer may specify a choice between three specializations: `chat`, `voice`, or `video`. Depending on conditions specified in the composition policies, such as type of terminal support in the Patient's Home, one of these three specializations will be chosen during dynamic composition of the service. In Figure 5.22, the PESM diagram has been extended to include the specialization of the collaboration use `pcr` of the call collaboration, to a choice between the collaboration use `pcr1` of the `chat` collaboration, collaboration use `pcr2` of the `voice` collaboration, and collaboration use `pcr3` of the `video` collaboration. This is done using the tool, by replacing the single collaboration with the three specializations, and replacing the original policy with three new policy rules, one for each specialization. We require that the specialized policy rules fulfill all the obligations of the policy rule that is being replaced. The resulting, revised global PESM diagram for the `aidedRegistration` collaboration is shown in Figure 5.22.

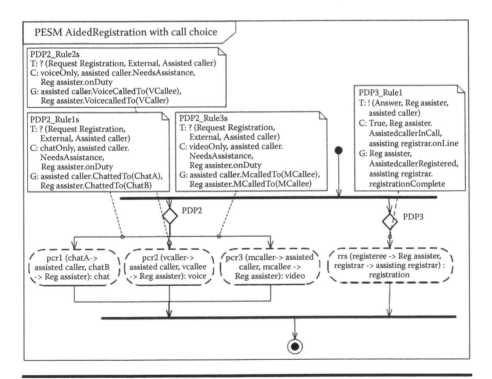

Figure 5.22   Global PESM diagram for the `aided Registration` collaboration with specializations of the `call` collaboration.

The three new rules, PRP2_Rule1s, PRP2_Rule2s, and PRP2_Rule3s shown in Figure 5.22 are all specializations of the policy rule PRP2_Rule1 identified by PDP2 of the global PESM diagram shown in Figure 5.7 of Section 5.4. The conditions of the condition part of the specializations imply the conditions of the condition part of the policy rule being replaced as each of the conditions of policy rule PRP2_Rule1 are also conditions of the three new rules, PRP2_Rule1s, PRP2_Rule2s, and PRP2_Rule3s. Similarly, the role goals of goal part of the specializations imply the role goals of the goal part of the policy rule being replaced.

In Figure 5.23, the specialization of the local PESM diagram for the **assisted caller** composite role is shown, as generated from the global PESM and composition policies using the transformation rules, during the reiteration of Step 5.

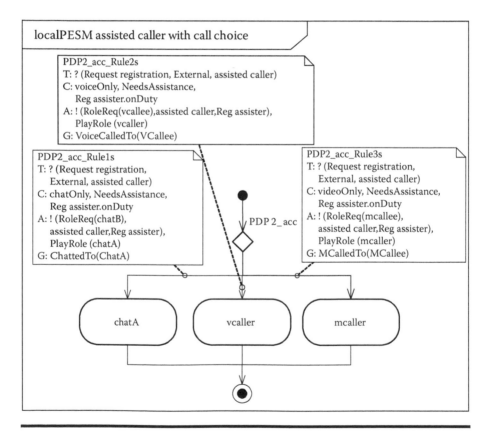

Figure 5.23 Local PESM diagram for the assisted caller composite role used in the aided Registration collaboration with specializations of the call collaboration.

5.7.2 Changing the Initial Specification by Adding Authentication

In this section, we explain how the methodology may be used to change an initial specification using the tele-consultation service as an example. Consider, for example, that after some period of use of the tele-consultation service, a security analysis recommends that authentication is required for direct registration to the electronic tele-consultation service desk. The analysis has determined that authentication is not required, for aided registration, as the patient speaks to the receptionist, and not directly to the electronic tele-consultation service desk. The specification of the service therefore needs to be adapted so that authentication is carried out before registration.

This is done by reiteration of Step 1, and the following steps. The collaboration diagram for the unilateral two-pass authentication (UTPA) is specified, and a collaboration use of UTPA is added to the tele-consultation collaboration as shown in Figure 5.24.

During a reiteration of Step 2, the sequence diagram for UTPA shown in Figure 5.25 is specified.

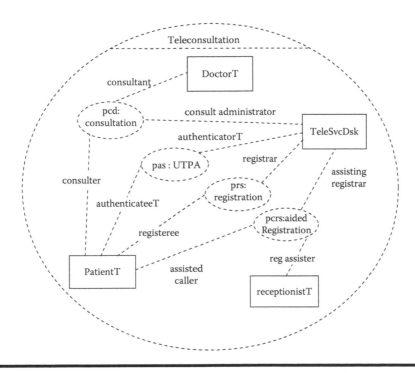

Figure 5.24 Collaboration overview diagram for the tele-consultation case with authentication added.

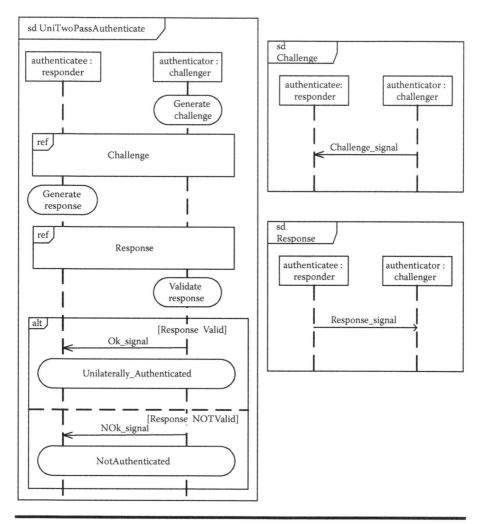

Figure 5.25 **UML 2.x sequence diagram for the UTPA collaboration.**

The global PESM diagram is modified during the reiteration of Step 3 by adding the pas collaboration use of UTPA as a node prior to the collaboration use prs of registration in the teleconsultation PESM with PDP inserted following the node, as shown in Figure 5.26. Additional policy rules are specified for PDP1, and policy rules are identified by the newly inserted PDP4. Existing policy rules may be changed, as exemplified by the changes to the condition part of PDP5_Rule1.

While we have chosen to exemplify the addition of authentication to the service using the UTPA collaboration, the approach and methodology allow for a choice of different authentication collaborations to be specified as described in Rossebø [5]. The UTPA collaboration may also be further specialized depending on protocol and algorithm to be used [5].

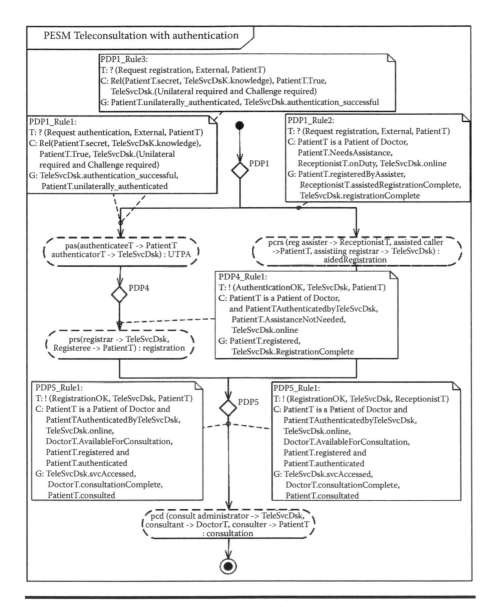

Figure 5.26 Global PESM for the tele-consultation service, with authentication added.

5.8 Discussion and Related Work

The collaboration-oriented approach presented in this chapter builds on the work on composing services presented in Sanders [2] and Sanders et al. [21]. The global PESM diagrams used in our approach are similar to the choreographies inspired by

UML 2.x activities presented in Castejón and Bræk [22] and further elaborated in Castejón et al. [23]. However, those choreographies describe static/nondynamic composition of collaborations, for which the choreography is hardcoded.

With our approach, dynamic coordination of the elementary collaborations and the conditions for choosing between alternative collaborations are governed by composition policies. Using policies for this purpose provides flexibility, in that the exact set of elementary collaborations involved in the composed service, and the ordering of these is determined first when the service is actually executed in the runtime system. Separating the specification of the policy rules from the collaborations, also allows for changes to be made at runtime by changing the policy rules without changing the service parts themselves. Our approach allows for dynamic adaptation by using policies to link together the separately specified collaborations. Adaptation is possible when triggered by an event or when policy rules are changed to allow for insertion of new behaviors.

In Castejón et al. [23] and von Bochmann [24], the authors focus on describing how UML collaborations can provide a framework for service modeling that enables service behaviors to be completely defined, and analyzed for problems. This provides a strong basis for analyzing problems that can arise in service composition that may lead to realizability problems such as race conditions. We have not focused on addressing such realizability problems, but rather have focused on the problem of enabling dynamic service composition.

In Castejón et al. [23] and von Bochmann [24], the authors have integrated the specification of which role plays the initiating role and which role plays the participating role in the specification of the subordinate collaborations in order to analyze behavior with respect to realizability, and to synthesize role behaviors. In Castejón et al. [23] and von Bochmann [24], there is no management layer present in the approach, and therefore the authors must make demands on the ordering of initiating and terminating roles in order to avoid realizability problems. Alternatively, the authors could have used synchronization messages. With our approach, however, the specification of which role plays the initiating role and which role plays the participating role is not integrated in the specification of the subordinate collaborations. The initiating party in our approach is the party that receives the trigger and has nothing to do with who makes the first move within the collaboration use. In our approach, the management layer takes care of coordination using role requests. This allows us to loosen the demands on the ordering of initiating and terminating roles in the specification of the subordinate collaborations, as the management layer takes care of the synchronization of roles.

We have chosen this approach because our focus is on enabling choice in dynamic service execution, including allowing for one collaboration to be replaced by another collaboration, regardless of which role internal to the collaboration sends the first signal. By decoupling the internal ordering of behavior from the choreography, we are able to enable such substitutions/choices between which collaboration should be executed next.

A UML 2.x activity is commonly used to model a WEB services choreography, capturing interactions from a global perspective [25,26]. The PESM is similar; however, it allows us to define the overall choreography in a dynamic way. Policies and rules are used in Chun [27] for discovering and finding compatibility to generate service flows in dynamic composition of Web services. In automatic WEB services composition, software agents perform composition based on predefined algorithms [28]. In rule-based service composition approaches, constructs are provided for specifying processes by means of sets of rules. Rules are used in the Web services composition approach to drive the process of service composition and govern service composition [29]. As in our approach, the rules are based on the so-called event–condition–action paradigm [13]. While there are similarities to our approach, we use policy rules to specify government of collaborative behavior in multi-initiative collaborative services as opposed to modeling business processes. In particular, the composition policies are designed to enable dynamic linking of collaborative behavior.

An area for which significant research has been done on composition is the area of aspect-oriented modeling (AOM) [30,31]. Both the AOM approach and our approach address cross-cutting concerns; however, we do this by modeling roles and using policy to bind them and define how they are to be composed, whereas in AOM, cross-cutting aspects are woven into so-called primary models using composition directives. Although composition directives and policies play similar roles in that they both define ordering and coordination of composition, the AOM approach is more prescriptive and static in nature. Our work is different from the AOM approach in that we use policy and dynamic role-binding. Policy rules declare what may be composed depending on trigger events and conditions, and the decision to compose may occur at runtime depending on the dynamic behavior state of the system, allowing for a choice between different behaviors. Furthermore, we do not use patterns, pointcuts, and advice to specify composition. However, we specify more explicitly where, under what conditions, and how composition should occur.

An approach to adaptive service composition with many similarities to our approach is presented in Funk [32]. Whereas they use a condition–action rule-based decision process, our policies are based on the event–condition–action paradigm. The decision system architecture is similar to our policy-based architecture, with a distributed management system. However, in Funk [32], the management is modeled more tightly integrated with the services than in our approach, where the policy management is specified separately from the individual services.

5.9 Conclusion

In this chapter, we have presented a three-layered modeling approach to dynamic service composition to address the need to be able to design and develop services so that they may be deployed and adapted dynamically. A methodology is provided explaining how to use the policy-enabled concept to model, design, and deploy dynamic composition of services. We have explained its use for a tele-consultation case.

We have presented an overview of the policy-based approach and we have explained how PESM diagrams and composition policies are used in the approach to specify the coordination of separately specified UML2.x collaborations. The PESM diagram allows us to emphasize the reusability of UML 2.x collaborations in service composition when combined with policy. We have demonstrated that the policy syntax can be used to specify composition policies for governing the choreography of elementary collaborations in dynamic composition of services. We have also demonstrated that local policy rules can be derived from global policy rules, and that local PESM diagrams can be derived from global PESM diagrams. We have explained how policy-enabled dynamic service composition can be deployed in the runtime system.

We provide methodology explaining how to use the policy-enabled concepts to model, design, and deploy dynamic composition of services. The methodology describes how to use the models to specify, design, and deploy services that may be adapted and managed more easily to suit the changing requirements of users and enable more rapid deployment of services.

Using policies and PESM diagrams makes the overall choreography flexible and adaptable, making this a promising approach to service composition. Policies are not just suitable for high-level abstraction/specification mechanisms, but also are a part of the deployment in the runtime system. The approach and methodology enables a large degree of variability in the services that the provider is able to offer to customers. Policy is also used to enable dynamic choice between services to be offered.

References

1. OMG. UML 2.2 Superstructure Specification, Formal/2009-02-02. Object Management Group, Needham, MA, USA, 2009.
2. R. Sanders. Collaborations, semantic interfaces and service goals: A way forward for service engineering. PhD thesis, NTNU, 2007.
3. J. E. Y. Rossebø and R. Bræk. Using composition policies to manage authentication and authorization patterns and services. In *The Third Int. Conf. on Availability, Reliability and Security (ARES 2008)*, pp. 597–603. IEEE Computer Society, 2008.
4. J. E. Y. Rossebø and R. K. Runde. Specifying service composition using UML 2.0 and composition policies. In *MoDELS 2008*, Vol. 5301, LNCS, pp. 520–536. Springer, Berlin, 2008.
5. J. E. Y. Rossebø. Dynamic composition of services—A model-based approach. PhD thesis, NTNU, 2009.
6. ITU-T. *Next-Generation Networks—Frameworks and Functional Architecture Models—General Principles and General Reference Model for Next Generation Networks*. ITU-T Recommendation Y.2011. Geneva, International Telecommunication Union, 2004.

7. J. E. Y. Rossebø and R. Bræk. Using composition patterns to manage authentication and authorization patterns and services. Avantel Technical Report 3/2008, NTNU, 2008.
8. R. Montanari, E. Lupu, and C. Stefanelli. Policy-based dynamic reconfiguration of mobile-code applications. *IEEE Computer*, 37(7):73–80, 2004.
9. J. Keeney. Completely unanticipated dynamic adaptation of software. PhD thesis, Trinity College, University of Dublin, 2004.
10. J. E. Y. Rossebø, R. Bræk, and R. K. Runde. Methodology for policy enabled dynamic composition of services. Technical Report AVANTEL 2/2009, Department of Telematics, Norwegian University of Science and Technology, 2009.
11. F. B. Engelhardtsen and A. Prinz. Application of stuck-free conformance to service-role composition. *The 5th Workshop on System Analysis and Modelling (SAM 2006)*, Kaiserslautern, Germany, June 2006.
12. J. Floch and R. Bræk. A compositional approach to service validation. In *Proc. 12th Int. SDL Forum (SDL 2005)*, pp. 281–297. Springer, Berlin, 2005.
13. J. Widom and S. Ceri (Eds). *Active Database Systems: Triggers and Rules for Advanced Database Processing*. Morgan Kaufmann, San Francisco, USA, 1996.
14. J. E. Y. Rossebø and R. K. Runde. Specifying service composition using UML 2.x and composition policies. Technical Report 378, Department of Informatics, University of Oslo, 2009.
15. Ø. Haugen and K. Stølen. STAIRS—Steps to analyze interactions with refinement semantics. In *UML 2003*, Vol. 2863, LNCS, pp. 388–402. Springer, Berlin, 2003.
16. R. K. Runde. STAIRS—Understanding and developing specifications expressed as UML interaction diagrams. PhD thesis, University of Oslo, 2007.
17. J. Whittle and J. Schumann. Generating statechart designs from scenarios. In *Proceedings, 22nd IEEE International Conference on Software Engineering (ICSE-00)*, pp. 314–323. Limerick, Ireland, 4–11 June 2000.
18. F. Seehusen and K. Stølen. Transformational approach to facilitate monitoring of high level policies. Technical Report All356, SINTEF Information and Communication Technologies, 2008.
19. F. A. Kraemer. Rapid service development for service frame. Masters thesis, University of Stuttgart, 2003.
20. F. A. Kraemer. Arctis and Ramses: Tool suites for rapid service engineering. In *Proc. of NIK-2007 (Norsk informatik konferanse)*. Oslo Norway, Tapir Akademisk Forlag, 2007.
21. R. T. Sanders, H. N. Castejón, F. Kraemer, and R. Bræk. Using UML 2.0 collaborations for compositional service specification. In *MoDELS 2005*, Vol. 3530, LNCS, pp. 460–475. Springer, Berlin, 2005.
22. H. N. Castejón and R. Bræk. Formalizing collaboration goal sequences for service choreography. In *26th IFIP WG 6.1 Int. Conf. on Formal Methods for Networked and Distributed Systems (FORTE'06)*, Vol. 4229, LNCS, pp. 275–291. Springer, Berlin, 2006.
23. H. N. Castejón, R. Bræk, and G. von Bochmann. Realizability of collaboration-based service specifications. In *Proc. of the Asia-Pacific Software Engineering Conference (APSEC 2007)*. IEEE Computer Society, 2007, Aichi.
24. G. von Bochmann. Deriving component designs from global requirements. In *Proc. of the Int. Workshop on Model Based Architecting and Construction of Embedded Systems (ACES-MB)*. Toulouse, France, 2008.

25. N. Kavantzas, D. Burdett, G. Ritzinger, T. Fletcher, Y. Lafon, and C. Barreto. *Web Services Choreography Description Language Version 1.0 (WSCDL)*. http://www.w3.org/TR/2005/CR-ws-cdl-10-20051109/, W3C (MIT, ERCIM, Keio), 2005.

26. G. Kramler, E. Kapsammer, W. Retschitzegger, and G. Kappel. Towards using UML 2 for modelling web service collaboration protocols. In *INTEROP-ESA'05*. Springer, Berlin, 2005.

27. S. A. Chun, V. Atluri, and N. R. Adam. Policy-based web service composition. In *Proc. Research Issues on Data Engineering (RIDE'04)*, pp. 85–92. IEEE Computer Society, Boston, MA, USA, 2004.

28. K. Fujii and T. Suda. Dynamic service composition using semantic information. In *Proc. Int. Conf. on Service Oriented Computing (ICSOC'04)*, pp. 39–48. ACM, New York, 2004.

29. J. Yang, M. P. Papazoglou, B. Orriens, and W.-J. v. Heuvel. A rule based approach to the service composition life-cycle. In *Proc. Web Information Systems Engineering (WISE'03)*, pp. 295–298. IEEE, Rome, Italy, 2003.

30. S. Clarke and R. J. Walker. Towards a standard design language for AOSD. In *Proc. Int. Conf. on Aspect-Oriented Software Development (AOSD'02)*, pp. 113–119. ACM, New York, 2002.

31. Y. R. Reddy, S. Ghosh, R. B. France, G. Straw, J. M. Bieman, N. McEachen, E. Song, and G. Georg. Directives for composing aspect-oriented design class models. In *Trans. on Aspect-Oriented Software Development I*, Vol. 3880, *LNCS*, pp. 75–105. Springer, Berlin, 2006.

32. C. Funk, A. Schultheis, C. Linnhoff-Popien, J. Mitic, and C. Kuhmunch. Adaptation of composite services in pervasive computing environments. In *Proc. of the IEEE Int. Conf. on Pervasive Services (ICPS'07)*, pp. 242–249. IEEE Computer Society, Istanbul, Turkey, 2007.

Chapter 6

Overview of Cognitive NGSDP Model: An Intelligent System in View of APIS (Applications, Performance, Intelligence, and Security)

Yong Zheng, Han-hua Lu, Ya-shi Wang, Li-juan Min, Shun-yi Zhang, and Yan-fei Sun

Contents

6.1 Overview

With the emerging developments of the Next-Generation Network (NGN) [1], IP Multimedia Subsystem (IMS) [2] is supposed to be the leading telecom solution to meet requirements of various multimedia telecom services and solve integration problems. The full implement of IMS is still a long way to go, while Service Delivery Platform (SDP) is considered as a forerunner to IMS. Therefore, IMS integrated with SDP is the directing orientation for telecom multimedia markets.

Actually, multimedia markets and telecom services vary from operator to operator. Several telecom operators and solution providers, such as Ericsson, Atos Origin, IBM, Huawei, and so on, have proposed their own SDP vista for the future development. However, different telecom operators almost agree on the same requirements of architecture. In this chapter, the third SDP architecture proposed by Moriana Model [3] is cited as an example for further explanation our cognitive SDP model, especially in view of performance and intelligence. The whole model can be depicted by Figure 6.1.

In this chapter, the Next-Generation Service Delivery Platform (NGSDP) [4] and the Cognitive NGSDP model [5–6] are introduced based on this model. Specifically, requirements and characteristics of Cognitive NGSDP are mainly

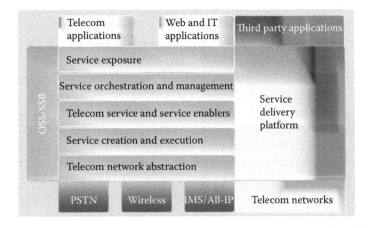

Figure 6.1 The third SDP architecture Moriana Model. (From The Moriana Group. *SDP 2.0: Service Delivery Platforms in the Web 2.0 Era.* **Moriana Group, 2008, http://www.morianagroup.com. With permission.)**

discussed in view of APIS (Applications, Performance, Intelligence, and Security), especially in view of performance and intelligence.

6.1.1 Next-Generation Service Delivery Platform

New requirements and new technologies come nearer with the rapid development of NGN and Web 2.0 [7] technologies. Accordingly, there are a new set of challenges the multimedia marketplace should face to address. Some of those challenges are indicated by Ericsson [8] as follows.

How can I launch new services faster? How can I attract consumers and create stickiness to increase revenues and reduce churn? How can I address partner to drive new streams of revenue monetizing on my assets? Or, put differently, how do I create an attractive proposition to engage a thriving developers' community to capitalize on the "long tail?"

How can I achieve reuse of common functionalities, like integration to network and business information capabilities, managing the multitude of services, partners, devices, customer segments and subscription and pricing models? How can I keep up with the quickly adopting new market and business situations? Last but not the least—how do I ensure world-class user experience?

Based on all those challenges, Web 2.0 technologies come into our sight. Our group defined NGSDP as a new service delivery platform based on SOA [9] and integrated with IMS in paper [4]. Also, with the development of semantic Web, a blueprint of new SDP which enables various device supports and allows personalized services within an open API platform is depicted by paper [6].

6.1.2 Requirements in View of APIS

Now that NDSDP is supposed to satisfy the increasing complicated telecom operations, what are its requirements and characteristics? This chapter aims to depict a bird view of these requirements and characteristics in view of APIS, while a brief introduction is indicated in this section.

The principle of APIS is proposed by Yong Zheng. Information technologies has benefited global development of information superhighway, especially information systems and service systems in various application areas, such as transportation, ticket business, stock market, government business, and so on. Telecom area is not beyond exceptions, including CRM, ERP as well as SDP. When it comes to research on these systems, APIS is a potential principle along with the public requirements and developments. Specifically, APIS can be described as follows:

- *Applications:* The most elemental one is to meet basic requirements of computing and informationizing. As a case in point, both computer architectures, such as CPU and GPU, and software systems, such as CRM and ERP, belong to these kinds of applications.
- *Performance:* The second one is how to evaluate and improve performance of these applications in order to cut down costs and best utilize and optimize applications to achieve diverse goals. These technologies can be drawn from parallel computing and QoS guarantees.
- *Intelligence:* The next one is how to make these applications smarter and more intelligent to operate flexibly and autonomously without unnecessary manual interventions. Artificial intelligence is obviously one of these representative research branches.
- *Security:* The last one is how to protect and guarantee normal operations of all these applications under specific defined rules. Areas of cryptology and data security are mainly oriented to this aspect.

More specifically, let us take NGSDP as an example. Then we can catch a sight of brief requirements of NGSDP in view of APIS:

The first and foremost, when it comes to application aspect, NGSDP is supposed to satisfy the basic requirements of SDP, including a horizontal platform, multimedia services supports, flexible service delivery, Web 2.0 integrations, and so forth. All these basic functions are the corresponding requirements for NGSDP in view of application.

Cognitive NGSDP was proposed by our group in view of aspects of performance and intelligence. Regarding actual development, SDP can be deemed as a distributed system which involved with distributed networks and software engineering. Therefore, performance problems brought by networks, software architecture, distributed middleware, and other minor aspects should be taken into consideration. Cognitive NGSDP is a NGSDP model with cognitive functions and characteristics [5–6] which includes self-awareness, self-organization, and even autonomy. All these self-functions are supposed to be the requirements for NGSDP in view of performance and intelligence. Besides, due to the description of NGSDP blueprint, how to flexibly support various user devices and how to provide an open API platform to enable the third-party development. Personalized services can be also deemed as requirements in view of intelligence.

Involving with distributed telecom networks, it is no denying that security is another aspect which cannot be ignored. Guarantees of both network security and software architecture should be taken account into basic requirements for NGSDP. Security problems along with Web service technologies are also increasing debated in actual software developments.

6.1.3 Brief Introduction of this Chapter

Requirements and characteristics of NGSDP in view of APIS are significantly introduced in this chapter. However, I entitled this chapter as "Overview of Cognitive NGSDP Model," which means that introduction of performance and intelligence aspects are emphasized to indicate in this chapter, while contents regarding application and security are relatively general descriptions.

- *Applications:* Indeed, application aspect spans various minor items. When it comes to NGSDP—a telecom system, we pay more attention on business services and operations, telecom frameworks, software engineering, and other aspects.
- *Performance and Intelligence:* Cognitive NGSDP is discussed in accordance with the consideration of aspects of both performance and intelligence. Cognitive Science and Cognitive Networks are introduced to further describe cognitive functions. Design principles and feasible solutions long with relevant theories and technologies are mainly discussed then for Cognitive NGSDP.
- *Security:* Due to that both distributed telecom networks and software engineering should be considered for security, they are described in this section for NGSDP.

Based on these introductions, we hope a whole clear impression on NGSDP, especially Cognitive NGSDP, can be made by our readers. Also, we hope our research on Cognitive NGSDP will contribute to further improvement in view of performance and intelligence for telecom operators and service providers.

6.2 Applications

Regarding basic requirements of NGSDP, different research groups, telecom operators and solution providers provide corresponding individual details. However, all these individual details can also represent the same vista of NGSDP. As indicated in paper [10], Atos Origin points out that a comprehensive and future-proof SDP supports all of the following:

- Off-the-shelf services. A broad catalogue of off-the-shelf services should be available that could be leveraged by the operator to cover most of its needs from the outset. For example, rich suite of services for both residential and enterprise segments contributing a new revenue stream, attracting new subscribers and increasing usage by existing subscribers. Next-Generation voice services (e.g., VoIP, Centrex, Unified Communications, etc.) should be supported.
- Extensive Protocol Support. Required in Convergent platforms, with the capacity to support traditional protocols—IN, MAP, SMS (SMPP, UCP), MMS, location, and so on—and new or pure IP protocols and standards such as SIP, Presence, Web Services, LDAP, CTI APIs, and so on, that is required for the development of different services. Convergent network adaptation supports hybrid networks allowing a smooth transition from IN services to rich, blended services while the network is evolved.
- Service Brokering for the combination of both internal and external services. Legacy services may have to be invoked in combination with new services in order to avoid their total replacement. For example, prepaid charging, number translation, MultiSIM, and so forth.
- This is used in transition processes to mitigate the impact of migration issues.
- Convergent charging. Should enable prepaid and postpaid charging and cost control, and should be available for any service. Services should be accessible from either the CS or PS network without the need to maintain two parallel service layer infrastructures.
- Web 2.0 integrations, SOA, mash-ups, revenue sharing, gaming, and so on. A Network API exposure layer should provide a robust interface with third parties while maintaining full system integrity.

6.2.1 Overview

Among APIS, the application aspect means basic and direct requirements of NGSDP, while performance and intelligence are advanced ones, and security is essential consideration. From the perspective of this chapter, when it comes to application, four aspects should be taken into consideration.

- *Telecom Services and Operations:* The ultimate goal of NGSDP is to help improve business services and further serve for better business operations and Quality of Experience (QoE) [11–12]. Detailed contents include IMS integrations and applications [13–17], telecom network infrastructures [18–19], rapid service creation environment [20–21], and actual converged applications, such as integration with Internet [22], supply chain management [23].

- *Telecom Frameworks:* It is well known that telecom services are always complicated and connected with each other, while telecom frameworks play a crucial role in telecom applications and operations. NGSDP should also better integrate with these frameworks, such as OSS/BSS [24] and Common Business Process Model (CBPM) in eTOM [25].
- *Software Engineering:* The most striking characteristic of NGSDP is to integrate with new Web technologies. Regarding software engineering, emerging Web technologies, such as Web 2.0, Web Service, and SOA [26–33], and actual implement approaches [34] combined with analysis methods [35] are discussed to make it clear.
- *Others:* Other aspects, such as Service Delivery Open Platform (SDOP) [36], user profile management [37], personalizable service discovery [38] and dynamic composition of personal network services [39] are introduced as supporting application requirements.

6.2.2 Telecom Services and Operations

SDP is supposed to serve for better telecom services and operations, especially oriented to multimedia markets. The first and foremost mission is to best integrate with IMS.

Regarding IMS integration, Ref. [13] suggests the use of SIP-based Micro Service Orchestration and Web service bus to seamlessly integrate IMS, WS, and the underlying services, while Ref. [14] focuses more on CPM (Converged IP Messaging)-IMS integration which enables implementers of CPM to reuse several existing platform modules and also minimizes the costs and maximizes the efficiencies. No matter what kind of approaches for integration, the basic requirements are obvious: First and foremost, integration should be implemented to work together for better multimedia services. Besides, best efforts should be made to reuse existing modules and components without unnecessary or repeated constructions. Additionally, how to maximize benefits and minimize disadvantages should be planned and realized.

Regarding costs, network infrastructures cannot be neglected. Toward NGN future, as Ref. [18] indicated, a strategy for evolving smoothly from modern networks to new network structure is essential in order to minimize the required investment during this transition phase.

Additionally, there is also one core part in the SDP model—Service Creation Environment (SCE) which coincides with the service creation and execution layer in Figure 6.1. Both Refs. [20] and [21] recommend establishing a template-based rapid SCE for SDP. Actually it aims to shorten and simplify the whole service creation lifecycle and cut down costs, especially for increasing new value-added telecom services requirements. From business perspective, the requirements of rapidness, flexibility, and robustness are significantly in need.

The most interesting and exciting point in NGSDP is that assorted technologies facilitate the third parties to offer various services more flexibly. Therefore, telecom-enterprise—Internet-combined services are involved in SDP. Reference [22] pointed out the requirements for SDP from the viewpoint of network operators, including time period shortening for application development, multivendor product supports, customization, and so on.

Reference [23] examines how utilization of SDP technology can potentially address some of the key issues facing the existing retail supply chain management. It reminds of another requirement of SDP, that is, SDP is supposed to help address existing business operation problems in addition to offering new services and functions. In other words, SDP is also expected to help fix current unhealthy architecture or existing issues.

From the perspective of telecom services and operations, SDP should be examined from a general view of telecom development tendency, not only for future evolutions, but also existing integrations and disadvantages.

6.2.3 Telecom Frameworks

Telecom frameworks undoubtedly play an important role in telecom operations. It usually involves with telecom architecture, various services, business process model, and so on. The actual designation, deployment, and implement of telecom systems are supposed to keep pace with telecom frameworks which guide a progress vista for telecom markets. Regarding construction of IMS and SDP, it should also be one of those considerations, especially when it comes to telecom modeling.

Reference [24] discussed "IMS/SDP ready" OSS/BSS as an example to use NGOSS principles in today's OSS/BSS projects. Several requirements indicated in the paper are proposed as follows [24]:

- Taking into account or interface all IMS/SDP architecture layers
- Manage multiservice ability
- Process complete order statuses from Customer Self-Service Portal/CRM to network
- Create and deploy services quicker than ever before
- Model Class of Service (CoS) at bandwidth level both during provisioning and assurance in order to manage IP capacity commercially
- Support E2E service-specific SLA/QoS guarantee and monitoring in accordance to the CoS
- Be inventory centric
- Use the NGOSS shared information data model
- Support flexibility and scalability
- Compliant with regulatory requirements such as emergency communication, security, and lawful interception
- Support integration or interoperability

Another example is IBM Common Business Process Model (CBPM). The horizontal SDP framework (metaSDP) provided by SDP transformation solution can work in concert with existing SDP silos. Thus, it can contribute to consolidating existing SDP and migrating toward complete converged SDP eventually. Reference [25] introduces the metaSDP CBPM which is developed by IBM China Research Lab in reference to domain engineering methods. According to the paper, the model refers to telecommunication industry standard TMForum and IBM telecom practice, including: the business process set of eTOM, the capacity set of telecom strategy process model, the architecture of IBM NG-SDP, and experiences of Global Business Services (GBS) in SDP real projects in China [25]. In other words, SDP integrated with telecom frameworks is put into practice via CBPM with its instantiation methods.

When it comes to SDP constructions, basic SDP requirements are in need to meet, while we should also view SDP from a higher perspective, such as long-vista guided by OSS/BSS or NGOSS principles and architectures. Due to the role of SDP in telecom operations, it should best integrate with telecom frameworks to fulfill and foster global telecom strategies.

6.2.4 Software Engineering

The most significant characteristic of NGSDP is to integrate with new emerging Web technologies, especially SOA and Web service. References [26–33] discussed several concerned converged Web technology problems which can be listed as follows.

- How to realize service requirements and business rules management in SOA architecture?
- How to better utilize Web service communication principles to address asynchronous, even-driven communication problems, and how to implement telecom communication models?
- How to deploy and manage lifecycle of telecom services by Web services? How to implement composite services considering SOA and Web service technologies?
- How to implement real-time Web service communications and also address relevant performance problems?
- How to combine distributed middleware with SOA and Web service principles to together serve for real practice?
- How to work together with current telecom network to perform best software performance?

All these questions are almost oriented to integration problems, including business integration, networks integration, and technologies integration. Before these actual implement of SDP, modeling and analyzing approaches should be taken into consideration first. Reference [34] proposed a model-driven approach which starts with establishing a model to describe business services from the perspective of business people. It aims to highlight the importance of modeling business specification of a service at all levels of granularity. Only in this way, business services can be designed, deployed, and delivered to maximize benefits from perspective from telecom operators and service providers. Additionally, analysis methods are also crucial. Reference [35] discussed a domain requirement model and relevant analysis method in software product line engineering for solution domains. It aims to utilize the relationships in the domain requirement model and follow the constraints in the configuration model, which enables domain requirements that are easy to navigate, update, communicate, and configure.

All in all, when it comes to software engineering, it involves with actual analysis, designation, modeling approaches, and technical implement. Any processes of this step will determine how the system will perform. Thus, both analysis sections and implement parts should be executed carefully and flexibly for further expanding and integrations.

6.2.5 Others

Some other proposals and requirements also pay to notice.

Reference [36] introduces the layered architecture of service delivery open platform (SDOP), and particularly describes the components and the functions of SDOP with its advantages, such as converged network, services and data resources, integration of a wide range of services ability to diverse and personalized business applications, service platform operation, and sharing of rich resources. It aims to integrate a wide range of basic service capacity, opening up a variety of network resources, on the third-party professional services providers/content providers and its business development to provide support and unified management [36].

With the rapid development of telecom market and various telecom services, the amount of customers is also increasing expanding. How to satisfy various customers, and how to better manage customer relationships are two key issues. Reference [37] describes a profiling engine that automatically learn user profiles (user preferences, interest domains and behaviors) by aggregating traces on converged SDP such as IPTV, mobile video, or IMS. User profile management and user preference surveys can further foster the development of business intelligence to better develop and deploy telecom services and service rules. Besides, as reference [37] indicated, such a profiling engine enables a wide range of personalized applications including targeted advertising, personalized portals, content recommenders, and social networking applications.

Assorted user device supports are one of the supports of SDP, while pervasive computing and systems arouse further requirements of SDP. Personalizable service discovery architecture in pervasive systems was proposed by Ref. [38]. It provides an approach to perform personalized service discovery based on a four-phase filtering process, including service type filtering, semantic filtering, contextual filtering, and privacy-based filtering, which can help find the best services matching the search criteria, so that QoE toward personalized services would be better satisfied based on this architecture.

Hybrid telecom services should be deployed and applied in various environments, while Personal Area Networks (PAN) and Wireless Sensor Networks (WSN) are not exceptions. Reference [39] addresses problems of integration of Personal Networks (PN) services into the future architecture of telecommunications and outlines a solution that coordinate services consisting of PN services and other regular IP-based services. It enables SN and PAN services integrations and also keeps interaction with IMS for new services creations, while expands telecom markets and allows IMS and SDP integrate with new added PAN and WSN services.

To sum up, these new applications are not strictly essential for NGSDP. However, they should also be considered when actually implementing NGSDP conceptual models. Both telecom strategies and technical details should be balanced for the construction of NGSDP. Through this way, NGSDP in view of applications can be fully viewed as a general required system.

6.3 Performance and Intelligence

Once basic application requirements are satisfied, attentions are supposed to be paid on performance and intelligence. Regarding performance, it aims to best utilize current resources and try to cut down costs and enhance efficiency at the same time. There are several influencing factors, including telecom networks, distributed middleware, communications, and interactions, involved in SDP's performance problems. When it comes to consideration of intelligence, a system should not only be robust, but also be intelligent, automatic and even autonomous without unnecessary manual interventions. In this section, Cognitive NGSDP proposed by our research group is introduced to address these consideration and problems.

6.3.1 Overview

Undeniably, it is human intelligence that separates us from animals; however, let us focus on the definition of artificial intelligence (AI). AI is the intelligence of machines and the branch of computer science which aims to create it. Textbooks define the field as "the study and design of intelligent agents," where an intelligent agent is a system that perceives its environment and takes actions which maximize its chances of success. John McCarthy, who coined the term in 1956, defines it as "the science and engineering of making intelligent machines" [40]. If the word "artificial" is focused, we can draw the conclusion that human-level intelligence is undoubtedly the most advanced intelligence in the real world, if human beings are considered as the smartest advanced animal in the world.

However, how can we extract pith of human intelligence and apply it into information technologies? Let us switch to our understanding of cognitive and adaptive functions. In my view, all human actions can be divided into two types: perception and decision making. Perception enables us to collect information and learn from our experience, as well as awareness of dynamic changes around us, while decision making is actually the crucial part of human living, because what we learn from experience and what we should do to act on the world are all reflected by decision making. Take ancient people for example; our ancients kept awareness of changes in environment and accumulated living experience from time to time. Therefore, based on all these experiences and knowledge, the ancients slowly learnt how to make best decisions to further act on the real world to make a living during that tough period. Another case in point, it is always the same case in how to establish a business. Questions like how to build companies, how to deliver and operate products or services, and how to attract customers or keep moving among competitors, can all be deemed as actions of decision making. If questions of "WHAT" can be deemed as actions of perception, the ones implying "HOW TO" can fully be considered as decision making. The more interesting is, people almost make decisions from our brains analyzing, while AI can contribute to decision making more scientifically, which can rely on traditional approaches, such as machine learning and data mining. The pith of it is how to collect information for perception and learn from useful experience or knowledge for our further decision making. Please notice that the definition of decision making here is not only a process of thinking and reasoning, but also a process of acting on the real world. We deem these powerful processes and functions as cognitive functions.

In this section, we are willing to introduce our core research contributions–Cognitive NGSDP. Cognitive Science and Cognitive Networks are introduced first to further explain and understand cognitive functions. After that, domain design patterns are discussed to better understand how we can make it work for practical applications. Then the whole Cognitive NGSDP, including the model, relevant theories and technologies, and feasible solutions, are finally described in this section. We hope our research on Cognitive NGSDP can contribute to actual performance and intelligence considerations for NGSDP.

6.3.2 Cognitive Functions

6.3.2.1 Cognitive Science

Cognitive intelligence comes from the discipline of cognitive science [41–45]. As indicated by Wikipedia, cognitive science is the study of mind or the study of thought. It embraces multiple research disciplines, including psychology, artificial intelligence, philosophy, neuroscience, linguistics, anthropology, sociology, and biology. It relies on varying scientific methodology (e.g., behavioral experimentation, computational simulations, neuro-imaging, statistical analyses), and spans many levels of analysis of the mind (from low-level learning and decision mechanisms to high-level logic and planning, from neural circuitry to modular brain organization, etc.). In the beginning, research on cognitive science focused more on human behavior learning, analyzing, and stimulating, while more recently, cognitive science can be applied to several application areas for practical usage, such as Language processing, perception and action, computation modeling, and so on.

At the same time, numerous emerging words, such as Cognitive Learning, Cognitive Modeling, Cognitive Networks, have been debated and researched to serve for real-world problems. From our group's perspective, cognitive intelligence can be considered as one of the most intelligent levels among human-level intelligence. And also, our previous research on Cognitive Networks contributed to our understanding and extraction of cognitive functions.

6.3.2.2 Cognitive Networks

Compared to a more intricate and complex development of network, traditional network is a level or layer-based static network. Increasing advanced requirements for network applications and higher QoE (Quality of Experience) from customers result in more difficulty in network operating and also monitoring and controlling of network behavior. It is how to adapt to various changes of dynamic network behavior and provide better QoE and QoS (Quality of Service) guarantees has turned out to be crux to be solved in order to switch roles of the current network to a distributed, dynamic, adaptive, and even autonomous one.

Based on such tendency, several concepts, constructions, and technologies of emerging networks are proposed and developed, such as Distributed P2P Network [46], Mobile Ad Hoc Network [47], Trustworthy Network [48], Cognitive Radio Network [49], and Ubiquitous Network [50]. The concept of Cognitive Networks is also produced under the background of this trend.

Perception is not a new concept which concentrates on the awareness of network business from the perspective of operators as well as customers. The specific technologies to implement the function of perception usually rely on data flow parsing and recognizing. Several basic level infrastructures and network architecture have been developed and put into practice, such as AON (Application-Oriented Networking) [51] from Cisco, and Acuity Architecture from Lucent.

Compared to perception, cognition can be considered as a concept of deeper perception, involving adaptive perception and relocation of assorted levels for network architecture and models. Cognitive Radio Network and Cognitive Packet Network [52] are two types of mature researches on cognition, while Cognitive Networks, especially based on wired, Internet and also NGN, is still under research and construction.

6.3.2.2.1 Differences and Relations between Cognition and Perception

Subtle are the differences between cognition and perception, and this chapter focuses on the deliberated ones to delve into the concept of Cognitive Networks.

From the perspective of processes and outcomes, perception focuses more on outcomes, while cognition more on an integrated process. Specifically, perception concentrates more on a series of approaches, including packet capturing, flow analyzing, resource distributing, business recognizing, and so forth, in order to more efficiently guarantee managements of resources, performance, SLA, and QoS. Compared to perception, cognition is a kind of deeper perception from assorted layers of the network model through approaches of flow or packet cognition, and so on. Cognition aims at not only awareness of business for management, but also resource distribution, business recognition, performance guarantees; thus, it enables operators more flexibly and smoothly to deploy, manage, detect, track, update, and amend the detailed network status. In the foreseen future, with the existence and development of Cognitive Networks, flows and protocols can be detailed tracked, parsed and classified, and various types of packets, including SIP packets, MPLS packets, RTP packets, and RTSP packets can be clearly recognized and classified to be routed, transferred or exchanged. Particular networks for multimedia transportation or QoS guarantees can also be logically divided from the whole network to be managed and controlled separately and flexibly.

From the perspective of depth of perception, cognition is a kind of deeper perception with a wider scope of perception and also a deeper understanding of network status. The original meaning of perception mainly focuses on the recognizing and parsing of packets, flows and protocols, in order to accurately recognize businesses and efficiently control the actual operations. However, cognition, as a deeper perception, is an integrated perception for all layers/levels of the network model in order to realize wholly controls and managements for operation, performance, security, guarantees, and also unexpected network faults or problems.

Furthermore, cognition is a process of feedback cycle. Reference [53] introduces such a feedback cycle which can be depicted by Figure 6.2. Based on certain environment, a series of sequenced processes of observing, orienting, deciding, and acting will ultimately act on the original environment to help it update its status information, those of which involves performance, security, guarantees, and other aspects of the whole system.

From the perspective of implementation, cognition is no longer a simple process of appending or expanding, instead, it is more likely to be a relocating, replanning, or redesignation for the network model. Scopes of cognition can be described by Table 6.1.

Comparing the differences between perception and cognition, their relation is more apparent. The concepts of cognition include ones of perception, but for a wider scope and deeper understanding. The connotation of cognition covers not

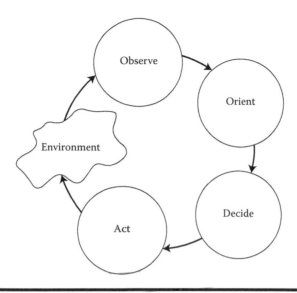

Figure 6.2   Feedback cycle of cognition. (Thomas, R.W., DaSilva, L.A., and MacKenzi, A.B. *First IEEE International Symposium on New Frontiers in Dynamic Spectrum Access Networks,* **© 2005. IEEE.)**

Table 6.1 Cognition Scopes

Layers	Scopes
Physical layer	Be self-aware and adaptive of topology, links and basic devices.
Network layer	Be self-aware and adaptive of data packets, flows and protocols.
Application layer	Be self-aware and adaptive of business services and network behaviors, and also related security aspects, such as access controls and intrusion detections.
Management layer	Be self-aware and adaptive of managements, including distribution, management, monitoring of network resources, management of network performances and securities, management and controls of SLA (Service Level Agreement), and also QoS guarantees.

only scopes of perception, but also ones of analyzing and decision making in order to meet the requirements of self-awareness, self-organization, and autonomy.

6.3.2.2.2 Definition of Cognitive Networks

There are no clear and unified definitions for Cognitive Networks. In accordance with scopes of Cognitive Networks analyzed above, further understanding and common meaning of Cognitive Networks are introduced in this section by investigating Cognitive Radio Networks and Cognitive Packet Networks as references.

The concept of Cognitive Radio Networks was proposed by Joseph Mitola III and Gerald Q. Maguire, Jr in 1999 [49]. It aims to collect and accumulate information through self-learning to be adaptive to changes in work environment and then make corresponding decisions or take actions. Specifically, approaches of spectrum shifting and sharing are applied to use spectrum resources more efficiently and also balance the resources distributions.

Compared to Cognitive Radio Networks, Cognitive Packet Networks [52] takes advantages of adaptive technologies to assign the best routing strategies according to user-defined QoS criterions and principles. The abilities of routing adjudging and decision making basically rely on the foundations of capabilities of self-awareness and self-learning.

From what has been discussed above, it can be concluded that self-awareness, self-learning, and adaptive capabilities are their significant common grounds; in other words, that is to say that these characteristics are also common grounds for cognitive functions. Based on these cognitive functions and functioning scopes, the definition of Cognitive Networks can be further introduced and discussed as follows.

Reference [53] defines Cognitive Networks as a kind of network including a process of cognition. Basically, Cognitive Networks can collect and be self-aware of information in certain environment, and automatically adapt to its changes by planning, decision making, and adjusting. Taking into account the end-to-end performance or efficiency, self-learning can be also carried out in the process of cognition in order to help collect more valuable information and also make further proper adjustments.

Reference [54] considers Cognitive Networks as a scheme to make up for the shortcomings of current networks and improve performances and status of networks through approaches of a series of active actions, including observing, learning, decision making, and also adjusting.

Reference [55] indicates positive agreements with views above and meanwhile, it also deems Cognitive Networks as a dynamic network which can be self-aware and adaptive for user requirements and dynamic changes in network environments. Cognitive Networks can also dynamically adjust network functions and even network topological structures; all those adjustments must strictly follow predefined business rules and corresponding network strategies.

Based on all information discussed above, our understandings of Cognitive Networks are proposed as follows: Cognitive Networks are a kind of intelligent and autonomous network, which can be self-aware and adaptive of information in network environments, including network resources, user behaviors, network problems, and so on. Furthermore, it enables network to adjust according to all these changes in network environments through self-functions and self-actions following predefined service levels and performance strategies. Consequently, network resources can be distributed and utilized more efficiently and flexibly, and the most important is that it is totally possible that Cognitive Networks can enable the whole processes to be nearly fully autonomous without unnecessary human interactions. The ultimate goals aim at flexible adjustments and improvements of network performance in order to offer more reliable guarantees of SLA, QoS, and QoE based on predefined business rules and service levels.

6.3.2.3 Cognitive Functions

With cleared indicated concept of Cognition Networks, its functions and requirements are inclined to be defined and located by a series of self-functions, such as self-awareness, self-organization, self-learning, self-optimizing, and so forth. According to definitions and understandings of this chapter, characteristics of Cognitive Networks are divided into basic and extended characteristics and basic ones can be expanded by extended ones. Three levels of characteristics can be depicted by Figure 6.3.

As depicted by Figure 6.3, characteristics of Cognitive Networks can be divided into three basic levels—self-aware, self-organization, and autonomy, all of which also include more detailed extended characteristics or functions. Detailed information on these three levels of characteristics is provided as follows:

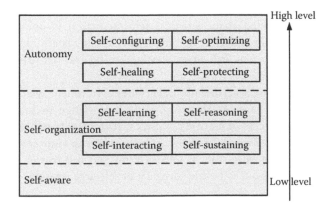

Figure 6.3 Characteristics of cognitive networks. (Zheng, Y., Lu, H.-H. and Sun, Y.-F. *The Second International Symposium on Intelligent Information Technology Application*, Vol. 3, pp. 137–140. © 2008 IEEE.)

6.3.2.3.1 Self-Awareness

Self-awareness is the basic characteristic of Cognitive Networks and also prerequisite and guarantee for characteristics of self-organization and self-autonomy.

Meanings of self-awareness can be drawn from several perspectives. From the perspective of the whole Cognitive Networks, self-awareness represents an ability of self-recognizing, modeling, controlling, and managing. From the perspective of functions of Cognitive Networks, self-awareness depicts a process of information collecting, filtering, reporting, and also feedbacking. While from the perspective of customers, self-awareness is a kind of intelligence which conceals detailed processes of perception and operation and makes customers likely to feel that it is self-adjusting and adaptive of operating.

6.3.2.3.2 Self-Organization

Self-Organization represents a kind of construction and organization mode from logical perspective of Cognitive Networks. It is more like a bridge connecting self-awareness and autonomy. There are two extended characteristics—self-learning, self-reasoning, self-interacting, and self-sustaining—included to fully complete the functions of self-organization.

Reference [55] indicates the meaning of self-organization from the perspective of biology. Actually, self-organization consists of a series of low-level components which are autonomous and operate independently following certain rules without unnecessary manual interventions. What is more important is that, these autonomous components can pose a positive or negative influence by changing their behaviors. Thus, the whole system can act as a biological system with self-feedback, self-modification, and self-progress. Meanwhile, the stability as well as the reliability of system is significantly improved and guaranteed.

Among those extended characteristics, self-interaction is one characteristic reflecting communication among components, while self-sustaining offers criteria for the stability of the system. Principles of self-learning and self-reasoning can be described by Figure 6.4.

In Figure 6.4, learning is carried out in certain environment system, which can also enrich knowledge base. At the same time, knowledge in knowledge base can also benefit the process of learning. Besides, reasoning based on knowledge, can directly act on the performance of the system under the same environment. To put it in a nutshell, the operation of self-learning and self-reasoning is another feedback cycle and turns out to be the core part of self-organization.

6.3.2.3.3 Autonomy

Autonomy is driven by business goals, service performance and efficiency, acting as an intelligent and automatic approach of guarantees of network performance and

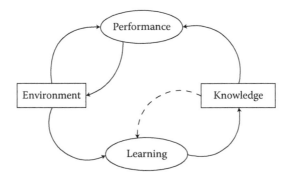

Figure 6.4 Functions cycle of self-learning and self-induction. (Zheng, Y., Lu, H.-H., and Sun, Y.-F. The 2nd International Symposium on Intelligent Information Technology Application, Vol. 3, pp. 137–140. © 2008 IEEE.)

security. Based on the characteristics of self-awareness and self-organization, it is established. There are four extended characteristics included in this part—self-configuring, self-optimizing, self-healing, and self-protecting.

Working principles of autonomy can be described as follows: Based on the characteristics of self-awareness and self-organization, when system status does not meet service requirements, decision making should be influenced by self-learning and self-reasoning. Then, the corresponding self-adjustments should be carried out to satisfy the requirements and meet end-to-end goals. All these actions should be autonomous, based on self-functions in this part.

Those four extended characteristics can be described from different perspectives. From the perspective of management, autonomy is a kind of ability of self-configuring and self-managing without unnecessary manual interventions. From the perspective of performance, autonomy is a kind of ability of self-optimizing which can help make decisions to optimize current status. From the perspective of security, autonomy is a kind of ability of self-protecting which can guarantee the whole processes of self-functions and can enable them to work normally preventing unexpected errors. From the perspective of stability and reliability, autonomy is a kind of ability of self-healing which is similar to immune systems of animals, enabling the whole system to be robust to face various small problems.

To sum up, characteristics of self-awareness, self-organization, and autonomy are layers correlated and also closely interacted. These three characteristics combined with corresponding minor self-functions can be deemed as cognitive functions.

6.3.3 Domain Design Pattern

All these cognitive functions can be implemented by a series of cognitive processes. The cognitive process can be summarized as a cognitive cycle—information collecting from context, knowledge formulating, and representation, machine learning and decision making, adjusting, and acting on a particular context for improvement. However, one system, such as SDP, is always a hybrid one involving with several infrastructures, including hardware, networks, protocols, and so forth. How to divide and manage system reasonably is supposed to be the most crucial problem when implementing.

6.3.3.1 Cognitive Domains

In order to collect and manage the whole system, all nodes' information should be collected and formulated as knowledge according to predefined rules. When it comes to large telecom networks, the amount of data is larger than it is imagined, not to mention real-time collecting and processing.

To reverse the situation, cognitive domains can be introduced to figure out the problem. According to network, physical and logical levels, the whole system, especially network environment, can be divided into several domain levels. Multilevel domains can be divided in accordance with the actual scale and overloads of telecom networks, while topper domain can be divided into minor domains as well. All these domains are cognitive domains taking charge of information collecting, nodes and domains communication, and information transferring and computing. Within the same lever, domains are similar to peer domains and are managed by topper domains. The highest level should contain a central domain server to deploy, dispatch, and manage communications and interactions among top cognitive domains.

6.3.3.2 Cognitive Domain Design Pattern

According to the principle of cognitive domain, the domain design pattern can be depicted by Figure 6.5. The whole system can be divided to several subcognitive domains, while individual cognitive domain is in charge of basic cognitive processes, mainly and especially involving with information collecting and transferring.

Individual cognitive domains deploy, dispatch, and manage their own nodes. Additionally, they are supposed to undertake basic computation functions, while topper domains will compute more to adjust and control lower domains, including network performance, resource distribution, dynamic best network routes, and so on.

Communications and interactions among same-level domains should be managed by domain servers. Every level contains at least one domain server to control and manage domain interactions, especially communications and decision making within the same-level domains.

6.3.4 Cognitive NGSDP

On the basis of cognitive functions and domain design pattern, our group integrated cognitive functions with NGSDP and then proposed Cognitive NGSDP model. In

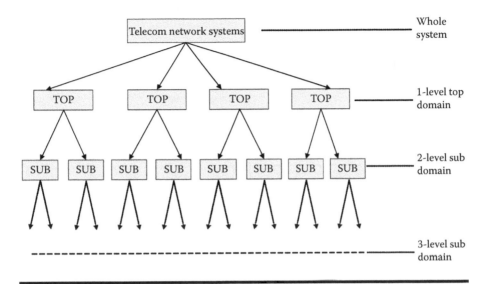

Figure 6.5 Cognitive domain design pattern.

this section, the model is introduced, while relevant theories, technologies, and feasible solutions are also discussed for further implementation of the network architecture of Cognitive NGSDP.

6.3.4.1 Model

The Cognitive NGSDP model was first proposed in Refs. [5,6], and it can be represented by Figure 6.6, which combines NGSDP model with cognitive functions.

As described in Ref. [5], the intelligent and cognitive SDP can be divided into three layers.

1. *Network layer:* The bottom layer is network layer which involves basic networks and the abstraction layer of NGSDP. The capability of self-awareness can be adopted here to recognize flow rates and collect related data. All these data will be the basis of advanced cognitive functions.
2. *Service/Business layer:* In this layer, services are created, deployed, and executed according to predefined logical rules. Service enablers are provided to control the processes of services and businesses. Functions of self-interacting, self-reasoning, or self-sustaining are needed to keep the whole process working correctly and orderly. The capability of self-organization can be added to enhance and protect the original functions.
3. *Management and Application layer:* Service orchestration, management, and exposure work here. All these jobs are expected to be more competent without unnecessary manual interventions. Some procedures should be processed

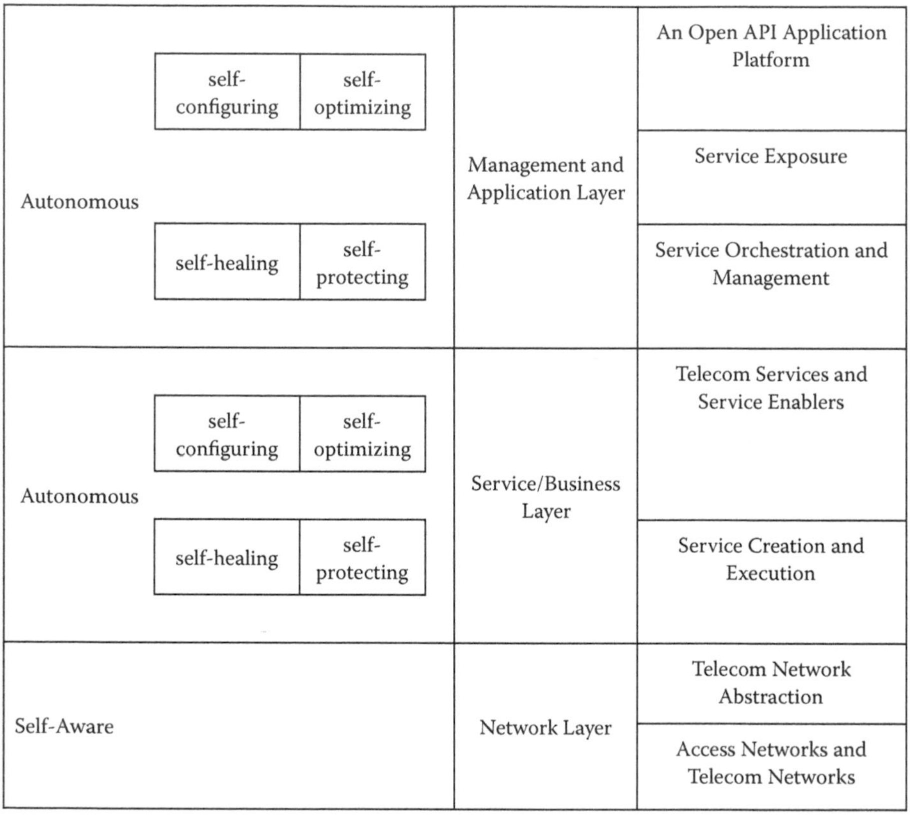

Autonomous	self-configuring	self-optimizing	Management and Application Layer	An Open API Application Platform
				Service Exposure
	self-healing	self-protecting		Service Orchestration and Management
Autonomous	self-configuring	self-optimizing	Service/Business Layer	Telecom Services and Service Enablers
	self-healing	self-protecting		Service Creation and Execution
Self-Aware			Network Layer	Telecom Network Abstraction
				Access Networks and Telecom Networks

Figure 6.6   An intelligent and cognitive service delivery platform model. (Zheng, Y., Lu, H.-H., and Sun, Y.-F. The 2nd International Symposium on Intelligent Information Technology Application, Vol. 3, pp. 137–140. © 2008 IEEE.)

autonomously and safely. The ability of being autonomous is the best choice to integrate with. For example, considering self-healing and self-protecting, the whole system would achieve higher fault-tolerance and be more robust with their supports. An Open API platform can also be deployed and working here, which is needed to be autonomous.

These three layers clearly integrate all layers of NGSDP with parts of cognitive functions. It does not mean that there is only one unique convergence pattern. There may be different combinations according to different requirements. But they have one characteristic in common, that is, all of them are intelligent and cognitive which will directly lead to an autonomous platform as a result. Unnecessary manual interventions will be replaced by cognitive analysis and intelligent processes. Meanwhile, the Open API application platform allows third parties and individuals to develop their applications according to flexible open rules, which will expand shares of service developers and providers.

6.3.4.2 Theories and Technologies

Cognitive NGSDP is just a conceptual model, while how to implement this model is what we must focus on. The core of Cognitive NGSDP is how to realize the cognitive functions. In this section, theories of perception and action are introduced to help implement cognitive processes, while machine learning and decision making contribute to actual design and implement approaches.

6.3.4.2.1 Perception and Action

The principles of perception and action root in research about human brain behaviors, especially human vision behaviors. It is said in Ref. [41] that, "Perception is the ability to take in information via the senses, and process it in some way. Vision and hearing are two dominant senses that allow us to perceive the environment. Some questions in the study of visual perception, for example, include: (1) How are we able to recognize objects? (2) Why do we perceive a continuous visual environment, even though we only see small bits of it at any one time? One tool for studying visual perception is by looking at how people process optical illusions. The image on the right of a Necker cube is an example of a bistable percept, that is, the cube can be interpreted as being oriented in two different directions."

Perceptions and actions form the core parts of cognitive processes. First and foremost, in order to further be adaptive and even autonomous, the requirement of being adaptive to dynamic contexts is essential without denying. Therefore, being adaptive means the system should own abilities of sensing dynamic changes in the context and taking actions to adjust and improve. This requirement totally coincides with the principles of perception and action.

As indicated, perception aims to sense dynamic changes, collect information, and even formulate understandable knowledge while action is in charge of intelligent learning, smart analyzing, and ultimately acting on the context.

6.3.4.2.2 Machine Learning and Decision Making

During perception and action, our research group focuses more on action. It not only involves taking actions, but also a series of preparatory works, including learning, analyzing, and decision making. It is almost similar to make commercial plans in industries. Before taking commercial actions, a company should make a decision, while before making a mature decision, a company should do surveys, learn from current or past successes and failures, and then scheme a suitable one for our own. Among numerous branches and minor application areas, all successes and failures can be deemed as experiences and then the whole process can be considered as a process of machine learning and decision making.

Machine learning is programming computers to optimize a performance criterion using example data or past experience. It also uses the theory of statistics in building mathematical models [56]. Therefore, machine learning can help to learn from context information and past experiences according to certain performance criterions and then further help improve performances.

Decision making can be regarded as an outcome of mental processes (cognitive process) leading to the selection of a course of action among several alternatives. Every decision-making process produces a final choice. The output can be an action or an opinion of choice [57]. Correspondingly, decision making can contribute to further actions, such as adjustment or improvement to implement adaptation to performance requirements.

The combination of machine learning and decision making will enable the whole system to perform intelligently to be adaptive to dynamic changes and finally help guarantee QoE and QoS, and also improve real-time performance. These are just the very requirements of Cognitive NGSDP taking into account performance and intelligence.

6.3.4.3 Feasible Solutions

Based on the requirements of cognitive functions and theories and technologies of perception and action as well as machine learning and decision making, our research group also proposed feasible solutions to implement the conceptual Cognitive NGSDP model. Regarding network architecture, we proposed the cognitive domain management scheme, while, in order to implement cognitive processes, active networks based on intelligent agents are introduced to figure out the problems.

6.3.4.3.1 Cognitive Domain Management

According to the principle of cognitive domain design pattern described in Section 6.3.3.2, the cognitive domain management scheme can be designed following its instructions.

From perspective of topology, it would be a kind of hybrid architecture combined with centralized style and distributed topology, which can be depicted as Figure 6.7. Domain servers are deployed to manage corresponding domain networks, while a central data server is prepared to collect information periodically. Dynamic network information is collected, transferred, and formatted to knowledge base for machine learning and decision making. Several network services and actions, such as QoS guarantees and route forwarding can be computed and then delivered smoothly. Taking into account the pressure of information transferring, the domain server is in charge of its own domain, while central data server serves for communications and

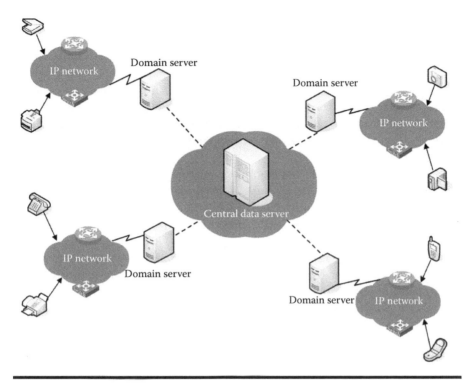

Figure 6.7   Topology of cognitive networks.

forwarding essential information among different domain servers. Intelligent mobile agents can be utilized to transfer information from end to end.

6.3.4.3.2 Active Networks and Mobile Agents

Active network is an approach to network architecture in which the switches of the network perform customized computations on the messages flowing through them. Those networks are active in the sense that nodes can perform computations on, and modify the packet contents [58]. Nodes in the active networking model differ from their passive counterparts in their ability to provide intermediate processing to the packets flowing through them, and hence offering these packets more services than the traditional store-and-forward services found in current routers [59].

Reference [60] deems mobile agents as programming entities which can act on networks following user-defined rules and principles. While Ref. [61] suggests that, "mobile agents are programs, typically written in a script language, which may be dispatched from a client computer and transported to a remote server computer for execution." Due to their mobility [62] and flexibility of mobile agents, they are widely used to solve assorted problems. According to Ref. [63], by using agents, network QoS parameters are controlled in an attempt to match the user's perceived QoE.

From information discussed above, active network can help collect information about network status and its nodes' own abilities of computations. Mobile agents can be flexibly written in a script language and also can be deployed in network environments to help act on networks.

Due to all these characteristics of active network and mobile agents described above, the combination of these two technologies has been researched to solve a multitude of problems.

Reference [59] focuses on the problem of mapping application sessions which require intermediate processing to available network resources and designs multi-agent-based system architecture for active network to meet these requirements.

According to Ref. [63], using agents, network QoS parameters are controlled in an attempt to match the user's perceived QoE. Thus, it develops an agent-based platform to map the quality of service to experience in conventional and active networks, which can enable QoS and QoE to be controlled and managed more flexibly and effectively.

To solve problems of network backward compatibility and standardization, Ref. [64] exploits the capability of mobile agent technology to construct active network-based architecture, and also presents an agent management/control mechanism to make it work well.

Reference [65] proposes a new kind of active network management model based on mobile agent to offer stronger management mechanism and strategy to make up for the shortcomings of SNMP Network and meet requirements more efficiently.

From the perspective of all these applications, the combination of these two technologies can help not only distribute network resources and guarantee QoS, but also manage network more flexibly and effectively. Furthermore, due to the abilities of information collection, mobility of mobile agents, the combination of these two technologies can help design a solution to realize the conceptual model of Cognitive Networks. The solution based on active network and mobile agents mainly contains three parts—active nodes, knowledge base, and mobile agents.

The solution model based on active network and mobile agents are described in Figure 6.8. The model is simplified and divided into four layers—physical layer, network layer, transport layer, and application layer. In physical layers, there are infrastructures, including basic and intelligent devices for networks. And in network layer, it is commonly designed or combined by active networks with essential active network nodes to collect and compute information, which can help realize the characteristics of self-awareness. Knowledge base is stored in the application layer to record useful information to help accumulate knowledge and make decisions. Mobile agents can travel through application layer, transport layer, and also network layer to help collect information, make decisions, and finally take action on the network. Characteristics of self-organization and autonomy can be fulfilled by interactions and operations among mobile agents. Due to

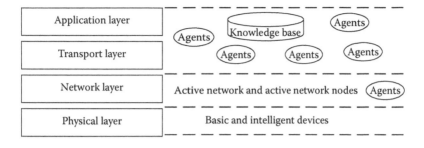

Figure 6.8 Solution model based on active network and mobile agents.

their mobility, operations related to organization and others can be flexibly defined and carried out by them.

All in all, as discussed above, cognitive domain management combined with active networks and mobile agents can help implement and construct the actual network architecture for Cognitive NGSDP, which directly provides a way to turn the conceptual model to reality. As a result, the cognitive process can be executed to exert cognitive functions on NGSDP.

6.4 Security

6.4.1 Overview

When a system is considered as a whole, security is one of the most essential issues which should be taken into account. Regarding NGSDP, there are also several aspects involved, including business processes, telecom networks, software engineering, and so on. In this section, business processes, telecom networks, and software engineering are discussed on the basis of introductions of published papers.

6.4.2 Business Processes

When it comes to business processes, a secured system should guarantee all services to be correctly deployed, managed, executed, and exceptions reported. What is still worth noticing is that, value-added applications provided by the third party should also be efficiently managed and protected to enable healthy telecom operations.

Reference [66] present an idea of charging validation in the service delivery platform to ensure correct execution and charging of complex third-party value-added applications by checking service charging records. Two methods, based on charging templates or service process model, are proposed to model application process and also automatically validate service charging records. SDP plays an important role in this process. It takes charge of service deployment, delivery, and management, as well as communications and negotiations between service providers and telecom operators. Also, in order to guarantee correct business processes, service execution, authentication and authorization, subscription management, and SLA control should also be managed by SDP.

This differs from common business security which focuses more on correct service operations. It emphasizes more on actual business collaboration taking into consideration the operator's guarantee of benefits. To maintain effective cooperation and also safeguard benefits from perspective of telecom operators is also a crucial aspect of security which should not be neglected.

6.4.3 Telecom Networks

Regarding network security, there are mainly two aspects—network robustness and protection from various attacks. The pith of Cognitive NGSDP is to implement adaption in order to further guarantee network robustness, while attacks detection and network protection should also be attached importance.

Reference [67] concerns Denial of Service (DoS) attacks involved in SIP standards, which should also be a consideration of IMS security. In order to protect the IMS network, the nonparametric Cumulative Sums (CUSUM) algorithm is proposed in this reference. Reference [68] pays attention on protecting IMS SDP from time independent attacks, including SQL injection and media flow attacks. All in all, both architecture modeling and network attacks should be considered when network security is concerned. Although research on network security has aroused attention, it is still a long way for us to proceed.

6.4.4 Software Engineering

Integration of Web technologies is one of the characteristics of NGSDP; meanwhile, it also produces new security concerns. As indicated in Ref. [69], Web service is regarded as the best candidate to design system interfaces. However, whether Web services technology can satisfy the telecom security requirements is still a question for telecom operators. WS-Policy, WS-Security, SOAP, and XML secure messages are all involved in Web service securities.

Additionally, Ref. [70] introduces a platform to figure out security problems. The Evros platform is built on three components, the Evros card, the Evros gateway, and the Evros management server. The Evros card is a laptop personal computer card with an embedded wireless wide area network modem, processor, battery, and nonvolatile memory. It independently establishes the Internet Protocol security (IPsec) tunnel that secures the remote access connection, and its WWAN interface makes the laptop reachable by the IT organization virtually anytime and anywhere. The Evros gateway is an enhanced secured access server that is deployed within the network perimeter. It hosts the enterprise end points of the IPsec tunnels and exploits the extended reachability of the laptop to improve the effectiveness of all remote management functions. The Evros management server is a software suite for the management of all Evros components [70].

6.5 Summary

IMS combined with SDP will lead to a new future for telecom operators and multimedia markets, while the actual constructions of NGSDP should take into consideration APIS—Applications, Performance, Intelligence, and Security. Cognitive NGSDP focuses more on performance and intelligence aspects, combining cognitive functions with NGSDP requirements. Principles of cognitive domain management and theories of cognitive science, machine learning, and decision making are utilized to contribute to the implementation of cognitive processes. Security problems, from business processes, telecom networks, and software engineering should all be considered to guarantee correct services execution, system security, and attacks prevention. With this overview of Cognitive NGSDP, we hope it can help understanding and actual progresses of SDP. And we believe SDP development and progress will lead to a new considerable future for both telecom industries and global customers.

Acknowledgments

Research on Cognitive NGSDP is supported by National Key Technology R&D Program in the 11th Five-Year Plan of China (No. 2007BAH17B04), National High-Tech Research and Development Plan (863) of China (No. 2009AA01Z212, 2009AA01Z202), Natural Science Foundation of Jiangsu Province (No. 2007BK603) and High-Tech Research Plan of Jiangsu Province (No. BG2007045). At the same time, the authors acknowledge the support of Fujian Fujitsu Communication Software Co., Ltd. (FFCS). With these supports, this study was successfully completed.

References

1. ITU. Definition of Next Generation Network. http://www.itu.int/ITU-T/study-groups/com13/ngn2004/working_definition.html
2. Agrawal, P., Yeh, J.-H., Chen, J.-C. and Zhang, T. IP multimedia subsystems in 3GPP and 3GPP2: Overview and scalability issues. *Communications Magazine*, IEEE, 2008, 46, 138–145.
3. The Moriana Group. *SDP 2.0: Service Delivery Platforms in the Web 2.0 Era*. Moriana Group, 2008, http://www.morianagroup.com
4. Lu, H.-H., Zheng, Y., and Sun, Y.-F. The next generation SDP architecture: Based on SOA and integrated with IMS. *The Second International Symposium on Intelligent Information Technology Application*, 2008, Vol. 3, pp. 141–145. Shanghai, China.

5. Zheng, Y., Lu, H.-H. and Sun, Y.-F. An intelligent and cognitive service delivery platform model. *The Second International Symposium on Intelligent Information Technology Application*, 2008, Vol. 3, pp. 137–140. Shanghai, China.

6. Zheng, Y., Lu, H.-H. and Sun, Y.-F. A cognitive SDP model: A telecom way to help social computing in communications and interactions. *The First International Symposium on Knowledge Acquisition and Modeling*, 2008, pp. 572–575. Wuhan, China.

7. O'Reilly, T. What Is Web 2.0? O'Reilly Network, http://oreilly.com/web2/archive/what-is-web-20.html, 2005.

8. Wilhelmsson, J. and Cegrell, C. *Towards the Multimedia Marketplace*. Ericsson White Paper, 2009.

9. Channabasavaiah, K., Holley, K., and Tuggle, E.M. Migrating to a service-oriented architecture. IBM Developer Works, 2003.

10. Atos Origin. A service delivery platform for the future. Atos Origin White Paper, 2009.

11. Patrick, A.S., Singer, J., Corrie, B., Noel, S., El Khatib, K., Emond, B., Zimmerman, T., and Marsh, S. A QoE sensitive architecture for advanced collaborative environments. *The First International Conference on Quality of Service in Heterogeneous Wired/Wireless Networks*, 2004, pp. 319–322. Dallas, Texas, USA.

12. Perkis, A., Munkeby, S., and Hillestad, O.I. A model for measuring Quality of Experience. *Proceedings of the Seventh Nordic Signal Processing Symposium*, 2006, pp. 198–201. Reykjavik, Iceland.

13. Shao, X., Chai, T.Y., Lee, T.K., Ngoh, L.H., Zhou, L. and Kirchberg, M. An integrated telecom and IT service delivery platform. *IEEE Asia-Pacific Services Computing Conference*, 2008, pp. 391–396. Yilan, Taiwan.

14. Saha, S. Deploying OMA converged IP messaging over IMS: Integration plan and architecture. *The Fourth International Conference on Wireless Communications, Networking and Mobile Computing*, 2008, pp. 1–6. Dalian, China.

15. Al-Hezmi, A., Friedrich, O., Arbanowski, S., and Magedanz, T. Requirements for an IMS-based quadruple play service architecture. *IEEE Network*, 2007, 21, 28–33.

16. O'Connell, J. Service delivery within an IMS environment. *IEEE Vehicular Technology Magazine*, 2007, 2, 12–19.

17. Magedanz, T. and Gouveia, F. IMS—The IP multimedia dystem as NGN service delivery platform, *e&i—elektrotechnik und informationstechnik*, Themenschwerpunkt: Vermittlungstechnik, e&i 123, Heft 7–8, 2006, pp. 271–282.

18. Kryvinska, N. and Strauss, C. Next generation service delivery networks: A strategic approach involving architectural planning and design, from a business perspective. *The Fifth Advanced International Conference on Telecommunications*, 2009, pp. 410–415. Venice, Italy.

19. Pavlovski, C.J. Service delivery platforms in practice [IP Multimedia Systems (IMS) Infrastructure and Services. *IEEE Communications Magazine*, 2007, 45, 114–121.

20. Jin, L., Pan, P., Ying, C., Liu, J. and Tian, Q. Rapid service creation environment for service delivery platform based on service templates. *IFIP/IEEE International Symposium on Integrated Network Management*, 2009, 117–120. Long Island, NY, USA.

21. Pan, P., Jin, L., Ying, C. and Liu, J.H. Template based rapid service creation environment for service delivery platform. *IEEE Network Operations and Management Symposium*, 2008, pp. 534–545. Salvador, Bahia.

22. Ohnishi, H., Yamato, Y., Kaneko, M., Moriya, T., Hirano, M., and Sunaga, H. Service delivery platform for Telecom–Enterprise–Internet combined services. *IEEE Global Telecommunications Conference*, 2007, pp. 108–112. Washington, DC, USA.

23. Hisatomi, M., Fukuda, K., Wilson, M., and Chujo, T. Application of service delivery platform for supply chain management. *IEEE Global Telecommunications Conference,* 2008, pp. 1–5. New Orleans, LO, USA.

24. Goestl, H. Using NGOSS Principles in today's OSS/BSS Projects NGOSS meets IMS/SDP. *The Twelfth International Telecommunications Network Strategy and Planning Symposium,* 2006, pp. 1–8. New Delhi, India.

25. Jin, L., Pan, P., Ying, C., Chen, X.Y. and Tian, Q.M. Common business process model for continuous SDP transformation. *IEEE Network Operations and Management Symposium,* 2008, pp. 546–559. Salvador, Bahia.

26. Carrero, M. Innovation for the Web 2.0 Era. *Computer,* 2009, 42, 96–98.

27. Callaway, R.D., Devetsikiotis, M., Viniotis, Y., and Rodriguez, A. An autonomic service delivery platform for service-oriented network environments. *IEEE International Conference on Communications,* 2008, pp. 327–331. Beijing, China.

28. Maes, S.H. Service delivery platforms as IT realization of OMA service environment: Service oriented architectures for telecommunications. *IEEE Wireless Communications and Networking Conference,* 2007, pp. 2883–2888. Kowloon, Hong Kong.

29. Auer, L., Strauss, C., Kryvinska, N., and Zinterhof, P. SOA as an effective tool for the flexible management of increased service heterogeneity in converged enterprise networks. *International Conference on Complex, Intelligent and Software Intensive Systems,* 2008, pp. 535–539. Barcelona, Spain.

30. Blum, N. and Magedanz, Th. Requirements and components of a SOA-based NGN service architecture. *E & I Elektrotechnik und Informationstechnik,* 125(7–8), 263–267.

31. Bo, C., Jie, G., Junliang, C., and Xiangwu, M. The communication model for real-time web service in telecom domain. *International Journal of Communication Systems,* 2009, 22, 773–787.

32. Bo, C., Xiaoyan, S., Xiangwu, M., and Junliang, C. Web services communication model in Telecom domain. *Ninth ACIS International Conference on Software Engineering, Artificial Intelligence, Networking, and Parallel/Distributed Computing,* 2008, pp. 747–752. Phuket, Thailand.

33. Sur, A., Skidmore, D., and Chakravarty, S. Web services based SOA for next generation telecom networks. *IEEE International Conference on Services Computing,* 2006, pp. 520–520. Chicago, IL, USA.

34. Nayak, N. and Nigam, A. Modeling business services for implementing on global business services delivery platforms. *The 9th IEEE International Conference on E-Commerce Technology and the Fourth IEEE International Conference on Enterprise Computing, E-Commerce, and E-Services,* 2007, pp. 577–583. Tokyo, Japan.

35. Tian, Q.M., Chen, X.Y., Jin, L., Pan, P., and Ying, C. Asset-based requirement analysis in telecom Service Delivery Platform domain. *IEEE Network Operations and Management Symposium,* 2008, pp. 815–818.

36. Hai-Hong, E., Mei-Na, S., Jun-De, S., Xiao-xiang, L., and Xiao-Qi, Z. A new service delivery open platform (SDOP) architecture. *IEEE International Symposium on IT in Medicine and Education,* 2009, pp. 404–409.

37. Aghasaryan, A., Betgé-Brezetz, S., Senot, C., and Toms, Y. A profiling engine for converged service delivery platforms. *Bell Labs Technical Journal,* 2008, 13(2), 93–103.

38. Frank, K., Suraci, V., and Mitic, J. Personalizable service discovery in pervasive systems. *The Fourth International Conference on Networking and Services,* 2008, pp. 182–187. Gosier.

39. Arbanowski, S., Lange, L., Magedanz, T., and Thiem, L. The dynamic composition of personal network services for service delivery platforms. *The Fourth IEEE International Conference on Circuits and Systems for Communications*, 2008, pp. 455–460. Shanghai, China.

40. Wikipedia. Artificial Intelligence. http://en.wikipedia.org/wiki/Artificial_intelligence

41. Wikipedia. Cognitive Science. http://en.wikipedia.org/wiki/Cognitive_science

42. Luger, G. *Cognitive Science: The Science of Intelligent Systems*. San Diego: Academic Press, 1994.

43. Port, R.F. and VanGelder, T. *Mind as Motion: Explorations in the Dynamics of Cognition*. Cambridge, MA: The MIT Press, 1995.

44. Bechtel, W. and Graham, G. (Eds.). *A Companion to Cognitive Science. Blackwell Companions to Philosophy*. Malden, MA: Blackwell Publishers, 1999.

45. Thagard, P. *Mind: Introduction to Cognitive Science*. Cambridge, MA: The MIT Press, 2005.

46. Chang, B.-J., Chen, C.-S., Liang, Y.-H., Lin, H.-D. A distributed P2P network based on increasing reliability and scalability for internet applications. *Proceedings of the 2006 International Conference on Wireless Communications and Mobile Computing*, 2006.

47. Badonnel, R., State, R., and Festor, O. Self-organized monitoring in Ad-Hoc networks. *Telecommunication Systems*, 2005, 30(1–3), 143–160.

48. Peng, X.-H. and Lin, C. Architecture of trustworthy networks. *Proceedings of the Second IEEE International Symposium on Dependable, Autonomic and Secure Computing*, 2006. Indianapolis, IN, USA.

49. Mitola III, J. and Maguire, Jr., G.Q. Cognitive radio: Making software radios more personal. *Personal Communications, IEEE*, 1999, 6(4), 13–18.

50. Nakata, M., Kushiro, N., Higuma, T., and Hibara, N. Ubiquitous network for building and home control with Ad-hoc wireless and plug-and-play mechanism. Ubicomp2005.

51. Bhuyan, L.N. Application oriented networking (AON): Adding intelligence to next-generation internet routers. *First International Conference on Wireless Algorithms, Systems, and Applications*, Xi'an, China, 2006. Berlin, Heidelberg: Springer, 2006, pp. 1–2.

52. Erol, G., Ricardo, L., Alfonso, M., and Zhiguang, X. Cognitive packet networks: QoS and performance. *Proceedings of the Tenth IEEE International Symposium on Modeling, Analysis, and Simulation of Computer and Telecommunications Systems*, 2002. Fort Worth, Texas, USA.

53. Thomas, R.W., DaSilva, L.A., and MacKenzi, A.B. Cognitive networks. *First IEEE International Symposium on New Frontiers in Dynamic Spectrum Access Networks*, 2005.

54. Thomas, R.W. *Cognitive Networks*. Blacksburg, VA: Virginia Polytechnic Institute and State University, 2007.

55. Mahmoud, Q.H. *Cognitive Networks: Towards Self-Aware Networks*. England: John Wiley & Sons Ltd, 2007.

56. Alpaydın, E. *Introduction to Machine Learning (Adaptive Computation and Machine Learning)*. MIT Press, ISBN 0262012111, 2004.

57. Wikipedia. Decision Making. http://en.wikipedia.org/wiki/Decision_making

58. Tennhouse, D.L, Smith, J.M., Sincoskie, W.D., Wetherall, D.J., and Minden, G.J. A survey of active network research. *IEEE Communications Magazine*, 1997, 35, 80–86.

59. Mostagir, M. and Decker, K. A multi-agent system architecture for active networks. *International Conference on Autonomous Agents and Multi-Agent Systems*, 2002. Bologna, Italy.

60. Milojicic, D.S., LaForge, W., and Chauhan, D. Mobile objects and agents. *Proceedings of the Fourth Conference on Object-Oriented Technologies and Systems*, 1998. Santa Fe, New Mexico.

61. Chess, D., Harrison, C., and Kershenbaum, A. *Mobile Agents: Are they a Good Idea?* Heidelberg, Berlin: Springer, 1997.

62. Borselius, N. Mobile agents and security. *Journal of Electronics and Communication Engineering*, 2002, 14, 211–218.

63. Siller, M. and Woods, J. Using an agent based platform to map quality of service to experience in conventional and active networks. *IEEE Proceedings-Communications* 2006, 153, 828–840.

64. Chen, W.-S. E. and Hu, C.-L. A mobile agent based active network architecture for intelligent network control. *Information Sciences*, 2002, 141, 3–35.

65. X. Huang, Y. Ma, and J. Zhang. Study of the active network management system model based on agent. *International Conference on Wireless Communications, Networking, and Mobile Computing*, 2008. Dalian, China.

66. Ren, L., Pei, Y.Z., Zhang, Y.B., and Ying, C. Charging validation for third party value-added applications in service delivery platform. *The Tenth IEEE/IFIP Network Operations and Management Symposium*, 2006, pp. 1–13. Vancouver, BC.

67. Rebahi, Y., Sher, M., and Magedanz, T. Detecting flooding attacks against IP Multimedia Subsystem (IMS) networks. *IEEE/ACS International Conference on Computer Systems and Applications*, 2008, pp. 848–851. Doha, Qatar.

68. Sher, M. and Magedanz, T. Protecting IP multimedia subsystem (IMS) service delivery platform from time independent attacks. *Third International Symposium on Information Assurance and Security*, 2007, pp. 171–176. Manchester.

69. Yu, X.-L., Chen, X.-Y. Fang, X Ding, X.-CH. Zhou, B., and Wei, B. Web services security in data service delivery platform for telecom. *IEEE International Conference on E-Commerce Technology for Dynamic E-Business*, 2004, pp. 374–377. Beijing, China.

70. Stiliadis, D., Francini, A., Kamat, S., Alicherry, M., Hari, A., Koppol, P.V., Gupta, A.K., and Skuler, D. Evros: A service-delivery platform for extending security coverage and IT reach. *Bell Labs Technical Journal* 12(3), 2007, 101–119.

Chapter 7

Service Innovation for Electronic Services

Christoph Riedl, Jan Marco Leimeister,
and Helmut Krcmar

Contents

7.1 Introduction

With an increasing importance of the service sector, the management of new service development (NSD) is becoming a key competitive concern for many companies (Gallouj and Weinstein, 1997; Johne and Storey, 1998; Fitzsimmons and Fitzsimmons, 2000; Johnson et al., 2000; Menor et al., 2002). Despite its importance, it is still not a very well-understood topic and ranks behind the research on new product development (Menor et al., 2002).

An increasing proportion of services are now electronic services delivered over the Internet. However, the systematic design of electronic services is not covered sufficiently in NSD literature. Yet, the importance and relevance of designing electronic services is demonstrated by examples of market success of services like Google, Amazon Web-services, or Salesforce.com. Moreover, a concerted research effort to address fields like "Internet of Services" and "Service Ecosystems" is forming (Janiesch et al., 2008; Riedl et al., 2009a,b). More general approaches such as Service Oriented Architectures and cloud computing are established and thus, exposing coarse-grained business components to simplify the assembly and deployment of business solutions built as networks of services (Beisiegel et al., 2005).

The purpose of this chapter is to derive a set of key attributes that distinguish electronic from nonelectronic services and their potential influence on NSD. These key attributes are then used as a framework for analyzing NSD literature with regard to their applicability to the development of electronic services.

To frame the object of interest a definition of electronic service is mandatory. Rust and Kannan (2003) define e-service as "the provision of service over electronic networks." Electronic networks include, but are not limited to the Internet. Other electronic environments such as mobile networks, ATMs, and self-service kiosks are also included by this definition. In business science literature, this usually refers to an Internet-based version of traditional services (Baida et al., 2004). This includes both, services that only use the Internet as a user-interface but where actual service fulfillment might include nonelectronic channels (e.g., online shopping), as well as services that are entirely delivered electronically (e.g., music download). The notion of e-services is not limited to the business-to-consumer domain but also encompasses the domains of business-to-business, government-to-public, and intra-organizational entities (Rust and Kannan, 2003). "Web-service" is a term used in computer science and is usually not found in business science. When used in business science, it either refers to the computer science definition or it simply refers to services delivered over the Web by way of e-service (Baida et al., 2004). In a computer science context, a Web-service is defined by Haas and Brown (2004) as a "software system designed to support interoperable machine-to-machine interaction over a network." Web-services have an interface described in a machine-processable format and other systems interact with the Web-service in the manner prescribed by its description using standardized messages.

For the purpose of this work, an electronic service will be defined as a *business activity of value exchange that is accessible through an electronic interface.* In that sense, an electronic service as it will be understood within the context of this work lies at the intersection of the business definition of a service (i.e., business activity of value exchange) and the technical implementation of a Web-service. Such a service is more than the pure technical implementation of a Web-service or another software implementation. The service has to implement a business activity that a user attributes value to. Yet, services delivered in a nonelectronic fashion, such as services offered by hospitality, are not within the scope of this work. An e-service following our definition may be provided through a single implementation of a Web-service or through a collection of Web-services that together form a new value-added service which is thus delivered through an electronic interface.

7.2 What Makes E-Services Different?

We argue that certain distinct characteristics of electronic services mandate a customized development process for these services as opposed to traditional new service development. Through an analysis of existing research related to electronic services we identified five key areas of difference: (1) the cost structure of services, (2) the high degree of outsourcing, (3) rapid development of new services, (4) the availability of transparent service feedback, and (5) the continuous improvement of services. The following sections motivate each area of difference.

7.2.1 Low Marginal Costs of Service Delivery

The economics of information have been recognized as dramatically different from the economics of physical items (Evans and Wurster, 2000). This leads to a unique cost structure both in comparison with physical products as well as other nonelectronic services.

The typical cost structure of an information technology supplier involves high fixed costs for developing the infrastructure and applications, and very low, sometimes near zero, marginal costs for actual service provision (Whinston et al., 1997; Bakos, 1998). Through the use of electronic intermediaries, the search and transaction costs are further reduced (Bakos, 1998). This further reduces variable costs of service provisioning and service use. Contrary to nonelectronic services that are sometimes very labor intensive (e.g., hospitality services), this difference should explicitly be addressed during service development.

7.2.2 High Degree of Outsourcing

Outsourcing is a standard concept that is being considered through making or buying decisions both in manufacturing and in services. In electronic services, outsourcing particularly plays an important role. First, since service provisioning occurs in the back-office and electronic services can easily be delivered from remote locations there is no need to collocate service production with the service consumption (Miles, 2005). Traditional services do not enjoy this opportunity, for example, through the need of attractive locations (think of a down-town café). Second, the high-degree of technical standardization achieved through various Web-service standards (Champion et al., 2002) and efforts to standardize Service Oriented Architectures (Beisiegel et al., 2005), this high-degree of outsourcing is accompanied by the necessary technical framework to make outsourcing of individual service components feasible. This is additionally fostered through the increased availability of high-speed networks and current developments in cloud computing (Böhm et al., 2009). The technical standardization allows the easy integration of other providers' components and services can be provided in a network of actors combining many service components (e.g., travel services integrating flight, hotel, local transportation, and other reservation services). Once these services are integrated through the development of appropriate interfaces infinite reuse of existing components with no further integration or assembly costs is possible. However, this can lead to complex value networks with different actors working together in a federated service environment. This leads to complex value constellation in distributed networks which are harder to manage with the increased number of involved actors (Vanhaverbeke and Cloodt, 2006).

7.2.3 Rapid Development of New Services

A differentiation strategy is difficult to attain as services can be copied easily and are not applicable to patent protection (Porter, 2001; Hipp and Grupp, 2005). Consequently, only continuous innovation can lead to economic success. However, these effects common to all service areas are magnified in the area of electronic services. Advances in electronic services are particularly rapid and low barriers of entry have been attributed to electronic services (*cf.* Evans and Wurster, 2000; Porter, 2001; Menor et al., 2002). This rapid development is further fueled by extremely fast technological progress and fast emerging of technologies. This fast technological progress not only creates opportunities for new service concepts, but also affects customers' expectations and preferences which require constant innovations (e.g., all the electronic services offered on the Apple iPhone store which was just created through the advances in mobile phone technology). Furthermore, the very nature of electronic services benefits radical innovation through major innovations and start-up businesses (Johnson et al., 2000; Menor et al., 2002).

7.2.4 Transparent Service Feedback

Through the electronic nature of service delivery, the interaction between a service consumer and the service itself becomes very transparent. A simple example of this effect is the monitoring of click-through-rates in online shops. This generates nearly a complete picture of customer interactions which a traditional shopping mall operator would dream of. This creates various opportunities for service design and innovation. Interactions between users and the service can be recorded and replayed. Thus, a service itself can gather information about what else users might want or need (Riedl et al., 2008).

The transparent nature of service feedback is also an option for new business models based on new licensing concepts. As the usage information is transparent to providers, billing is not only possible based on the actual use, but also on the value generated for the customer. For example, instead of charging for a Customer Relationship Management (CRM) service based on concurrent users, charging based on the actual revenue generated through the CRM service would be possible.

7.2.5 Continuous Improvement and Deployment

Unlike software being sold over the counter, electronic services are no longer restricted to a scheduled release cycle where changes, improvements, and bug fixes require months to be integrated into the service (termed "perpetual beta" by some authors, *cf.* Morris, 2006; O'Reilly, 2007). Rather, services are developed in the open with tight integration of service users or even by the users themselves. For example, Google services like search and many of the online applications are constantly updated. There are no distinct releases with version numbers assigned to the service instance currently offered. Rather, improvements slip into the market almost unnoticed. The innovation process is full of small cycles that allow a service to be improved almost instantly. Additionally, as services are delivered through a global delivery system, there are no local differences in the services offered and the new version is instantly available to all users. This would be very hard to implement for nonelectronic services where physical facilities would need to be upgraded and personnel to be trained.

This has two fundamental effects on the development of new electronic services. First, the benefits of perpetual beta and continuously improvements can be used to upgrade services with the improvements instantly visible to all users. Second, service providers have to make sure that improvements are visible to users and are valued as such.

7.3 State-of-the-Art Review of Developing Electronic Services

7.3.1 Analyzed Aspects

Based on the key differences between electronic and nonelectronic services and their impact on the new service development process, we developed an analysis framework. This framework has been used to review existing literature on their suitability for guiding the development of new electronic services. The following list presents our analysis framework.

■ Are there defined methods and processes to guide the development of new services?
■ Are electronic services explicitly covered by the method?
■ Are all phases of the innovation process included in the method or just selected aspects such as idea generation or implementation?
■ Does the approach pay special attention to the IT service specific cost structure?
■ Is a high degree of outsourcing and modularization supported?
■ Does the method provide support for very fast cycles and immediate deployment?
■ Does the method integrate aspects of continuous improvement through transparent feedback?
■ Does the method include a step to look for existing components to reuse to take advantage of low search costs and standardization to shorten time to market and reduce fixed costs?
■ Is special attention paid to complex value constellation in distributed networks (based on outsourcing)?

The following sections will review the literature in the area of new service development with regard to their prescriptive support for designing and developing services. Special attention will be paid to those aspects distilled above that are unique to electronic services compared to nonelectronic services.

7.3.2 Analyzed Literature

A systematic literature review has been performed. An initial search using the key words "new service development" or "NSD" on online databases ScienceDirect and EBSCOhost has been performed to cover a broad range of high-quality, peer-reviewed publications. The review time period was from 1997 to 2008 as NSD research received significant attention during this time (Zhou and Tan, 2008). The initial search returned over 300 articles. Accounting for duplicate results and after a preliminary scan of the article's abstracts the number of articles to be included could be substantially reduced. Reasons for excluding articles where, among others, a different understanding of e-service that related more to information system adoption or articles that refer to NSD literature or use NSD methods but do not contribute to extend NSD research itself. Moreover, several cross-referenced articles and books not found in those databases have been included and further extended by a comprehensive review of relevant academic journals that we expected to have published articles on NSD. Finally, 42 relevant journal and conference articles as

Table 7.1 Overview of Topics Covered in the Analyzed Articles

NSD Research Theme	Frequency
Types of innovation	10
Antecedents/success factors	10
Process models/methods	13
Generic (literature reviews)	9

well as books and book chapters have been included in the review. The literature on NSD focuses mainly on success factors and the development of (process) models (Zhou and Tan, 2008) as well as a large set of summary and review-based articles. The topics covered in the analysis and the number of articles that predominantly deal with this topic are shown in Table 7.1. A similar distribution of main themes covered in NSD research has also been reported by Zhou and Tan (2008). Table 7.2 gives an overview over the publications by year.

7.3.3 New Service Development

NSD involves the development of service offerings such as financial services, health care, telecommunications services, leisure and hospitality services, information services, legal and educational services as well as many more (Johne and Storey, 1998). Contrary to new product development (NPD) which is regarded as a base for much research in this area, new service development stresses core differences between products and services: intangibility, heterogeneity, simultaneity (Fitzsimmons and Fitzsimmons, 2000). Despite a growing body of knowledge, our understanding of new services development processes for especially electronic services is still limited (Menor et al., 2002).

Table 7.2 Publications by Year

Year	Frequency	Year	Frequency
1997	4	2003	5
1998	2	2004	2
1999	0	2005	4
2000	10	2006	3
2001	1	2007	4
2002	3	2008	4

7.3.3.1 Types of Service Innovation

The first set of articles tries to bring structure to the types of innovations found in services by proposing typologies of service innovation. Edvardsson and Olson (1996) suggest that service innovation includes the development of (1) a service concept (what customer needs are satisfied), (2) a service system (the resources necessary to deliver a service), and (3) a service process. These three areas make service innovation a complex and multidimensional undertaking (Essen and Conrick, 2008). Johnson et al. (2000) suggest six categories to structure service innovation (Table 7.3).

Other types of service innovation noted are, for example, the new combinations of existing services or the combination of customer coproduction with new service characteristics or competencies (Gallouj and Weinstein, 1997; van der Aa and Elfring, 2002). Hipp and Grupp (2005) identified four patterns of key factors

Table 7.3 A Typology of New Services

New Service Category	
Radical Innovations	
Major innovation	New services for markets as yet undefined; innovations usually driven by information and computer-based technologies
Start-up business	New services in a market that is already served by existing services
New service for the market presently served	New service offerings to existing customers of an organization (although the services may be available from other companies)
Incremental Innovation	
Service line extensions	Augmentations of the existing service line such as adding new menu items, new routes, and new courses
Service improvements	Changes in features of services that are currently offered
Style changes	Modest forms of visible changes that have an impact on customer perceptions, emotions, and attitudes, with style changes that do not change the service fundamentally, only its appearance

Source: Adapted from Johnson, S. et al. 2000. *New Service Development: Creating Memorable Experiences*, Fitzsimmons, J. and Fitzsimmons, M. (Eds.), pp. 1–32. Thousand Oaks, CA: Sage Publications.

influencing service innovation: knowledge intensity, network basis, scale intensity, and supplier dominance. Especially the network-based innovations seem to match most electronic services due to their reliance on technological systems for information and communication processing. Menor et al. (2002), moreover, argue that the nature of electronic services especially benefit radical innovations (major innovation and start-up business).

Barras (1990) argued that IT-based service innovation follows a pattern that is different from that found in manufacturing. He claims that in the early life-cycle phase of a service "technology push" is the main driving force whereas in the later phases incremental process innovation through "demand pull" is the driving force. In the latter phase pressures by users increasingly force service providers to differentiate themselves leading to differentiated products and product innovation. To account for this fact and the specifics of service industries, Barras proposed a reverse product life-cycle (RPC) model for services. The reverse product life-cycle model suggests that innovation takes place in three phases: improved efficiency, improved quality, and new services phase (Barras, 1990). Other articles also discuss the specific influence of IT innovations on service innovation. Especially the process innovation aspects achieved through the use of IT in back-end service provisioning and automation potentials are notable (Miles, 2005). However, these types of innovation are not specific to e-services as IT is a technology to be applied to the generic information-processing activities of services (Miles, 2005). He concludes that a study of IT's influence does not reveal much about the dynamics and processes of innovation.

None of the studies cited above explicitly addresses electronic services but are taken from diverse industries. However, there is an established hypothesis that innovation patterns in services are less sector-dependent, and that every type of innovate can be found within each individual service industry. In particular, there is no specific industry to offer electronic services *per se*. Moreover, knowledge insensitivity does not necessarily imply that the service is delivered electronically. For example, many financial services, though highly IT-based, are not electronic. Yet, an electronic ticket reservation service offered by an airline is. As Miles (2005) notes "some online information services originated form in-house data management services, for example, from publishing firms."

7.3.3.2 Antecedents of Success

Related to the different types of innovation a substantial part of the literature addresses the question what antecedents of NSD success are (de Jong and Vermeulen, 2003). Generic antecedents include strategic fit, skilled front-line employees, high-involvement teams, clear project structure, formal processes, top management support, and product champions (de Brentani, 2001; de Jong and Vermeulen, 2003; Vermeulen and van der Aa, 2003). Stevens and Dimitriadis (2005) report, that NSD is successful especially when learning has been achieved during the development process. Furthermore, two evolutionary stages of "manage key activities" and "create a climate for continuous innovation" have been identified (de Jong and Vermeulen, 2003). In an analysis of the antecedents of NSD success, IT systems and process structure have been found to have a positive impact on the speed of NSD (Froehle et al., 2000). None of the studies explicitly addressed electronic services. It can be assumed that these antecedents are generic enough to play an important role for electronic services as well, but specific aspects of e-services have not been studied so far. In an analysis of the antecedents of NSD success, IT systems and process structure have been shown to have a positive impact on the speed of the NSD process (Froehle et al., 2000). As NSD speed is of particular importance for e-services, this is a valuable contribution. A notable exception is Menor et al. (2002) who did not study e-service antecedents of success but propose that the aspect of external newness is especially salient as electronic services are often replications of services already known to customers but that are now offered in an electronic way.

7.3.3.3 Processes

With regard to traditional services NSD can be seen as a rather complete method covering all phases of the service life cycle. There are in particular wide set of process models defined for the development of new services. In a comparative study of existing NSD literature, Johnson and Menor (1997) proposed a basic model of four phases: design, analysis, development, and launch. Although models included in the literature review did not match precisely and different phases were more detailed in some models and more succinct in others, these four phases were found in all.

More recently, Johnson et al. developed a new NSD process based on four broad stages and 13 detail tasks to produce and launch a new service (Johnson et al., 2000). The model emphasizes the nonlinearity of the NSD process through a continuous cycle as well as the importance of enabling factors: teams, tools, and organizational culture (see Figure 7.1).

On a very generic level, Bessant and Davies (2007) suggest that organizations have to manage four phases in the innovation process: search and scan their environment to pick up signals for potential innovation, strategically select those ideas that the organization will commit resources to, implement the innovation, and finally reflect on the previous phases to achieve organizational learning.

An issue also commonly addressed in NSD is "design for delivery" (Bullinger et al., 2003). As many services are highly labor intensive (e.g., hospitality services) the motivation to optimize new services for efficient delivery is high. As electronic services follow a reversed cost structure, these approaches are not suitable for e-service development.

Another common distinction in service development is that between a "front-office" and "back-office" (e.g., Metters and Vargas, 2000). Yet, sole focus on

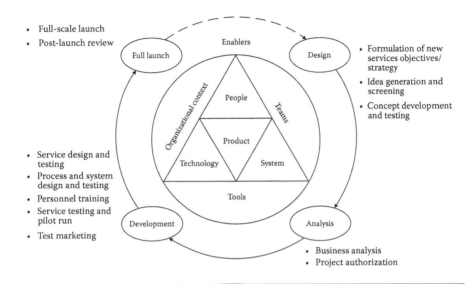

- Full-scale launch
- Post-launch review

Enablers

Full launch

Design

People

Teams

Organizational context

Product

Technology   System

Tools

- Formulation of new services objectives/ strategy
- Idea generation and screening
- Concept development and testing

- Service design and testing
- Process and system design and testing
- Personnel training
- Service testing and pilot run
- Test marketing

Development

Analysis

- Business analysis
- Project authorization

Figure 7.1   NSD process cycle. (Adapted from Johnson, S. et al. 2000. *New Service Development: Creating Memorable Experiences.* **Fitzsimmons, J. and Fitzsimmons, M. (Eds), pp. 1–32. Thousand Oaks, CA: Sage Publications.)**

"back-office" operational efficiency is not enough and has been neglected with many e-services. As argued by Riedl et al. (2008), perceived quality measures have to be taken into account to address satisfaction issues commonly addressed in "front-of-fice" design. Moreover, Johnson et al. (2000) note that different NSD processes are necessary for different types of innovation. In particular, they identify incremental service innovations, radical service innovations, and technology-driven services as key differences that should be used to choose the appropriate NSD process and propose this as an avenue for future research. With regard to the perpetual beta aspect of electronic services, this result might be useful in guiding the selection of a specific process that is designed especially for incremental innovations (de Brentani, 2001).

Froehle and Roth (2007) propose a framework for new service development that integrates both process- and resource-oriented approaches. The resource-oriented practices focus on cultivating and developing the intellectual, organizational, and physical resources that support NSD capabilities. The process-oriented practices focus on planning, defining, and executing the actual stages of the service development (see Figure 7.2). Their belief and motivation for this integrated view is that both resource and process capabilities are required for successful service development.

Pavitt (2005) acknowledges the fact that services have to be continuously improved and a continuous mapping of service artifacts to market needs and demands is necessary. However, there is no consideration for the vast transparent feedback available in e-services and the very fast cycle times.

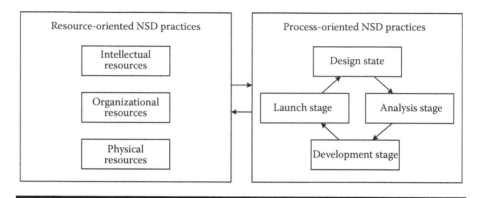

Figure 7.2 The Resource-Process framework of new service development. (Adapted from Froehle, C. and Roth, A. 2007. *Production and Operations Management* **16(2):169–188.)**

7.3.3.4 Generic and Organization-Related Issues

Syson and Perks (2004) address network issues in NSD. They conclude that interactions are critical for NSD and that the incorporation of disparate perspectives is beneficial (i.e., they increase creative potential) and that the network perspective helps incorporate relevant resources and actors. However, the very nature of services (intangibility, heterogeneity, and inseparability) brings considerable complexities to the exchange processes of NSD. As services are copied easily, the development of a network approach to NSD could provide firms with a source of competitive advantage. They do not, however, address the management of the resulting complex value networks.

7.4 E-Service Development and Open Innovation

How can innovations be developed in open, networked, and dynamic systems and markets such as those most electronic services are found in? Outside the core area of New Service Development a new research stream called *open innovation* is making progress (Chesbrough, 2003). The tools and methods proposed by open innovation seem to address the problems that are encountered when developing electronic services (e.g., fast development) and might also help to exploit the new opportunities (e.g., transparent feedback).

Three core process archetypes in open innovation have been identified: the outside-in process, the inside-out process, and the coupled process (Gassmann and Enkel, 2004). The outside-in process enriches a company's knowledge and innovation base through the integration of external knowledge sources, particularly the knowledge sources of customers and suppliers, to increase its innovativeness. The inside-out process exploits a company's unused inventions in different markets and a managed trade of intellectual property, for example, through licensing. The coupled process is a combination of both the outside-in and the inside-out processes intended to maximize the benefits of both approaches.

These three archetypes are achieved through various means of perspectives on opening the innovation and development including: (1) globalization of innovation, (2) outsourcing of R&D, (3) early supplier integration, (4) user innovation, and (5) external commercialization of innovations (Gassmann, 2006). These open innovation processes lead to inter-firm cooperation and development of ecosystems of networked firms sharing technology and trading intellectual property (West et al., 2006; Stathel et al., 2008). The benefits of customer integration, user-driven innovation and open innovation are well recognized and established in research (e.g., Surowiecki, 2005; von Hippel, 2005; Chesbrough, 2006; Gassmann, 2006; Ogawa and Piller, 2006; Lakhani and Panetta, 2007; West and Lakhani, 2008). However, these results have not yet been linked to NSD and thus escaped the systematic literature review.

This research argues that an open-innovation paradigm rather than a closed-innovation paradigm is necessary for successful innovation of electronic services. This is due to their heavy reliance on reuse, their reliance on new business models, and knowledge leveraging as services are implemented as software (Gassmann, 2006). As open innovation is geared toward systematically integrating external ideas and influences into the internal innovation process, a systematic approach becomes available for the integration of the transparent feedback generated by service usage of electronic services.

To support such an open innovation process in the networked environment in which electronic services operate it would be necessary to provide a central, shared innovation repository through which the diverse actors like service provider, customer, and aggregator can interact (Riedl et al., 2009a). Thus, a duality of an open and networked structure for the delivery of many electronic services and an open and networked model for the development of these services is created.

Taken together the special requirements posed by developing new electronic services, the environment of highly networked services, and the move to an open innovation model offer the potential for new and improved service innovation approaches that are until now little understood.

7.5 Conclusion

NSD is a rather complete method describing key processes and tasks. Moreover, it covers all phases of the life cycle from design, analysis, and development to launch as is apparent from the wide collection of process models that have been reported. Especially noteworthy is the cyclic model of Johnson et al. (2000). However, the design of electronic services is not explicitly covered except in articles offering basic definitions of e-services. A notable exception is the article by Menor et al. (2002) that also shows gaps that exist and points to research challenges.

While several research results exist that can point in the direction of successful development of new e-services, there are certain gaps in NSD research with regard to key attributes of electronic services and their influence on NSD. In particular, current NSD methods are not well equipped to address the rapid nature and specific cost structure found in electronic services. Moreover, current NSD methods are not well suited to fully exploit the various advantages offered by electronic services over non-electronic services. These are in particular the transparent feedback generated by service usage and potentials for continuous improvement and rapid deployment of service changes.

This research only focuses on core NSD literature. However, there are other streams of research that might be suitable to target some of the key issues. We propose in particular that an open innovation process should be taken toward the development of new electronic services. Future research should address how the research gaps identified in this analysis of NSD literature can be addressed by other streams of research and how open innovation could be specifically used for the innovation of electronic services. To address the potentials of increased outsourcing and also increase the speed of development NSD processes for electronic services could also be extended to include a specific step of searching for existing service components that can be reused. Literature on mash-ups could serve as initial guidance in this area. Furthermore, there are potentials of participatory development of new e-services and open innovation (e.g., Leimeister et al., 2009; Riedl et al., 2009a) that could provide useful for fast and successful development of new services.

In summary, the research on electronic services in general and the development of these services in particular is, despite their increasing importance, still very limited. This research provides an initial basis in elaborating on the key aspects that distinguish nonelectronic from electronic services and points to gaps in the literature that could be addressed by future research.

Acknowledgment

This research received funding from the German Federal Ministry of Economics and Technology (BMWi) under grant code 10MQ07024. The responsibility for the content of this publication lies with the authors.

References

Baida, Z., Gordijn, J., and Omelayenko, B. 2004. A shared service terminology for online service provisioning. In *Proceedings of the Sixth International Conference on Electronic Commerce (ICEC '04)*, New York, NY: ACM Press, 1–10.

Bakos, Y. 1998. The emerging role of electronic marketplaces on the Internet. *Communications of the ACM* 41(8):35–42.

Barras, R. 1990. Interactive innovation in financial and business services: The vanguard of the service revolution. *Research Policy* 19(3):215–237.

Beisiegel, M., Blohm, H., Booz, D., Dubray, J.-J., Colyer, A., Edwards, M., Ferguson, D. et al. 2005. Building Systems using a Service Oriented Architecture. Whitepaper, SCA Consortium.

Bessant, J. and Davies, A. 2007. Managing service innovation. In *Innovation in Services— DTI Occasional Paper No. 9*, Bessant, J., Davies, A., Tether, B., Howells, J., Voss, C., Zomerdijk, L. and Massini, S. (Eds.), pp. 61–95. UK: Department of Trade and Industry.

Böhm, M., Leimeister, S., Riedl, C., and Krcmar, H. 2009. Cloud computing—outsourcing 2.0 or a New business model for IT provisioning? In *Application Management*, Keuper, F., Oecking, C. and Degenhardt, A. (Eds.), Wiesbaden: Gabler.

Bullinger, H., Fähnrich, K., and Meiren, T. 2003. Service engineering—Methodical development of new service products. *International Journal of Production Economics* 85(3):275–287.

Champion, M., Ferris, C., Newcomer, E., and Orchard, D. 2002. Web Services Architecture, W3C Working Draft 14 November 2002, http://www.w3.org/TR/2002/WD-ws-arch-20021114/ (accessed November 1, 2009).

Chesbrough, H. 2003. The era of open innovation. *MIT Sloan Management Review* 44(3):35–41.

Chesbrough, H. 2006. *Open Innovation: The New Imperative for Creating and Profiting from Technology*. Boston: Harvard Business School Press.

de Brentani, U. 2001. Innovative versus incremental new business services: Different keys for achieving success. *Journal of Product Innovation Management* 18(3):169–187.

de Jong, J. and Vermeulen, P. 2003. Organizing successful new service development: A literature review. *Management Decision* 41(9):844–858.

Edvardsson, B. and Olsson, J. 1996. Key concepts for new service development. *Service Industry Journal* 16(2):140–164.

Essen, A. and Conrick, M. 2008. New e-service development in the homecare sector: Beyond implementing a radical technology. *International Journal of Medical Informatics* 77:679–688.

Evans, P. and Wurster, T. 2000. *Blown to Bits: How the New Economics of Information Transforms Strategy*. Boston: Harvard Business School Press.

Fitzsimmons, J. and Fitzsimmons, M. 2000. *New Service Development: Creating Memorable Experiences*. Thousand Oaks, CA: Sage Publications.

Froehle, C. and Roth, A. 2007. A resource-process framework of new service development. *Production and Operations Management* 16(2):169–188.

Froehle, C., Roth, A., Chase, R., and Voss, C. 2000. Antecedents of new service development effectiveness: An exploratory examination of strategic operations choices. *Journal of Service Research* 3(1):3–17.

Gallouj, F. and Weinstein, O. 1997. Innovation in services. *Research Policy* 26(4):537–556.

Gassmann, O. 2006. Opening up the innovation process: Towards an agenda. *R&D Management* 36(3):223–228.

Gassmann, O. and Enkel, E. 2004. Towards a theory of open innovation: Three core process archetypes. In *Proceedings of the R&D Management Conference (RADMA)*. Sessimbra, Portugal July, 8–9.

Haas, H. and Brown, A. 2004. *Web Services Glossary*. W3C Working Group Note 11 February 2004, http://www.w3.org/TR/ws-gloss/ (accessed November 1, 2009).

Hipp, C. and Grupp, H. 2005. Innovation in the service sector: The demand for service-specific innovation measurement concepts and typologies. *Research Policy* 34(4):517–535.

Janiesch, C., Ruggaber, R., and Sure, Y. 2008. Eine Infrastruktur für das Internet der Dienste. *HMD* 261:71–79 (in German).

Johne, A. and Storey, C. 1998. New service development: A review of the literature and annotated bibliography. *European Journal of Marketing* 32(3/4):184–251.

Johnson, S. and Menor, L. 1997. Integrating service design and delivery: A proposed model of the new service development process. Paper presented at the annual meeting of the Decision Sciences Institute, San Diego.

Johnson, S., Menor, L., Roth, A., and Chase, R. 2000. A critical evaluation of the new service development process: Integrating service innovation and service design. In *New Service Development: Creating Memorable Experiences*, Fitzsimmons, J. and Fitzsimmons, M. (Eds.), pp. 1–32. Thousand Oaks, CA: Sage Publications.

Lakhani, K. and Panetta, J. 2007. The principles of distributed innovation. *Innovations: Technology, Governance, Globalization* 2(3):97–112.

Leimeister, J. M., Huber, M., Bretschneider, U., and Krcmar, H. 2009. Leveraging crowdsourcing—Theory-driven design, implementation and evaluation of activation-supporting components for IT-based idea competitions. *Journal of Management Information Systems* 26(1):197–224.

Menor, L., Tatikonda, M., and Sampson, S. 2002. New service development: Areas for exploitation and exploration. *Journal of Operations Management* 20(2):135–157.

Metters, R. and Vargas, V. 2000. A typology of de-coupling strategies in mixed services. *Journal of Operations Management* 18(6):663–682.

Miles, I. 2005. Innovation in services. In *The Oxford Handbook of Innovation*, Fagerberg, J., Nelson, R. and Mowery, D. (Eds.), pp. 433–458. New York, NY: Oxford University Press.

Morris, J. 2006. Software Product Management and the Endless Beta, http://jimmorris.blogspot.com/2006_08_01_jimmorris_archive.html, 2006-08-30 (accessed November 01, 2009).

Ogawa, S. and Piller, F. 2006. Reducing the risks of new product development. *MIT Sloan Management Review* 47(2):65–71.

O'Reilly, T. 2007. What is Web 2.0: Design patterns and business models for the next generation of software. *Communications and Strategies* 2007:17–37.

Pavitt, K. 2005. Innovation processes. In *The Oxford Handbook of Innovation*, Fagerberg, J., Nelson, R. and Mowery, D. (Eds.), pp. 86–114. New York, NY: Oxford University Press.

Porter, M. E. 2001. Strategy and the internet. *Harvard Business Review* 79(3):62–78.

Riedl, C., Böhmann, T., Leimeister, J. M., and Krcmar, H. 2009a. A framework for analysing service ecosystem capabilities to innovate. In *Proceedings of 17th European Conference on Information Systems (ECIS '09)*. Verona, Italy.

Riedl, C., Böhmann, T., Rosemann, M., and Krcmar, H. 2008. Quality aspects in service ecosystems: Areas for exploitation and exploration. In *Proceedings of International Conference on Electronic Commerce (ICEC '08)*. New York, NY: ACM Press, 1–7.

Riedl, C., Böhmann, T., Rosemann, M., and Krcmar, H. 2009b. Quality management in service ecosystems. *Information Systems and e-Business Management* 7(2):199–221.

Rust, R. T. and Kannan, P. 2003. E-service: A new paradigm for business in the electronic environment. *Communications of the ACM* 46(6):36–42.

Stathel, S., Finzen, J., Riedl, C., and May, N. 2008. Service innovation in business value networks. In *Proceedings of XVIII International RESER Conference*. Stuttgart.

Stevens, E. and Dimitriadis, S. 2005. Managing the new service development process: Towards a systemic model. *European Journal of Marketing* 39(1):175–198.

Surowiecki, J. (2005). *The Wisdom of Crowds: Why the Many are Smarter than the Few.* London: Abacus.

Syson, F. and Perks, H. 2004. New service development: A network perspective. *Journal of Services Marketing* 18(4):255–266.

van der Aa, W. and Elfring, T. 2002. Realizing innovation in services. *Scandinavian Journal of Management* 18(2):155–171.

Vanhaverbeke, W. and Cloodt, M. 2006. Open innovation in value networks. In *Open Innovation: Researching a New Paradigm*, Chesbrough, H., Vanhaverbeke, W. and West, J. (Eds.), pp. 258–281. New York, NY: Oxford University Press.

Vermeulen, P. and van der Aa, W. 2003. Organizing innovation in services. In *Service Innovation*, Tidd, J. and Hull, F. M. (Eds.), pp. 35–53. London: Imperial College Press.

von Hippel, E. 2005. *Democratizing Innovation.* Cambridge: MIT Press.

West, J. and Lakhani, K. 2008. Getting clear about communities in open innovation. *Industry and Innovation* 5(2):223–231.

West, J., Vanhaverbeke, W., and Chesbrough, H. 2006. Open innovation: A research agenda. In *Open Innovation: Researching a New Paradigm*, Chesbrough, H.,Vanhaverbeke, W., and West, J. (Eds.), pp. 285–307. Oxford: Oxford University Press.

Whinston, A., Choi, S., and Stahl, D. 1997. *The Economics of Electronic Commerce.* Indianapolis: Macmillan Technical Publishing.

Zhou, Q. and Tan, K. C. 2008. A bibliographic analysis of the literature on new service development. In *Proceedings of the International Conference on Management of Innovation and Technology (ICMIT '08).* IEEE, 872–877.

Chapter 8

Service Orchestration in the IP Multimedia Subsystem

Richard Spiers, Richard Good, and Neco Ventura

Contents

8.1 Introduction

In the past couple of years, a wave of new innovative services has appeared on the Internet. These services, including Skype, YouTube, Facebook, and Twitter are driven largely by a user-centric model of service provisioning, where the end user is responsible for producing most of the content or the inherent value of these services. For example, Facebook had more than 390 million users that were adding 2 billion photos and 14 million videos every month by October 2009 [1]. This emerging generation of services, named the Web 2.0 generation, relies on content that is generated by the end user as well as enabling third-party application developers to act as service providers. Facebook has more than 350 thousand applications developed by more than 1 million developers [1]. These Internet services are being provided "over the top," that is, they do not rely on the network operator beyond providing access to the Internet. Free VoIP services on the Internet are eroding the operators voice revenues and the so-called Internet "App Stores" are allowing mobile developers to completely bypass the network operators' walled gardens. End users can now buy applications to use on their phones without paying the network operator (beyond the existing connectivity fees). In 2007, the network carriers' portal Web sites were visited by 57% of all UK mobile Internet users. By 2008, the figure dropped to 22% as users turn to other sources for their mobile content and services [1]. Network operators have recognized the challenges presented by the Web 2.0 revolution, and have begun the transition toward Service Delivery Platforms (SDPs). These SDPs are designed to allow the rapid prototyping of new and innovative services that can take advantage of the higher bandwidth available on the current access networks, as well as the enriched multimedia capabilities of today's telecommunication devices. It is hoped that these SDPs will allow the network operator and third-party developers the opportunity to provide a myriad of new services, opening up new revenue streams and increasing the Average Revenue Per User (ARPU). These SDPs are largely based on top of a Next Generation Network (NGN), which is defined by the International Telecommunication Union (ITU) as "a packet-based network" that makes use of "multiple broadband, Quality of Service (QoS) enabled transport technologies and in which service related functions are independent from underlying transport-related technologies" [2]. In particular, the IP Multimedia Subsystem (IMS) has been standardized by the Telecommunications and Internet converged Services and Protocols for Advanced Networks (TISPAN) as a core part of a NGN, and forms the main focus of this chapter.

During the development of the IMS, two core principles have been established. The first one is the realization that the network architecture needs to be revamped to fully utilize the bandwidth available via new access technologies, allowing multimedia communication sessions over a noncircuit-based network. Secondly, proprietary closed signaling protocols inhibit the growth of new services and stifles innovation. The IMS uses an open, human readable protocol for the control signaling that is deployed widely on the Internet. This lowers the entry barrier for third-party service developers, allowing them to use the experience gained from the Internet in the development of similar services over the network operators' network as well as reducing the need for specialized knowledge about telephony protocols.

However, one drawback to the expected influx of developers and services is that research needs to be completed on how to handle interactions (both negative and positive) between these new services. Service conflict detection and resolution is not a new issue. Previous network architectures, such as Intelligent Networks (IN), have suffered from service conflict detection and resolution issues [3]. A summary of techniques used to detect and resolve these conflicts in previous architectures can be found in [4]. The IMS will be no different unless mechanisms are designed and developed to handle these interactions. The situation is expected to become increasingly complicated as services will no longer be provided solely by the network operator but also by their partners and third party developers. This area has not been standardized yet, unlike the main core of the IMS framework. A study concluded by the Third Generation Partnership Project (3GPP) in September 2008 highlighted this area as requiring further critical research [5]. This chapter presents the current state of service orchestration in the IMS as well as future developments in the area. First, the existing IMS facilities for service orchestration are introduced and limitations discussed. Second, the requirements for an improved service orchestration system are presented. This is followed by a discussion on the current state of research in the area. The existing research focuses mainly on determining where to locate a conflict detection and resolution mechanism, as well as how these conflicts can actually be resolved. The chapter is concluded with some recommendations for future work.

8.2 IP Multimedia Subsystem

Traditionally, network operators had to develop single point solutions for each service that they offered the end user. This led to a duplication of effort, with vastly complicated Operations and Business Support Systems (OSS/BSS). Multiple interfaces had to be developed, often proprietary, between the services and the existing charging and support systems. The IMS aims to reduce this duplicated effort (and thus, reduce the CAPEX and OPEX of the operator) by running each service over a single platform. It has been designed to allow the rapid prototyping of new, innovative services that can be enriched through the addition of any kind of multimedia. This is enabled through the use of a packet-based network and common functions such as user authentication and charging which can be reused by any service. The result is a single IP-based network and the operator no longer has to maintain several different platforms and networks, greatly reducing the cost of maintenance. This also increases the speed at which new services can be developed and tested as each service no longer requires its own mechanisms to handle charging, authentication, and so on.

The IMS has a dedicated control layer, which is handled separately from the multimedia traffic. The media traffic can follow a different path from the control traffic, allowing for low-latency multimedia sessions, while the control signal traffic is passed securely through the network operators network. It is also access agnostic, as the service functionality is separated from the access technology. The IMS can be used with a variety of fixed line or mobile access technologies.

8.2.1 Architecture and Signaling

As mentioned previously, the IMS uses an open, human readable control protocol that is popular on the Internet. The Session Initialization Protocol (SIP) is used to establish, modify and tear-down any form of multimedia session [6]. It is very flexible, and can be used to create sessions for IPTV, video calls, conferencing, and so on. It is used in conjunction with the Session Description Protocol (SDP), which allows the multimedia to be fully described separately from the session control information. Together, these protocols allow the session details to be negotiated before the session is established, or renegotiated during a session. This greatly enhances compatibility between different devices and implementations as each side can provide information about their capabilities (e.g., supported video codecs) in order to determine a set of matching capabilities to be used for that particular session.

There are four main functional elements in the IMS core. These are the Proxy Call Session Control Function (PCSCF), the Interrogating Call Session Control Function (ICSCF), the Serving Call Session Control Function (SCSCF), and the Home Subscriber Server (HSS). It is important to note that these are logical elements, and are not tied to a single piece of hardware. For example, one can have multiple SCSCFs in order to scale the network or to provide redundancy through a fail over mechanism. The Call Session Control Functions (CSCFs) are SIP servers that perform a variety of tasks. The PCSCF is the functional node that handles the initial and subsequent communication with the end user. It acts as a SIP Proxy, forwarding messages from the end user to the other functional elements. It has some security functions such as ensuring that the SIP message is compliant with the relevant protocols, as well as optimization functions such as the compression and decompression of SIP messages that travel over slow access networks. The SIP messages are small compared to the media traffic that is sent, thus this compression serves to reduce the end-to-end delay or total round trip time rather than the amount of bandwidth consumed. As the PCSCF is the point of entry to the IMS network for the end user, it is included in the signaling path and plays a role in every message that the end user sends (even if that role is to forward messages to the other nodes).

The SCSCF is of particular interest to the discussion contained within this chapter. It handles the registration of the end user, and contains detail on any sessions currently underway. It communicates with the HSS to load information about the end user, which includes the initial Filter Criteria (iFC) which are used to trigger services. These iFCs will be discussed in further detail in the next section, but for now it is suffice to say that the SCSCF decides where to send the end user's SIP message, that is, the SCSCF decides which services get triggered for a particular request. The ICSCF acts as a point of contact for other IMS networks, and forwards messages to the local end user on behalf of a foreign network. It does not play a crucial role in service triggering, and as such will not be discussed further.

The HSS contains information about the end user, including the necessary information to authenticate the user as well as the end user's profile. It performs many of the same functions that the Home Location Register (HLR) and the Authentication Center (AUC) provide in GSM networks.

All the elements described so far are involved in the control and operation of the network. The actual services are provided by SIP Application Servers (AS). These AS can operate in a variety of modes, in order to provide different types of services to the end user. These modes are the SIP Proxy mode, the SIP User Agent (UA) mode (i.e., it acts as an endpoint), or the Back-to-Back User Agent (B2BUA) mode (i.e., it acts as two SIP user agents joined together). A service such as voice-mail would normally be implemented in the SIP UA mode, whereas a click-to-dial service would normally be implemented as a B2BUA SIP AS. The SIP protocol is used by the AS to communicate with the SCSCF and end user, but the AS uses the Diameter protocol to communicate with the HSS in a safe and secure manner. The Diameter protocol is an extendable protocol that has been designed to mainly provide authentication, authorization, and accounting (AAA) services [7]. The core IMS network uses Diameter to provide these functions (Figure 8.1).

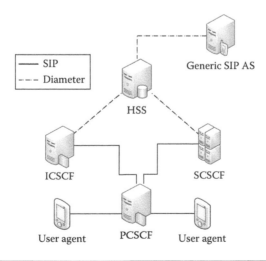

Figure 8.1   The standard IMS architecture.

8.2.2 Service Triggering in the IMS

All nonuser originating services in the IMS are hosted on AS. An example of a user originating service is a normal call placed by the end user. If a more advanced service is required, it is performed by the AS. The SCSCF directs user's requests to different AS depending on certain conditions. Filters containing these conditions are compared against any incoming SIP requests, which are forwarded to the appropriate AS if a match is found. These filters are the iFC. The iFCs are vital to the operation of the network, and are stored on a per user basis in the HSS. Only SIP requests that create a dialogue or are a stand-alone request trigger these filters. Thus, these filters would never be examined on the termination of a session due to a SIP BYE request.

The SCSCF retrieves the end user's profile from the HSS when the user registers on the IMS network. This user profile contains the iFCs. It is not necessary to discuss the technical structure of the IFCs in detail; it shall suffice to mention one important field. This is the priority field, which indicates the order in which the iFCs are evaluated. This directly controls the order in which services can be used by the end user. It is possible to have more than one matching iFC for any particular SIP request. The matching iFC with the highest priority determines which AS receives the request. If the first AS does not change the SIP Request-URI or terminate the dialog, the request can be forwarded on to the AS with the next highest priority matching iFC. While these priorities can be modified by the network operator, which would require the HSS to send the changed iFCs to the SCSCF, they are considered to be largely static.

For example, consider the following scenario. A service chain could include a Destination Barring service, an Advertising service, and another service such as a Video on Demand (VoD) service. The Destination Barring service would block all sessions to a particular destination or service (e.g., block all calls to a black list of numbers). The Advertising service could play an advertisement to the end user, and then finally the end user's request would be served by the VoD service (Figure 8.2).

It is clear that the order in which services are triggered is critically important. For example, if a particular AS terminates SIP dialogues and has a very high priority, it will prevent any of the other services being triggered. As the number of iFCs increase (i.e., as the number of services offered by the network or a third party increases) the delay incurred in the establishment of a session increases, as the SIP request may be passed through a chain of services.

8.2.3 Limitations

The current mechanisms for service invocation in the IMS have a number of limitations. First, there is no external information used when determining which service to invoke. Only information contained in the original SIP message is used. External information such as the status of the end user, the time of day, or the outcome of

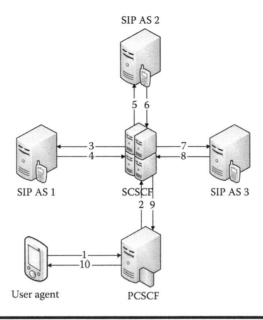

Figure 8.2 Example of chaining services in the IMS.

previous decisions cannot be included in this process. Second, handling multiple user interactions is not trivial, as each user has their own profile with an individual set of iFC, which can result in a different chain of services being triggered. This creates a problem when multiple users wish to utilize a service together. As each user's service chain and thus signaling paths are different, their messages will not be routed in the same fashion. This introduces a variable delay in the establishment of a session, as well as potentially losing some control signaling as the different services pass or modify the signaling that flows through them.

Currently, the iFCs are only evaluated for initial SIP requests. This needs to be expanded to allow services to be triggered on certain responses as well [8]. There is a feature called "record-routing" which allows an AS to become part of the signaling path so that it can inspect the message flow for a particular response. However, this is not ideal when multiple AS is involved, as the signaling path and hence, the overall delay would increase. Not allowing services to be triggered on responses such as a SIP REINVITE message inside an existing SIP dialogue limits the flexibility and type of services that can be offered. For example, the service orchestration system should allow a service to be triggered when the end user receives a notification from a present AS about a change in some monitored state, or it should be possible to trigger another service when the current service ends (i.e., the SCSCF receives a BYE request destined for, or from, the end user).

There are no mechanisms to detect when services interact with each other. These interactions can be positive (such as building an improved service out of several services), or they can be negative (when the services offer competing or incompatible features). There is no way of handling situations where services operate successfully in an independent fashion, but cannot operate correctly when they are triggered together. The 3GPP recently completed a "Study on Architecture Impacts of Service Brokering" [5]; Service Brokering is another term used for service orchestration, and a Service Broker will have largely the same functionality as a Service Capability Interaction Manager (SCIM). While there can be differences between the usage of these two terms, they are close enough to be used interchangeably for this chapter's purposes. The 3GPP identified several areas that need investigation as well as proposing different architecture alternatives that can provide a starting point for further research. These proposed alternatives will be discussed in the next section. An important area not discussed in the report was the ability of the end user to have a greater form of control over their services. In the future, the end user should be able to personalize and control their services. They should be able to build the final blended service from a range of service providers.

8.3 Service Orchestration System Design

The idea of a function to organize and control service capabilities was first raised by the 3GPP in 2001 [9]. It was not defined in detail, and was placed in the application layer as a stand-alone entity. It was envisaged that it would not use standardized data to control the services, but would have interfaces between the SCSCF and the other AS. The call for this functional element was made after gaining experience from the difficulties arising in service interaction management from previous network architectures [4]. This entity was then moved to become part of a SIP AS, where it remained dormant for several years. It eventually fell under the management of the 3GPP Work Group SA2 (Architecture), and renamed the SCIM. The 3GPP started a study in June 2006 to investigate this function and by September 2008 it had determined that many enhancements were needed.

8.3.1 Combined Service Examples

In order to speed up the development of new services, the Open Mobile Alliance (OMA) develops a set of "enablers" or service capabilities. For example, the OMA provides a Presence enabler that can be used to add presence to any existing application [10]. These enablers (including email notification, instant messaging, etc.) have been designed in a modular fashion to allow the easy creation of rich combinational services. It is predicted that in the future the SCIM will control these interactions. A few examples of these combinational services are provided below to show the flexibility and potential of these service interactions. They are not meant to be deployed real-world services, but rather example concepts to provide the reader with a reference framework for the following sections.

8.3.1.1 Presence-Based Simultaneous Ring

An IMS user can register several different identities and several different devices with the IMS network at the same time. Through the use of SIP forking, it is possible to ring these different devices at the same time when trying to establish a call with the given user. Through the use of a presence capability, it is possible to enhance this service by only ringing the device that indicates the user's presence as available and not away. Thus, not all his devices start ringing at the same time. If there is no answer at the first device, it can proceed to ring all of the devices.

8.3.1.2 Enhanced Call Control

It is possible to enhance the normal call barring or call forwarding services through the use of location or presence data. If the end user sets his presence to "busy," then all incoming calls could be barred or forwarded to his voice mail. Depending on the end user's location, certain calls could be handled differently. For example, if the end user's location is currently set to "Home," then calls from work related numbers could be sent to voice mail and he could instead receive an email notification of the call (with a text transcription of the voice mail message).

8.3.1.3 Rich Media Content Distribution

Goveas et al. [11] suggest a blended service where the end user receives a combination of offers depending on the end user's location and device capabilities. If the user's device is capable of receiving a multimedia stream, they receive a trailer for the latest movie release. If they are not capable of receiving the media stream, they can receive a poster for the movie instead. A virtual discount coupon can be sent to the end user if their location indicates that they are near a movie theater [11].

8.3.2 Service Conflict Detection and Resolution

The two main problems encountered in designing a system to handle service orchestration is detecting the various interactions between services, as well as resolving any conflicts that arise. A system needs to be put into place that can determine when services are in conflict. Conflict resolution that takes place before the services are invoked is known as offline service resolution, whereas conflict resolution that happens during run time is known as online service resolution. An offline service resolution requires knowledge of all possible service interactions that can occur, and can be handled in several different ways. A popular approach is to build acyclic graphs or state tables indicating all the possible state changes/service flows that can occur. However, this is not a scalable approach as the graphs or state tables grow larger and more complex as additional services are developed. In addition to this, any time a service is changed or added, these flows need to be updated which increases the administrative overhead of the system. Resolving all these potential conflicts offline or ahead of run-time implies that the time to detect conflicts is greatly reduced during run-time. A drawback to offline conflict detection resolution is that there might be unpredictable behavior from a service which had not been covered by the offline detection resolution. This could lead to a nondeterministic operation of the system resulting in conflicts.

While both online and offline conflict detection systems require information about each particular service, online systems do not need to keep a record of all the possible interactions between the different services. Online systems should be able to detect a potential conflict as the session is established, and trigger the necessary recovery mechanisms in real time. This means that there is less maintenance required whenever a new service is added. Another benefit is that the real-time analysis should be able to handle any unexpected interaction as it happens. This real time analysis does increase the time to establish sessions as more processing time is needed to evaluate the required services.

Both online and offline systems have their advantages, and are not mutually exclusive. Both approaches should be considered in the design of a service conflict detection and resolution system, and examples of each of these systems will be discussed further on.

Regardless of whether offline or online conflict detection is used, some mechanism to resolve conflicts needs to exist. This mechanism can either disable the service with the lowest priority, or manipulate the session so that the conflict does not occur. This manipulation can take the form of modifying the SIP request to fall within acceptable bounds for either service. For example, if a particular service modifies a header that another service requires to function correctly, the conflict resolution system can restore the original value before invoking the second service. This needs to be tightly controlled however, and requires significant intelligence to identify which parameters are core to a service's operation.

8.3.3 Requirements

Several requirements for the success of a service orchestration system can be identified:

- *Should not require large modifications to existing architecture.* Any enhancements to the core of the IMS network should not require large modifications to the existing architecture or to existing services. It should allow for backwards compatibility with existing services, while providing a smooth transition should developers wish to utilize additional functionality.
- *Flexible enough to handle new services.* As many services will be designed by a wide range of developers on an ongoing basis, the SCIM should be able to adapt to the addition of new services without requiring large amounts of maintenance.

■ *Support existing control protocols.* SIP is used extensively throughout the IMS as the control protocol, while Diameter is used to convey information in a more secure manner. As the SCIM will need to interface with the SCSCF as well as the AS it is necessary to support SIP. The SCIM will also need to pull information from the HSS and to access information stored in the user's profile. This information can be used to determine what services the end user is allowed to access, as well as the manner in which these services should be accessed.

■ *Intelligent conflict detection and resolution.* Negative service interactions have been a problem in numerous network architectures, and the IMS is expected to exacerbate this situation by greatly increasing the number of services and developers. The SCIM should be able to make intelligent decisions based on external factors in addition to the pre-existing conditions contained within the SIP request. This will allow richer and more flexible services.

■ *Service access management.* It should be possible for the service orchestration function to authorize or control the access between different enablers or service capabilities. Network policies should be used to prevent unauthorized access to particular services' capabilities. Only authenticated and authorized use of the network's services should be allowed, even between different services.

■ *Minimize session setup delay.* The SCIM should take advantage of offline conflict detection in order to reduce the processing time needed to establish sessions. It should not get involved in the signaling path unless it is necessary. For example, the SCIM should not be invoked for an ordinary voice call between two parties.

8.3.4 Existing Research

The study concluded by the 3GPP proposed three categories of network architectures for Service Brokers (SBs). These are the centralized, distributed, and hybrid approaches. The centralized architecture has one Service Broker (SB) that communicates with and controls multiple AS. In this approach, the SCSCF treats the SB as another AS. The SB can be integrated with the SCSCF or it can be a stand-alone entity. The interfaces between the SB, the SCSCF, and the AS are all SIP-based interfaces. The distributed approach contains multiple SBs that each communicate with a single AS (Figure 8.3). Here, the SB can be integrated into the AS or it can be a stand-alone entity. As in the previous architecture, the SCSCF treats the SB as an AS, and the interfaces between the SCSCF and AS are SIP based. The SCSCF needs to ensure that all the appropriate messages are passed in a chain from SB to SB, until they have all finished executing their services. With the hybrid approach, the number of SBs and their connections to the various AS are not defined as strictly. It is possible to have a SB that acts both as a centralized and distributed SB

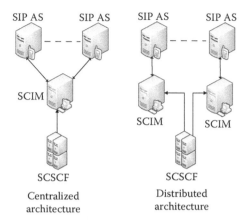

Figure 8.3 Difference between centralized or distributed SCIM architectures.

(in terms of how it connects to the AS). The following sections discuss the approaches taken by existing research in this area (Figure 8.3).

8.3.4.1 Architecture

Gouya et al. [12] discuss a centralized hybrid approach, where the SCIM is placed outside the SCSCF and it is viewed as an AS that communicates with the individual service capabilities [12]. They also propose a second approach that can be classed as a distributed hybrid architecture. In this architecture, they have multiple SCIMs and multiple SCSCFs, with each SCIM integrated into a single SCSCF. They claim that their centralized network architecture is not scalable as each SCIM creates a bottle neck, whereas their distributed approach is scalable as they place each SCIM within a SCSCF. However, in their centralized architecture they have one SCSCF talking to one SCIM, and therefore, they can have the same number of SCIMs in both architectures. Thus, the only real difference between their architectures is the placement of the SCIM within the SCSCF or integrated with the AS.

In terms of managing the fault tolerance of the system, they claim that having the SCIM integrated with the SCSCF is a better approach as each SCIM/SCSCF has a smaller impact when it fails. However, if the SCIM fails, the SCSCF will fail and vice versa as they are now a single functional entity. If this integration did not take place (i.e., the SCIM was a separate entity but still connected to the SCSCF) the network could carry on operating, and only services requiring a SCIM would fail. The SCSCF will carry on handling basic session establishment. Thus, it is more robust to have the SCSCF and SCIM as separate entities [13]. By integrating the SCIM and SCSCF it is possible to reduce the complexity of the SCIM as it can now take advantage of the existing diameter connection to the HSS, and does not need its own interface. This approach means that each SCSCF needs to be modified to include the SCIM, and it is no longer possible to deploy a SCIM in an existing IMS network without additional work. As the IMS architecture stands, the SCSCF does the most work on the control signaling as it is responsible for passing every message through the network and determining where to route the initial requests. Therefore, any steps that increase its load need to be carefully examined.

Nakajima et al. [13] agree with the placement of the SCIM outside the SCSCF. They propose that while placing the SCIM inside the SCSCF would also fall within the 3GPP-proposed standards, there is a scalability benefit and increase to failure tolerance to be gained by placing it outside as a stand-alone function. They state that the "SCSCF is considered to be busy handling basic calls. For scaling, the SCSCF has to concentrate on handling basic calls by delegating the proposed functions to the SCIM" [13]. In terms of failure impact, they classify the different types of calls or sessions into three distinct groups. These three groups are: basic calls, emergency calls, and enriched services. If the SCIM is separated from the SCSCF, and the SCIM fails, only enriched services will fail to operate. Most importantly, the emergency calls will continue to operate successfully, together with the basic calls. If a combined SCSCF/SCIM fails, then all three groups of sessions will fail. The latest IMS standards have introduced another element called the Emergency Call Session Control Function (ECSCF) to provide a dedicated entity to deal with the establishment of emergency calls, and thus, the concern about the emergency calls being affected by the combined SCSCF/SCIM falls away. However, it is still preferred to only have enriched services failing with a SCIM failure. In both scenarios, if the SCSCF fails, all groups of sessions will fail as there would no longer be a functional entity passing messages from the end user to the intended destination (whether that is the SCIM or another user). Therefore, it is better to separate these functions when one considers the potential impact that a component failure will have on the network. Another consideration discussed is that the AS can be provided by third parties, and therefore, should not be trusted implicitly [13]. As they will be talking to the SCIM or SCSCF, the exposure of the network to these third-party service providers should be limited. If they manage to overload their point of contact with the network (i.e., the SCIM or SCSCF/SCIM), any damage dealt to the network's operational ability must be limited. Thus, from a threat management point of view, it makes sense to separate the SCSCF/SCIM functionality.

By defining the SCIM as a separate functional entity to the SCSCF, it no longer has to operate as a SIP Proxy (the SCSCF is standardized as a SIP Proxy). Thus, the SCIM can operate in another mode, such as a B2BUA. B2BUAs are allowed to create and terminate SIP dialogues, implying that they are no longer restricted to operating within the SIP boundaries of a proxy. They can use their own timers to control when to terminate or establish sessions, and do not have to rely on the timeout mechanisms built into SIP.

Qi et al. [14] follow a more analytical approach to defining their architecture. They investigate the performance implications of having a centralized architecture, and suggest splitting some of the functionality into two separate entities. They suggest having a module that is integrated with each AS, as well as a central node that informs this module of the next destination for service requests. The core idea they propose is to allow the AS to pass messages between each other without having to send the message back to the SCIM/SCSCF. By bypassing the SCIM and SCSCF their simulations show a large reduction in the load on the SCSCF, as well as decreasing the session setup delay. The benefits of this model are related to how long it takes each service to execute its logic, as the benefits arise from determining the next hop for a service request in parallel with the execution of the current service's logic. The number of signaling messages increases in order to communicate this information between the central node and the modules integrated with the AS. Further work is needed with this architecture to investigate how it handles faults. If a particular AS malfunctions and does not forward the request correctly or times out, the service chain is broken. The recovery mechanisms for this approach are only as complex and intelligent as each module that is integrated to the AS allows.

Xia et al. [8] discuss Lucent's proprietary approach to a SCIM. The major difference with this architecture is that they allow other services to interact with the SCIM through various protocols in addition to SIP. They propose allowing services to communicate with the SCIM through protocols such as HTTP. No matter which protocol is used to communicate with the SCIM externally, it would translate it to SIP to forward on to the internal IMS network and the relevant AS. This combined approach does not seem to be aligned with the current standardized practice of having a gateway function that only handles interoperability, such as the gateway to the Parlay X API [15].

While all the architectures discussed so far have focused on the placement of the SCIM within the overall hierarchy of the network, Goveas et al. [11] place more emphasis on defining the internal structure of the SCIM. They discuss aspects relating to the "Policy Engine," the "Data Federator," the actual "Service Broker" as well as a "Charging Agent." It is important to mention that the Data Federator introduces new interfaces through which data can be fetched from various sources, such as the HSS or a XML Document Management Server (XDMS). This information would then be used to make intelligent decisions regarding the triggering of services and their interactions.

8.3.4.2 Functionality

Regardless of the approach taken in the layout of the network architecture, several functions have been identified as being necessary for a SCIM. Before the SCIM can control any service interactions or AS, it needs to have some form of knowledge about these services. It also needs to have some mechanism to cope with the detected conflicts.

Almost all the approaches covered here use the standard diameter interface with the HSS to download user information (the user's IMS profile). However, we are more interested in how these systems handle the information about each service.

Gouya et al. [12] propose using a "formal model of all the services [the SCIM] may trigger. This formal model would [make] explicit the decomposition of a service into service capabilities and the interactions needed between service capabilities and the end-users." This would be classed as an offline conflict detection mechanism, and while it has a high level of control over the service interactions, it is not ideal because as the number of services increase this model would need to be updated. The extension of this system to perform in a dynamic way has been identified by Gouya et al. [12] as an area requiring future work. They state that "increased diversity in the domain of next generation services" demands that the "extension of the service capability interaction rules must be performed in a dynamic way to facilitate the introduction of new services" [12].

Nakajima et al. [13] make a distinction between two main types of services to make detecting and resolving conflicts easier. The separating factor between these two groups is the amount of call control that the service exhibits. Where there is a large amount of call control the services are classed as "Connection Control Applications." An example of a Connection Control Application would be a "Click-to-Dial" application. This is where the end user visits some mechanism, for example a Web page that allows him to establish a voice call. He would click on a link on a page, and would interact with an AS that would connect him to the called party. Another call control service is Third Party Call Control (3PCC), where the actual call is controlled by a third party that is not necessarily included in the call. If there is not a large degree of call control exhibited by the service, it is classed as a "Coordination Application." Examples of these services include services such as call redirection, call blocking, and so on. By making this distinction, Nakajima et al. [13] can detect conflicts as only one service from the first group is allowed per session. Multiple service interactions from the second group are allowed, as they enrich the service and do not terminate the call or session.

Qi et al. [14] store information about the possible service interactions and the details of the AS in XML documents. They also store policies which allow the network operator to control these interactions.

Another offline conflict detection and resolution mechanism has been proposed by Goveas et al. [11]. Here each combined service is stored as an acyclic graph. Each service capability is represented by a node, and different routes through the graph indicate different combined services. They define rules which identify, along which edge the overall service flows (i.e., determining the next service capability to trigger from the current one). These rules take advantage of context-specific data as well as the known information about each service. These routes or service flows are defined statically by the developer of the service before the service provisioning actually takes place, and thus, limits the flexibility of the service. This approach leads to difficulties when trying to integrate services offered by multiple different third parties.

An interesting online conflict detection and resolution system is proposed by Kolberg et al. [16]. Here each service is wrapped in a "cocoon," which sits in front of the AS. Each SIP message that is destined for the AS is instead received by this cocoon, which saves a copy of the message. It sends the message to the service processing logic unchanged. When the service has finished executing, it may have made some changes to the SIP message. The original message is compared against the message returned by the service, and a header is injected into the SIP message. This header is called the "ConType" header, and contains information about which service was activated, and any changes that were made to the SIP message. This allows it to detect any conflicts on subsequent service interactions, as each cocoon knows what parameters it requires or modifies. If a cocoon receives a message with a preexisting ConType header, it examines it to detect any conflicts before passing the message on to the service. If a conflict is detected one of the services needs to be disabled. This leads to a problem with this approach, as the lower-priority service (i.e., the service that should be disabled in case of conflicts) could have been the first service to activate, and has hence already spent time processing the message and executing the service logic. This results in the service being terminated, and a SIP 380 Alternative Service message is returned to the originating user. In this message the ConType header is returned, indicating which service needs to be disabled. The original signaling flow for the session set up would then be repeated with this ConType header injected into the original request. When the SIP request reaches the cocoon of the disabled service, the cocoon would detect that it should be disabled, and would pass the message on to the next service without triggering its own service. This leads to many redundant signaling messages, as the entire session setup procedure would have to be repeated for every conflict detected. Another drawback to this approach is that every service needs to know the priority of every other service, so that each cocoon can know when to disable its service and allow another one to take over. Thus, every time a new service is configured or added, each service needs to be updated with the new list of priorities.

While the work done by Kolberg et al. [16] uses the SIP signaling protocol, it is not specific to the IMS architecture. Chiang et al. [17] extend this solution to the IMS platform by placing the cocoon functionality inside the SCIM. The SCIM caches the original SIP request so that it can act on the end user's behalf in reestablishing the session when service conflicts are detected. The functionality of the ConType header is largely replicated, although it is given a new name in the Service-Indication header. This means that the redundant signaling does not have to travel over the access network, which will reduce the session setup time considerably as well as reducing the bandwidth used. However, it does not eliminate this overhead entirely as it still takes place over the internal core IMS network. Services that will play no part in the final session are still included in the initial session setup, and the session setup is repeated for every combination of services that result in conflicts.

8.3.5 Open Research Areas

While progress has been made on the offline mechanisms for detecting service conflicts, an efficient real-time online resolution mechanism that does not incur penalties due to redundant signaling is still required. In both online and offline detection systems, information about the services and their interactions are needed. It is proposed that there should be a mechanism allowing the automatic exchange of this information whenever a new service comes online. This would require a standardized way of capturing all the succinct details of a service in a manner that is compatible with third party developers. Third-party developers could build this into their new services, such that their own service would communicate its needs and requirements to the SCIM. This information could include the details of what SIP headers the service reads and/or modifies any external conditions such as time of day and the capabilities provided by the service such as audio or video streams. Another detail that could be recorded is whether the service is a call control service or a service that does not need further interaction (such as an email notification service). This would allow the SCIM to handle these services in parallel with the call control services, speeding up the session setup.

An area of importance that the 3GPP have left for further study at the moment is one of allowing the user more control in their choice of service integration. It is envisaged that roughly the same functionality can be provided by multiple third parties, and the user should be able to choose which service capabilities make up their final blended service. The 3GPP mention that a framework that allows the "users to personalize and control their services" needs to be developed. "The architecture should allow end users to personalize and control how applications work together when there are multiple choices of integration available" [5]. This framework would need to have a user interface, as well as functional entities that can store the policies that would govern this interaction. These policies would need to be expressed in a form that the SCIM can understand, and a mechanism that handles the transfer of these policies to the SCIM needs to be investigated.

Any modifications to the session setup delay need to be examined closely. A test bed implementation of the above-mentioned solutions should be examined to determine the impact they have on key performance indicators such as the session setup delay. If the session setup delay grows too large, the quality of experience for the end user will be degraded.

8.4 Conclusion

This chapter has studied the currently available mechanisms for service conflict detection and resolution in the IMS, which were found to have several limitations. It has examined the different architectures possible to deploy a Service Broker, and found that there are numerous advantages to placing the SCIM outside of the SCSCF and AS. The requirements for a successful service orchestration system running over the IMS platform were determined. The IMS has the potential to offer the end user an enhanced, richer communication experience that is better than currently deployed services on the Internet. However, there are still challenges facing the development and deployment of these services. By using service capabilities or enablers, it is possible for the developer to leverage existing work to create innovative services that users will want to use. There is a need for further research in the area of real-time conflict resolution mechanisms as well as a standardized way to represent service information. Both offline and online conflict detection systems have their advantages, and it is recommended that future research in this area focuses on a system that utilizes both offline and online conflict detection. User-centric services deployed by "over-the-top" providers are currently performing well, and network operators will have to match this level of service if they wish to compete with enriched multimedia services available on the Internet.

References

1. Morgan Stanley, Economy and Internet Trends, Web 2.0 Summit, October 2009. [Online]. Available: http://www.morganstanley.com/institutional/techresearch
2. Next Generation Networks Global Standards Initiative (NGN-GSI). [Online]. Available: http://www.itu.int/ITU-T/ngn/definition.html
3. S. Tsang and E. Magill, Detecting feature interactions in the intelligent network, *Feature Interactions in Telecommunications Systems*, 236–248, 1994.
4. M. Calder, M. Kolberg, E. H. Magill, and S. Reiff-Marganiec, Feature interaction: A critical review and considered forecast, *Comput. Netw.*, 41(1), 115–141, 2003.
5. 3GPP, *23.810 v8.0.0, On Architecture Impacts of Service Brokering, (Release 8)(SA2)*, September 2008.
6. J. Rosenberg, H. Schulzrinne, G. Camarillo, A. Johnston, J. Peterson, R. Sparks, M. Handley, and E. Schooler, SIP: Session Initiation Protocol, RFC 3261 (Proposed Standard), Internet Engineering Task Force, June 2002, updated by RFCs 3265, 3853, 4320, 4916, 5393. [Online]. Available: http://www.ietf.org/rfc/rfc3261.txt
7. P. Calhoun, J. Loughney, E. Guttman, G. Zorn, and J. Arkko, Diameter Base Protocol, RFC 3588 (Proposed Standard), Internet Engineering Task Force, September 2003. [Online]. Available: http://www.ietf.org/rfc/rfc3588.txt
8. N. Xia and W. J. Zhai, Study on IMS Service Broker, in *Proc. Third International Conference on Convergence and Hybrid Information Technology ICCIT '08*, Vol. 2, November 2008, pp. 340–343.
9. *The IMS lantern: Standardization: SCIM and service broker*. [Online]. Available: http://theimslantern.blogspot.com/2007/05/standardization-scim-service-broker.html
10. *Open Mobile Alliance, Presence Simple Version 2.0*. Draft Enabler Release, March 2008.
11. R. Goveas, R. Sunku, and D. Das, Centralized Service Capability Interaction Manager (SCIM) architecture to support dynamic-blended services in IMS network, in *Proc. Second International Conference on Internet Multimedia Services Architecture and Applications IMSAA 2008*, December 2008, pp. 1–5.
12. A. Gouya, N. Crespi, and E. Bertin, SCIM (Service Capability Interaction Manager) implementation issues in IMS service architecture, in *Proc. IEEE International Conference on Communications*, Vol. 4, June 2006, pp. 1748–1753. Istanbul, Turkey.

13. M. Nakajima, M. Kaneko, M. Hirano, and H. Sunaga, SIP servlet dialogue handling for NGN Service Capability Interaction Manager, in *Proc. APSITT Information and Telecommunication Technologies 7th Asia-Pacific Symposium*, April 2008, pp. 41–46. Bandos Island.
14. Q. Qi, J. Liao, X. Zhu, and Y. Cao, DSCIM: A Novel Service Invocation Mechanism in IMS, in *Proc. IEEE Global Telecommunications Conference IEEE GLOBECOM 2008*, December 2008, pp. 1–5. New Orleans, LA, USA.
15. Open Service Access (OSA), *Parlay X Web Services, (Release 7)*, June 2007.
16. M. Kolberg and E. H. Magill, Managing feature interactions between distributed sip call control services, *Computer Networks*, 51(2), 536–557, 2007, feature Interaction. [Online]. Available: http://www.sciencedirect.com/science/article/B6VRG-4KXVB79-1/2/d2ba17665d6f030e94ce37a2444efc63
17. W. Chiang and C. Tseng, Handling feature interactions for multi-party services in 3GPP IP multimedia subsystem, in *Proc. International Conference on IP Multimedia Subsystem Architecture and Applications*, December 2007, pp. 1–5. Bangalore, India.

Chapter 9

Design and Implementation of an IMS-Based Testbed for Real-Time Services Orchestration and Delivery in Heterogeneous Networks

Luying Zhou, Lek Heng Ngoh, Teck Yoong Chai, Teck Kiong Lee, Xu Shao, and Joseph Chee Ming Teo

Contents

9.1 Introduction

New and innovative communication services are the latest buzzword for both telecommunication companies and network providers. In the future IP (Internet Protocol)-centric climate, all wired and wireless communications in the form of Telephony, Fax, Email, Internet access, Web services, Voice over IP (VoIP), Instant messaging, Videoconference sessions, Video on demand (VoD), and so on can be carried in "packet frames." With such convergence anticipated, operators need a new architectural framework that embraces flexibility and expandability in introducing new services as they continuously emerge and evolve, without forcing customers to frequently change carriers.

To avoid developing services individually for each specific application, a service platform is needed that presents standard interfaces to allow applications to access and orchestrate these services. Such standard platforms have been developed over the past years, for example, Parlay for telecommunication networks [1], and IP Multimedia Subsystem (IMS) for IP-based fixed and mobile networks [2–4]. These platforms are composed of service functions and standard interfaces for applications to discover, access, and orchestrate those service functions.

The IMS is a specification of an environment where these types of services can be rapidly developed, deployed, and delivered with relative ease in a standardized fashion as compared to current vendor-based solutions. IMS is built for services and applications, providing operators the capability to manage rich multimedia services across both next-generation and traditional (with limited functionalities) networks. In addition to deriving added value from their own networks, operators can further benefit by opening up these networks to host third-parties' services and applications. IMS is standard based and one of the key features is the use of open interfaces and functional components in both hardware and software systems to support real-time interactive services and applications.

IMS was introduced by the 3rd-Generation Partnership Project (3GPP) initiative to control and provide multimedia services over 3G mobile networks, but it is now adopted in the next-generation network (NGN) standardizations. The key technology behind IMS is Session Initiation Protocol (SIP) [5]. Many of the important interfaces between IMS components use SIP as the underlying signaling protocol.

IMS specification began in 3GPP Release 5 as part of the core network evolution from circuit switching to packet switching and was refined by subsequent Releases 6 and 7 [6]. Till today, IMS standards are still evolving. Both the International Telecommunication Union (ITU) [7] and the European Telecommunications Standards Institute (ETSI) [8] are also involved in the IMS standardization efforts.

We aim to build our testbed on the principle of Service-Oriented Architecture (SOA), with an emphasis for real-time network services. SOA is an architectural style that guides all aspects of creating and using business processes packaged as services throughout their lifecycle, as well as defining and provisioning the IT infrastructure that allows different applications to exchange data and participate in business processes, regardless of the operating systems or programming languages underlying those applications. A number of SOA platforms and Web-based solutions are focused mainly on business process issues. For our development work, we need a platform on which an assortment of experimental tools and products may be deployed and allowed to interact in real-time, thus, allowing for rigorous, transparent, and replicable testing to be achieved. For our testbed platform, IMS technology is applied to form the core network.

9.2 IMS-Based Testbed

A system overview of our IMS-based testbed is shown in Figure 9.1, and a more detailed description of the testbed components, especially, of a service orchestration module and its application, was presented in [9]. At the core of this testbed is the "Open IMS Core" from Fraunhofer Institute FOKUS, Germany [10]. The underlying physical network is an Optical Ethernet network with Ethernet Passive Optical Network (EPON) access technology [11] and WiFi wireless network. The Optical Network Control and Management Plane [12] is the network-specific interface to reconfigure the optical network and provision bandwidth with the ability to support carrier-grade Ethernet transport services, for example, QoS (Quality of Service).

9.2.1 Open IMS Core

The Open IMS Core project of the Fraunhofer Institute FOKUS was officially launched in November 2006. The Open IMS Core is an open-source implementation of 3GPP's IMS standard. It consists of Call Session Control Functions (CSCFs)

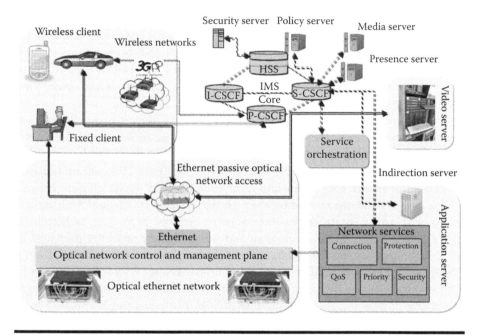

Figure 9.1   IMS-based testbed system overview.

and a Home Subscriber Server (HSS). The Open IMS CSCFs (Proxy, Interrogating, and Serving) form the central routing elements for IMS-based signaling, while the HSS manages user profiles and associated routing rules.

The Open IMS Core interfaces with multiple servers such as Security, Policy, Media, Presence, Video, and Application Servers, so that fixed and wireless clients alike can establish multimedia sessions with them in a standard manner. Lastly, the Core is linked to a SIP-based service orchestration module [13] where the composition logic is stored and executed. This module provides the possibility of rapid service creation and delivery in response to market needs.

9.2.2 Presence Service

Our testbed consists of a SIP-based Presence Service [14] that provides presence functionalities to the rest of the application servers using standards developed by the Internet Engineering Task Force (IETF), Open Mobile Alliance (OMA), 3GPP, and ETSI. This service is linked into the IMS core and application servers via the native SIP presence protocol and service building blocks (SBB), respectively.

In contrast to the HSS database that contains user information that is short term and transient in nature (e.g., login status and subscription profile), the presence service manages state information which are more static in nature (e.g., resource authorization rules and policy).

The Presence Service actually comprises three separate but interrelated servers. This means that each of these servers can be used independently or combined to provide the necessary integrated functionalities.

The functions of each of the servers are briefly described below:

- The XML document management Server (XDMS) is responsible for handling the management of network user XML documents such as presence authorization rules, static presence information, contact and group lists (see "resource list server" next), policy data, and so forth.
- The Resource List Server (RLS) handles subscriptions to Presence Lists. It creates and manages back-end subscriptions to all resources in the Presence List. The list content is retrieved from the XDMS.
- SIP Presence Server is responsible for accepting, storing, and distributing presence information on behalf of resources. It obtains the required resources data such as authorization rules from the XDM server, and manages subscriptions and generates notifications.

9.2.3 Indirection Service

In its simplest form, the Indirection Service translates a channel name into the video server address from which the client could stream the video content. The Indirection Service can be invoked through SIP protocol, either directly by the client or as part of a chain of service invocations. In either case, the service should behave exactly the same way. The channel mapping table may contain information about the video contents, for example, video formats and bit rates, to aid media capability negotiation. The negotiation process, however, may be handled by a separate service making use of the information obtained from the Indirection Service. In our testbed, the Indirection Service is implemented as a SIP servlet, hosted on Mobicents Platform [14]. Mobicents SIP Servlets is an open-source implementation of the SIP servlet's open standard [15]. The implementation of the Indirection Service is general enough to run on any platform that supports the standard, for example, Sun's SailFin [16].

9.2.4 Network Service

The Network Service, implemented as SIP servlet in Mobicents, packages and exposes proprietary network functionalities such as bandwidth management and network protection as reusable service components which can readily be combined with other service enablers to provide multimedia services with managed QoS. The network functionalities, though vendor-specific in the way they are configured and controlled, are exposed in a generic and standard way via the IMS framework, as shown in Figure 9.2. The Network Service is a suite of services that abstract the

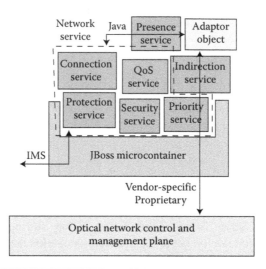

Figure 9.2   Network service exposed as standard IMS service.

network functionalities so as to allow users to request managed optical Ethernet connections with dedicated bandwidth and other properties such as user authentication and data integrity.

The Network Service includes the following basic service modules:

- *Connection Service:* Supports end-to-end and multipoint LAN connections. Through this service, connections may be established, modified (e.g., nodes join or leave a multipoint connection), or released.
- *QoS Service:* Enables the specification of quality of service parameters for individual/multiple traffic flows or connections (e.g., bandwidth and availability guarantees).
- *Priority Service:* Allows multiple traffic flows to share a link or a switch port, and assigns different transmission priorities to individual traffic flows.
- *Protection Service:* Provides different types of path protection, being shared or dedicated protection, full protection, or partial protection.
- *Security Service:* Provides user authentication and data encryption/decryption service for the established connections.

9.2.5 Security Service

The security server SECS is implemented in our improved IMS-based architecture as shown in Figure 9.3 to provide security services. SECS provides security services such as two-party key management, group key management, data origin authentication, data integrity and confidentiality, and privacy of client identity. The current

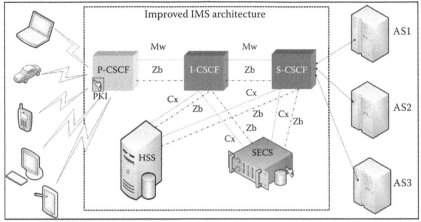

Mw : SIP-reference point among different CSCFs

Cx : Reference point for S-CSCF and I-CSCF to acquire information from
HSS and SECS, in which the employed protocol is the diameter protocol

Zb : Security interface in 3GPP network domain security

PKI : Public key infrastructure

AS : Application server

Figure 9.3   Authentication and data security server.

IMS security standards only address two technical specifications, namely access security [17] and network domain security [18]. Access security is concerned with the mutual authentication between the client and the IMS core via the IMS Authentication and Key Agreement (AKA) protocol [19]. Access security is also concerned with securing the first hop signaling between the client and P-CSCF through the security features of the SIP protocol. Network-domain security is concerned with providing security within the network domain for communications between the IMS components, namely HSS and the P-CSCF, I-CSCF and S-CSCF via the Cx interface. However, both access security and network domain security do not address the security of the applications and services provided. Our security service can provide secure video multicast service using the IMS architecture, the video multicast service can request for the group key management and data confidentiality service from the security server. The group key management service will generate a secure group key that will be used by the video server to encrypt the video data before multicast and by the IMS-based clients to decrypt the received video data (i.e., to provide data confidentiality). The group key management service will also ensure that the group key is changed whenever new IMS-based clients join the secure video multicast group or whenever clients leave the secure video multicast group. This ensures that new clients are not able to watch videos that are multicast before they join the group and leaving clients are not able to continue watching multicast videos after they have left the group.

9.2.6 Service Orchestration Module

The main purpose of this module is to enhance the basic functions provided in the current IMS's S-CSCF (the S-CSCF 'sits' on the path of all signaling messages, and can inspect every message). Under this new arrangement, the SIP-based service orchestration module is the first module to be contacted by the I-CSCF after it is contacted by the user terminal. I-CSCF will also provide the relevant user-related information. It decides both the order and to which application server(s) (both IMS-based and Web Service (WS)-based) messages will be forwarded in order to provide the relevant services. It accomplishes this by sending the messages via the S-CSCF module.

There are two objectives to be achieved by this new service orchestration module:

1. *Chaining of SIP-based services:* This function allows for the setting up of a SIP session that involves two or more application servers (AS). For example, a "Do Not Disturb (DND)" service can be added and placed in front of the "Press to Call (PoC)" service so that before a PoC service is engaged, an instance of the corresponding session description protocol will be generated. This session description protocol will then be used to direct the setting up of a PoC service with the requirement to first consult the DND service. Depending on the outcome of initializing DND, a PoC may be set up subsequently. To accomplish this function, the service orchestration component is expected to interact with the S-CSCF according to dynamically generated session description protocol information.

2. *SIP- and WS-based service chaining:* In contrast to (1), the chaining of services now involves some of those services offered at WS level and others at SIP level. For example, to realize a "blended" service that allows a viewer to answer his call while watching IPTV (Internet Protocol Television). This will engage the IPTV AS implemented in WS- and SIP-based equipment involving accepting and answering voice calls. Whenever a call arrives, the orchestration component will determine the appropriate AS to notify the user, and pass along the ID of the caller. Depending on the decision of the viewer to either accept or reject the call, the appropriate signaling action at the SIP level will be activated accordingly. Notice that all this while, all session-related information is kept in the orchestration components and not at the IPTV AS. To accomplish this function, appropriate static mapping between the appropriate user-interface component and the SIP-level session trigger information must be provided.

9.3 IMS-Based Testbed Experimentation

We have conducted experiments on the IMS-based testbed to evaluate performance of the design for various applications. We evaluate the effectiveness of the service orchestration module on composing new service, and show the multimedia applications over EPON network are supported through the IMS enabled service plane architecture. We also show that the service orchestration can be achieved through smart phone over wireless mobile network.

9.3.1 Service Orchestration

To demonstrate the power of service orchestration, a new composite service is created in the SIP-based Service Orchestration Module, which combines Indirection and Network Services. The usage scenario, as shown in Figure 9.4, is as follows: After successful logon to the IMS, a user would select an IPTV channel from a list shown on the client software, which will then be streamed from the video server with the most appropriate network QoS settings. The configuration of the Indirection Service is similar to that of the University of Cape Town's IMS-based IPTV [20] with the following additional contributions from us:

■ The video data is transmitted over a managed network, which can be configured through our Network Service based on the actual bandwidth requirement of the streaming session.

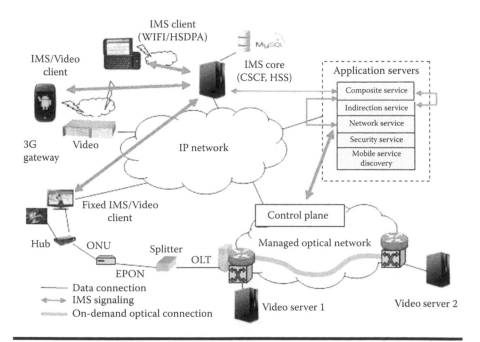

Figure 9.4 Mobile service orchestration over IMS testbed.

- The video data transmission between the video client and the video server is encrypted to provide proper security and privacy. The encryption and decryption modules are built into the client and server.
- A composite service, which combines two independent services, that is, the Indirection and Network Services, is provided by means of service orchestration in our SIP-based Service Orchestration Module.

The experiment is based on the following open-source software. Though the code base is from diverse sources, they integrate well due to good standard compliance.

- Open IMS Core from the Fraunhofer Institute FOKUS [8].
- UCT IMS client from the Communications Research Group at the University of Cape Town [20].
- VLC media player and video server from VideoLAN project [21].
- Mobicents Platform [13].
- Mobicents Sip Servlets and Presence Server [13], which run on Mobicents Platform.

Our composite service invokes the constituent Indirection and Network Services as part of the steps in its orchestration logic to serve videos through managed optical Ethernet connections with dedicated bandwidth. The orchestration logic is composed in Java using the SIP servlet API. We implemented the SIP servlet message handling methods to invoke the desired actions in response to each type of message. The message exchange between various parties is shown in Figure 9.5 (in Section 9.3.3, more specific message exchanges for wireless mobile application). When a SIP INVITE is received, the message is relayed to Indirection and Network Services in turn. Both the outgoing INVITE messages are associated to the incoming INVITE message in a back-to-back relation. The Indirection Service would reply with SIP OK carrying the access information related to the requested channel, for example, server address, codec, and so on. The Network Service then interacts with the Optical Network Control and Management Plane using proprietary signaling to set up the network as required. If both INVITE messages are successful, as indicated by the receipt of SIP OK messages from Indirection and Network Services respectively, SIP OK will then be sent to the requester along with the video information (copied from the Indirection Service's SIP OK), based on which the requester can start streaming video.

9.3.2 Multimedia Services over EPON Access Network

EPON, specified by IEEE 802.3ah, enables multiple ONUs to share a symmetrical 1 Gbps downstream and 1 Gbps upstream bandwidth resources. As a low-network-cost and high-data-rate network, EPON is an attractive solution to broadband access networks for business and residential users. In order for

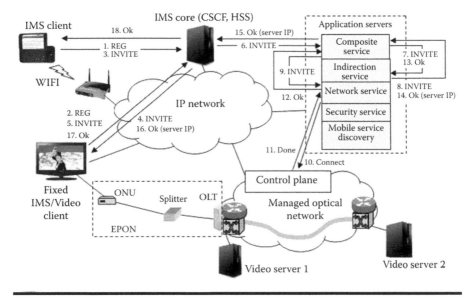

Figure 9.5 Signaling sequence for mobile phone initiated service.

multimedia service to be efficiently supported on the EPON network for subscribed users, the EPON should be able to allocate and guarantee the downlink
bandwidth resource for these multimedia services, that is, resource on EPON is
allocated according to user's service requirements. Furthermore, to allow users to
access subscribed multimedia services, such issues as user authentication and
authorization, requested service discovery/location, and service usage charging,
which are related to service management, must also be addressed. It is obvious
that the problems of resource allocation on EPON and service management are
related and should be addressed in an integrated manner to provide efficient multimedia services. We apply the IMS service plane enabled network architecture
and develop a resource allocation method that, by integrating service plane functionality with EPON network traffic flow control scheme, provide efficient service access and management, and bandwidth resource allocation for multimedia
services over EPON networks.

The implemented system architecture consists of a service plane that provides
the required service and resource management functions, and the EPON network
with a traffic controller that classifies traffic flows into different service priority
queues, and an inter-module communication mechanism that facilitates the information exchange between the service plane core and the EPON controller. We
developed an EPON network resource management module in the service plane
for service request admission and EPON resource control. This module utilizes
users profile information available from the HSS in the IMS core, such as subscription and service priority to optimize the EPON networks resource allocation

for multimedia traffic, both unicast and multicast traffic, and communicates with other modules in the service plane, such as billing module, to apply different price plans. With this module, the EPON resource controller will get instant user service information and apply it in making resource allocation decision based on user's service agreement and service priority. The resource management module performs user's service request administration control and resource sharing under the circumstance of differentiated traffic priority and traffic congestion, and provides guaranteed performance for prioritized services. The resource management module maintains a record of users' average resource usage for subscribed services and the service type, for example, unicast or multicast, requested by the users in the EPON network.

In the experiment, the traffic admission decision made in resource management module is applied to EPON OLT to schedule the downlink packet transmission. After the requested service is granted, the resource management module sends control message to EPON controller to inform the user's traffic type and priority. In the OLT traffic controller design, the subscribed multicast packets are put in one logical queue in OLT to avoid duplicate packet delivery, unicast packets are put into different unicast queues based on their service priorities. Packets in different queues have different priorities to be transmitted. Traffic of subscribed services has higher priority than that of unsubscribed services, which is transmitted in a best-effort manner. The resource allocation scheme controls the downlink bandwidth resource allocation and provides a fair EPON resource usage among the users in the case of users competing for the resource, and is able to differentiate user's service priority and charge users for different service and resource usage accordingly.

9.3.3 Multimedia Services over Wireless Mobile Network

This experiment illustrates the use of smart-phone to initiate and orchestrate the playback of online video program on-the-go, and to a display device of choice. Using the IMS-based mobile service platform, an Android phone authenticates and interacts with the IMS core network to subscribe a video channel from the program listing. The orchestrated service shows selected videos streamed to a remote client for display via quality-of-service-guaranteed connections established via an IMS-based optical network service plane.

Figure 9.5 shows the steps and signaling sequence for the wireless mobile experiment.

The detail steps are as follows:

Step 1: The Android phone sends a SIP REGISTER message to the IMS Core to register to the IMS.

Step 2: The TV (video client) also sends a SIP REGISTER message to the IMS Core to register itself to the IMS.

Step 3: The Android phone sends a SIP INVITE message to the IMS Core. This SIP INVITE message contains the request for a particular IPTV Channel, say Channel 1, to be shown on the TV (video client). It could be noted that IPTV Channel 1 is residing on VideoServer2 which is connected to the Managed Optical Network. The TV (video client) is also connected to the optical network.

Step 4: The IMS Core, upon receiving the SIP INVITE message in Step 3, will send the SIP INVITE message to the TV (video client).

Step 5: Upon receiving the SIP INVITE message in Step 4, the TV (video client) will invoke the IPTV service (Channel 1) and send a SIP INVITE message to the Application Servers (AS) via the IMS Core to request for IPTV service.

Step 6: The IMS Core will forward the SIP INVITE message from the TV (video client) to the AS. Note that the SIP INVITE message contains the IPTV Channel 1 request.

Step 7: The Composite Service at the AS receives the SIP INVITE from Step 6 and will relay it to Indirection and Network Services in turn. The Composite Service provides application routing function, that is, to invoke the right services in sequence according to various dynamic conditions such as user preference and privilege.

Step 8: The Composite Service routes the INVITE message to the Indirection Service where the access information related to the requested channel, for example, server address, codec, and so on, is looked up in a database and added to the message.

Step 9: The Indirection Service passes on the INVITE message. The Composite Service, playing the role of application router, intercepts the message and determines from the composition logic that the next service to be invoked is the Network Service. The Composite Service will then relay the message to the Network Service.

Steps 10 and 11: The Network Service will interact with the Optical Network Control and Management Plane using proprietary signaling to set up the network as required for that particular video channel.

Steps 12–14: Upon successful network setup, the Network Service will send a SIP OK message back along the same route with the video information (copied from the Indirection Service's INVITE message), based on which the requester can start streaming video.

Step 15: The Composite Service at the AS sends a SIP OK message containing VideoServer2 IP address to the IMS Core (to be routed to the TV (video client)).

Step 16: The IMS Core sends the SIP OK message (with VideoServer2 IP address) to the TV (video client) which will then connect to the IPTV Channel 1 (located in VideoServer2) via the Managed Optical Network.

9.4 Integrated Telecom and IT Service Delivery Platform

After talking about the IMS-based platform, we discuss the possibility of integrating IMS with Web Service (WS). IMS and WS are SOA architectures proposed and promoted separately by the telecommunications industry and IT industry, respectively. As a result, some overlap exists between the two technologies and respective modules. We propose a Service Delivery Platform (SDP) that can inter-work between IMS and WS. The proposed SDP can leverage and blend telecom and IT services to reach a larger user base and provide more interesting/complex services. Despite frequent signs that the telecom industry is on a collision course with the IT industry, the potential for cooperation between them will benefit both segments of the market. Conventionally, IT vendors are strong in WS while telecom vendors are strong in IMS. Generally, they target different user groups by satisfying dissimilar requirements. Therefore, integrating IMS and WS will provide new opportunities for them to reach an even larger user base and create synergies, in terms of new and exciting service offerings. For example, by combining an IMS-enabled "push-to-talk" service with WS-based map service, a restaurant can make use of this "blended" service to advertise and offer instant-order services to its prospective customers over the World Wide Web.

The main challenge of this proposed inter-working layer is to harness multiple applications and to provide a set of combined or blended services. This problem is gaining importance as we head toward a fully functional IMS deployment and also to support competitive multimedia services from the Internet [13].

9.4.1 Proposed Service Delivery Platform for Inter-Working between IMS and WS

SIP-based Micro Service Orchestration and Web service bus are used to seamlessly integrate IMS, WS and the underlying services. The design objectives of the SDP for inter-working between IMS and WS are

- To be transparent to services it controls. For example, it enables users to select, invoke, and compose services and services can be added, deleted, and moved without knowing the underlying network architecture.
- To enable network service blending, for example, fast and effective way to combine (compose) two or more services (both local and remote), and handle SIP messages originated from services and AS under its control, and routing of SIP messages among services and AS.
- To ease Internet service mash-up, for example, availability, accessibility of services via WS SOA with effective SOAP-SIP/ISC mapping "intermediaries" (e.g., JSR-116/289).

- To enhance IMS service triggers and to enhance IMS initial filter criteria (iFC) with customization and interactive user triggers and controls.
- To inter-work with legacy services, for example, Customized Applications for Mobile networks Enhanced Logic (CAMEL) [22] and Open Software Architecture (OSA) based inter-working via SIP "intermediaries" (e.g., message adaptation, charging event generation, etc.).
- To provide policy management by following unified service and resource management and the IETF Common Open Policy Service (COPS) framework.

Figure 9.6 shows the architecture of the proposed SDP architecture for inter-working between IMS- and WS-based services. The architecture is a collection of different services from IMS (lower half of Figure 9.6) and WS applications (upper half of Figure 9.6), linked (indicated by solid arrow lines) to the proposed SIP-based microservice orchestration layer by standardized signaling protocols such as SIP and DIAMETER [23]. Notice that both the legacy and current server technologies can be fitted into the proposed architecture. For example, CAMEL [22], Weblogic in [24] and Parlay services in [25] are modules between the module of WS bus as well as the linked modules of the SIP-based microservice orchestration layer. Similarly, both policy manager and home subscriber server have interface (indicated as dotted lines) to the WS service bus as shown in Figure 9.6. The end goal for the proposed solution is to achieve effective blending of both

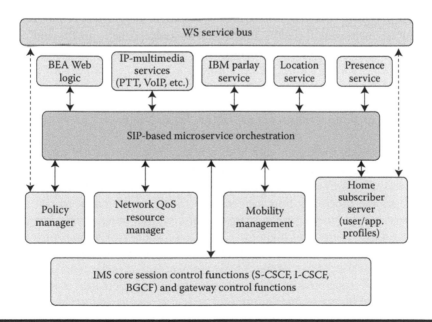

Figure 9.6 **Inter-working between IMS and WS components via SIP-based microservice orchestration module.**

WS- and IMS-based services. The relevant WS and IMS inter-working components of Figure 9.6 are briefly explained below, and the sequence of execution involving the proposed SIP-based microservice orchestration is illustrated in the next subsection.

- *Interface with WS-Based Servers:* Building of SIP-aware applications using the API available in programming environments such as JAIN and JSLEE does not represent a general-purpose solution to interconnecting ASs to IMS. This is because such an approach will generally make ASs network aware. In addition, such an approach will not be viable to inter-work IMS with legacy systems implemented based on CAMEL or OSA definitions. A preferred solution is to define a set of basic interface functions (AKA IMS's service capabilities interface manager—which is not completely defined in the current IMS standard) between the WS-based ASs and SIP-based Micro Service Orchestration layer.
- *SIP-Based Micro Service Orchestration Module:* The main objective of this module is to enhance the basic function provided in the current IMS's S-CSCF (the S-CSCF sits on the path of all signaling messages, and can inspect every message). Under this new arrangement, the SIP-based microservice orchestration module is the first module to be contacted by the I-CSCF after it is contacted by the user terminal. I-CSCF will also provide the relevant user-related information. It decides both the order and to which application server(s) (both IMS-based and WS-based) messages will be forwarded in order to provide the relevant services. It accomplished this by sending the messages via the S-CSCF module.

9.4.2 Basic Operations

The basic operations of the proposed SDP are listed as follows:

- *Invocation of services:* As with Web services, there are two key roles in SOA: service requestor and service provider. The requestor application invokes services offered by one or more provider applications by sending request messages and processing response messages. Some service providers can also be service requestors: they aggregate capabilities from other service providers to construct composite, higher-level services. Today's SOA implementations commonly use HTTP, resulting in synchronous communications—the requestor waits for a response before proceeding. Synchronous exchange is used to invoke services which are expected to complete in a relatively short period of time—seconds or minutes. It is also possible to have asynchronous service interactions, in which the service requestor does not wait for a response but expects it to be delivered later. For example, invoking a business process to request a quote for a custom passenger aircraft, involving many partners or human interaction in the process.

■ *Brokering between services:* A service broker is an intermediary between a service provider and a service requestor—for service location, brokered trust arrangements, and so on. A service broker facilitates communications between a service requestor and a service provider by allowing the requestor to discover a service interface previously registered by the provider. As such, the service broker is essentially a clearing house for service interfaces. In addition, the service broker may effectively act as a secure boundary that allows the domain owner to provide controlled access to internal services. In some cases, the service broker can also serve as a messaging broker and route messages to different services.

■ *Service orchestration:* Service orchestration is a centralized mechanism that describes how diverse services can interact. This interaction includes message exchange, business logics, and order of execution. Service orchestration introduces state and flow control to create dynamic relationships between services that are determined at runtime, rather than the construction phase. A business process specifies the potential execution order of operations from a collection of services, the data shared between these services, which partners are involved and how they are involved in the business process, joint exception handling for collections of services, and other issues involving how multiple services and organizations participate. Breaking the opaqueness of diverse services means specifying constraints on how the operations of collections of services and their joint behavior can be used. The description of the sequence of activities that make up a business process is given by service orchestration. Rules provide a powerful business logic representation in many modern information systems. Built-in integration for rule engine support separates orchestration logic from business policy and reduces the effort of adapting to changing business needs.

9.5 Conclusions

In this chapter, we presented an IMS-based testbed designed to provide a platform for research in real-time services integration and orchestration. The components used within the testbed were developed based on open-source software. SIP-based service orchestration module is reported and discussed along with service-oriented system functionalities like mobile client authentication protocol, optical network connection management, and EPON OLT data management. We showed the service orchestration through wired and wireless devices over the IMS-based testbed. We also proposed an integrated Telecom and IT service delivery platform that facilitates Telecom services and IT service combination and blending.

References

1. Parlay URL: http://www.parlay.org
2. IP-Multimedia Subsystem, Third Generation Partnership Project (3GPP), http://www.3gpp.org/article/ims
3. G. Camarillo and M. Garcia-Martin. 2006. *The 3G IP Multimedia Subsystem (IMS) Merging the Internet and the Cellular Worlds*. New York, NY: John Wiley & Sons, 2006.
4. IP Multimedia Subsystem, SIP Center, http://www.sipcenter.com/sip.nsf/html/IMS+IP+Multimedia+Subsystem
5. J. Rosenberg et al., SIP: Session Initiation Protocol, Internet Engineering Task Force (IETF) RFC 3261, June 2002, http://www.ietf.org/rfc/rfc3261.txt
6. Technical Specifications and Technical Reports for a UTRAN-based 3GPP system, 3GPP TS 21.101, http://www.3gpp.org/ftp/Specs/html-info/21101.htm
7. Telecommunication Standardization Sector (ITU-T), International Telecommunication Union (ITU), http://www.itu.int/ITU-T/
8. European Telecommunications Standards Institute (ETSI), http://www.etsi.org/
9. T. Lee, T. Chai, L. Ngoh, X. Shao, C. Teo, and L. Zhou, An IMS-based Textbed for real-time services integration and orchestration, *Proc. IEEE Asia-Pacific Services Computing Conference (IEEE APSCC)*, Singapore, December 2009.
10. Open IMS Core, Fraunhofer Institute FOKUS, http://www.openimscore.org/
11. L. Zhou, T. Y. Chai, X. Shao, C. V. Saradhi, V. Kumaran, Y. Wang, and C. Lu. iOPEN testbed for dynamic resource provisioning in metro ethernet networks, *Proc. of IEEE Create-Net Conf Testbeds and Research Infrastructures (IEEE Tridentcom)*, March 2006.
12. L. Zhou, T. Y. Chai, M. Kirchberg, T. K. Lee, L. H. Ngoh, X. Shao, and Y. K. Yeo. A service-oriented optical ethernet network design and implementation, *Proc. 7th International Conference on Optical Communications and Networks (ICOCN)*, December 8–11, 2008, Singapore.
13. Xu Shao, Teck Yoong Chai, Teck Kiong Lee, Lek Heng Ngoh, Luying Zhou, and Markus Kirchberg, An integrated telecom and IT service delivery platform, *2008 IEEE Asia-Pacific Services Computing Conference (IEEE APSCC 2008)*, December 9–12, 2008, Yilan, Taiwan.
14. Mobicents, http://www.mobicents.org/
15. The Java Community Process Program, JSR #289, http://jcp.org/en/jsr/detail?id=289
16. SailFin Project, https://sailfin.dev.java.net/
17. *3G Security: Access Security for IP-based Services, 3rd Generation Partnership Project, Rel. 8*, 3GPP TS 33.203, V8.0.0, September 2007.
18. *3G Security: Network Domain Security; IP Network Layer Security, 3rd Generation Partnership Project, Rel. 8*, 3GPP TS 33.210, V8.1.0, September 2008.
19. *3G Security: Security Architecture, 3rd Generation Partnership Project, Rel. 8*, 3GPP TS 33.102, V8.0.0, September 2008.
20. UCT IMS Client, http://uctimsclient.berlios.de/uctiptv_ advanced_howto.html
21. The VideoLAN Project, http://www.videolan.org/project/
22. 3GPP TS 02.78, Release 96, *3GPP specification*, http://www.3gpp.org/ftp/Specs/html-info/0278.htm
23. P. Calhonu et al., Diameter Base Protocol, *IETF RFC 3588*, September. 2003.
24. Application Notes for the Kagoor VoiceFlow 200 with Avaya Converged Communication Server (CCS)—Issue 1.0, *Avaya Solution and Interoperability Test Lab*, 2004.
25. IBM IMS Vision for Network Operators, *White Paper, IBM*, http://www-03.ibm.com/industries/telecom/doc/content/bin/ibm_imS_Vision_2006_v2.doc

Chapter 10

Personalization Paradigm in Service Delivery Platforms

Sofiane Abbar, Mokrane Bouzeghoub, and
Stéphane Lopes

Contents

10.1 Introduction

Service delivery platform (SDP) refers to a set of components that provide service delivery architecture (such as service creation, session control, and protocols) for a type of service. In spite of some standardization effort (e.g., TeleManagement Forum (TM Forum)) and deployments (e.g., Teligent SDP, HP SDP 2.0), there is not yet a consensus for the definition of SDPs and components composing them. Indeed, some vendors use the term SDP for their application servers, whereas others include a whole portfolio of products and components they offer.

However, since SDP aims at enhancing services provided to end users by simplifying their interactions and taking benefits from Customer Relationship Management's (CRM) experience to provide them with the most relevant contents and services, personalization is a key issue in such architectures.

In our vision, personalization encompasses a set of techniques that make applications sensitive to users' profiles and contexts. The user profile groups information that characterizes the user himself (e.g., preferences, interests, demographic data, etc.) while the context groups a set of features that describe the environment within which the user interacts with the system (e.g., location, temporal information, devices, temperature, etc.).

Personalization paradigm aims at adapting applications as much as possible to the user preferences and to the user context. Adaptation may concern several aspects, such as system reconfiguration, communication protocols, data sources selection, query reformulation, data layout, and user feedback handling. Data personalization refers to the set of techniques that allow one to provide the user with the most relevant data, depending on his domain of interest, his data quality requirements, his location at the querying time, the time at which the data are required, the media used to supply these data or any other constraint related to data pricing, user privacy, or business policy.

Works on personalization techniques concern, generally, a very few dimensions among these, and the proposed solutions are encapsulated as part of systems and applications features, thereby reducing their capabilities of extension and evolution. Actually, there is no effort devoted to the definition and production of personalization services which are generic enough to be used in many applications. Then, the personalization process is approached in different ways, depending on the applications and on the technologies used.

In information retrieval systems, the personalization is considered as a machine learning process based on user feedback [14,49]. The user is fully involved in the query evaluation which is conducted as a stepwise refinement process where the user can decide at each step which data he likes and which data he dislikes. From the log of this behavior, user profiles are elaborated and introduced as new filtering rules refining the results of further queries.

In database systems, personalization is considered through two viewpoints: query language extension and user query expansion. Query language extensions, such as SQL/f [38] or PreferenceSQL [30], enable expressing within each query user preferences. Query expansion consists in the user query enrichment using preferences given in the user profile [31,34].

Both approaches deal with user preferences characterizing the desired data. They are not concerned by the adaptation to the user environment. However, the need of considering the user mobility and omnipresence [13] has imposed new considerations such as user location, the media used for the interaction, and many other features grouped in the concept of user context. A *context* is a set of features that describe the environment in which the user interacts with the information system, while a *profile* is a set of features characterizing user needs in terms of data and quality of these data.

Many context-aware applications, such as smart-homes [47] or context-sensitive search engines [27], are able to adapt their processing and their services to the user context. Even if all applications agree with the importance to have a profile and a context, there is a lack of consensus in the definitions of these concepts. There are as many profile and context definitions as application domains and technologies. Thus, classifying, organizing, and structuring the knowledge describing the user and the context are the key elements to have a global vision on data personalization.

The goal of this chapter is to provide a formal definition of a personalized access model (PAM) with its underlying notions of profile, context, and contextualization. The definition of such a PAM is driven by the following requirements:

◾ Define profile and context meta models that are generic enough to be adapted to a wide range of SDP applications and that are open to integrate specific knowledge not included in the initial modeling

- Define a set of services that can be used to personalize existing SDP applications or to build new personalized applications
- Propose a generic PAM process that can be deployed over several personalization scenarios, depending on architectural issues
- Allow partial or full usage of this process such that each application can use part or entire knowledge defined by the meta model

The chapter is organized as follows. Section 10.2 gives a global view of our meta model that captures the semantics of profile and context. Section 10.3 describes the PAM and its personalization services. Section 10.4 presents the PAM application to context-aware recommender systems (CARS). Section 10.5 concludes this chapter with the summary of the main concepts introduced and further research on PAM and CARS.

10.2 Personalization Meta Data: Profiles and Contexts

Profiles and contexts are the main concepts used by modern applications (e.g., e-commerce, recommender systems, and publish/subscribe systems) to adapt content delivery services to the users' needs and preferences. While the two terms are intensively used in many papers and products, their definitions are specific to each application and the relationship between the two remains unclear. Few applications are profile based while others are claimed to be context aware. Some others deal with contextual or situational preferences. Although there is some standardization effort (e.g., P3P, CC/PP, CSCP, UAPref) and application deployments (e.g., iGoogle, MyYahoo), these concepts need to be clearly defined both in their semantics and usage. This chapter is a step forward in this concern. It aims at providing disjoint definitions of the two concepts through a multidimensional meta modeling.

10.2.1 Profile Modeling

A profile is a user model which contains all knowledge that characterize this user, regardless of any application or any technology environment. A profile may cover, for a given domain of interest, all what the user likes or dislikes. A user profile can be seen as a set of organized prescriptions that allow applications to filter and rank any content they deliver to the user. A profile may be either a user property (external profile) or a service provider property (hidden profile). In the former case, the user may impose privacy constraints on some elements of his profile. In the latter case, the user may require to know which data a system knows about him. A user can have several profiles, depending on his various domains of interest. These profiles act as ubiquitous objects that facilitate users' access to massive data and optimize providers' selection of target customers.

The user profile definition varies from one application to another, depending on how application designers perceive personalization, regardless of how users want personalization. As a first example, the CASPER project [14] defines a user profile as a sample of searched documents associated with their relevance with respect to user preferences. Another example [18] defines a profile using a set of utility functions over a domain of interest: two complementary clauses, DOMAIN and UTILITY, enable to define, respectively, the domain of interest of the user as a set of abstract subjects and the relative importance of these subjects as a set of equations. In [33], the user profile is defined with respect to a database schema as a list of weighted predicates that represent the most frequent selection predicates that are supposed to appear in user queries. Each predicate weight, defined between 0 and 1, emphasizes its relative importance with respect to others.

To go beyond these specific definitions, we consider a profile as a mutidimensional structure, each dimension being characterized by a set of attributes, possibly organized into abstract entities, whose values can be user defined or dynamically derived from users' behavior. These attributes are generally ranked and organized as sets of preferences that will serve to drive query compilation, query execution, and data delivery. Preference expressions may be of several kinds: introducing partial or total orders within profile predicates or keywords, emphasizing on some attributes or predicates by assigning relative weights, imposing the subset of predicates that should be fully or partially satisfied (top K predicates), discriminating between query results (top K results), adapting data delivery to the context (interaction media, spatio-temporal position), and so on. The next subsections introduce, respectively, the main dimensions of our profile model as well as the various ways preferences are specified within these dimensions.

10.2.1.1 Multidimensional Modeling of Profile

The multidimensional view of a profile is not new. Several attempts have already introduced various dimensions such as personal data [10], data delivery preferences [24], data quality [16], and security preferences [20]. We capitalize on these approaches and generalize them to achieve a certain level of genericity where existing profile models are simple instances of our model, and where each new personalized application can easily elaborate its own profile model. The ultimate goal is to provide a framework where personalized applications based on different profile models can interoperate with each other, offering the user an expanding personalization environment. A brief description of each dimension of our model is given below.

Personal data dimension groups attributes and preferences related to the user himself, that is, his identity, demographic data, professional data, health-care data, and so on. In some information systems, personal data will be used to filter query results with respect to the age of the user, his gender, or his area of work. In others, personal data are useless for information filtering itself but they are still useful as an "exchange currency" between the user and the information provider. This is the case in many e-commerce applications and Web-based systems that collect personal data for statistics purpose or for publicity dissemination.

Domain of interest is the central dimension of the user profile. It groups all attributes and preferences related to general needs of a given user. The domain of interest may describe the user's expertise or qualification in a specific field as well as the main object types he is interested in. It can be defined in different ways depending on the application needs (Figure 10.1). For instance, in information retrieval, the domain of interest is usually described by a set of possibly weighted keywords [39] or ontology graphs [22], while in database field, it is commonly described by a set of predicates [32] or expressions in a given formalism such as Horn clauses [21] or utility functions [18]. In some applications, the domain of interest is represented by the history of the user interactions with the system. This includes examples of elements to which are associated the actions performed by the user on them. Knowledge defined in the domain of interest is mainly used to reformulate user queries, either by term substitution, by complementing the queries with new selection predicates, or by introducing orders between predicates.

Data quality dimension is one of the most important issues in data personalization. Most of the user preferences relate to data accuracy, data freshness, data consistency, and so on. Data quality does not only concern data values, but also data sources (e.g., confidence, update frequency) and data derivation process (e.g., response time, reliability). Attributes and preferences of the quality dimension can be used by the query processor or more generally by the data management system to filter the accessed data sources when their number is significantly high [35] or to find a good balance between data freshness and response time [16].

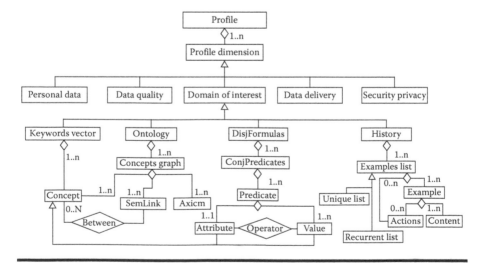

Figure 10.1   Domain of interest dimension.

Data delivery dimension describes different preferences related to user interface modalities (e.g., presentation style, query results layout), temporal accessibility (e.g., moment at which queries are issued and results notified), query results size (e.g., first N objects), and so on. Delivery modalities may depend on the media used: the same query results will not be presented in the same way depending on whether the media is a laptop, a Personal Digital Assistant (PDA), or a mobile phone. Delivery preferences are closely related to the context in which the user interacts with the application.

Security and privacy dimension describes security rules and constraints that can be applied either to the data resulting from queries, to the queries themselves, to the user identity, or to the whole profile as a sensitive knowledge. Security dimension mainly refers to privacy policies as described in different standards such as P3P [CDE + 05] for example. Privacy preferences may force personalized applications to not keep track of some user knowledge (e.g., identity, banking data).

Dimensions are described by ground attributes or complex attributes (called subdimensions) over which preference rules are defined. Due to the constraint of space, this chapter will not provide a detailed description of all attributes and subdimensions of the profile model. Figure 10.1 gives a flavor of the domain of interest dimension; detailed description can be found in [3]. The next subsection adds a little more emphasis on different types of preferences that the profile model allows one to define.

Preferences were subject to many researches during the last 10 years [19,33,41]; they may be of several kinds: *Example-based preferences* consist in a set of content examples [14]. These examples can be put in positive (POS) and negative (NEG) sets representing respectively liked and disliked contents (e.g., *POS–set* = {<*movie. genre=action*>}). *Weight-based preferences* known also as quantitative preferences [32] consist in assigning scores to the objects (e.g., <*movie.genre=action*>, 0.8). *Order-Based preferences*, called qualitative preferences [19] rely on binary-order relations in order to express preference over two objects (e.g., (*movie.genre=action*) $>_p$ (*movie. genre=drama*)). *Clause-Based Preferences* are specified through a set of predefined clauses [30] such as AROUND(A, v) clause, where the user prefers objects for which the value of the attribute A is close to v (e.g., AROUND(*movie.duration, 2 h*)).

Customizing the Profile Model to a Specific Usage: The profile model we have defined before is generic enough to cover most of the knowledge needed by a wide class of personalized applications. However, a given profile instance does not necessarily contain all model elements (dimensions, subdimensions, attributes, and preferences). Consequently, before using such a generic model, some customization might be done to derive a submodel (or a view) that best fits the need of a specific application domain. Such customized model becomes the reference model which is then instantiated with real user profiles. Customization of the generic profile model (GPM) is done by a set of restructuring operations such as hiding a model element, defining a new model element, and renaming a model element. These operations constitute the first set of model management operations associated to the GPM.

10.2.2 Context Modeling

In many personalized applications, providers aim at not only satisfying users' preferences, but also at adapting their applications to the users' environments such as the device with which they interact, as well as to the moment or the location where the interaction happens. All these features are grouped into a context entity that complements in the provider side users' profiles.

The advent of mobile and pervasive computing needs has pushed multimedia services and content providers to consider the context of users in personalization and delivery processes. This consideration gave rise to context-aware applications able to adapt their services to users' contexts such as smart home intelligent agents [47] and context-aware information retrieval [27] for which the context refers usually to spatial and temporal information of the interaction session as well as users' recent activities.

Dey [23] defined the context as being any information that can be used to characterize the situation of an entity. An entity is a person, place, or object that is considered relevant to the interaction between a user and an application, including the user. Although context modeling has been addressed in many researches and commercial products, its semantics differs from one reference to another. The most known context model is the key-value model [8,40,42]. In this approach, the context is formalized as a conjunction of predicate of the form <attribute, value>. Stefanidis et al. [42] model the context with a set of predefined contextual parameters—a parameter consists in one or more predicate. In [28], the context (situation) is modeled in ER-model. Authors proposed a context-meta model based mainly on three dimensions: time, locality, and influences. The last one may concern human influences (e.g., physical state, mood) as well as surrounding influences (e.g., weather, luminosity). The context is also modeled through ontologies [46]. The ontology-based model enables sharing and reuse contextual information in groupware, and ubiquitous computing systems; further, it allows performing reasoning on these models. There exist many other formalisms for modeling context such as XML based, object oriented, and logic based; an exhaustive survey of these modeling approaches can be found in [43].

By synthesizing these proposals and after clarifying our understanding of the context, we have proposed a multidimensional model, clearly separated from the profile model but related with an explicit and well-defined relationship called contextualization. The following subsection gives an overview of our context model, while the next section details the contextualization concept.

Multidimensional Modeling of Context: As for the profile, identifying and organizing contextual information are key issues in context modeling. The composite capabilities/preference profiles (CC/PP) standard [48] for composite capability/preference profile defines the context as being device features. CC/PP is RDF based, it consists in three components (dimensions): terminal hardware, software, and browser. Each component consists in a set of predefined attributes (e.g.,

TerminalHardware.displayWidth). The comprehensive structured context profile (CSCP) [17] is a more flexible standard for context modeling. CSCP is also RDF based, but unlike CC/PP, CSCP has no fixed hierarchy of dimensions and attributes: it allows expressing a more complex and general context. In our proposal, we identified five important dimensions: temporal, spatial, equipment, user state, and environment. Figure 10.2 gives a general overview of the context model. Detailed description can be found in [3].

The temporal dimension groups attributes related to the temporal aspect of the user interaction. It allows personalizing an application with respect to the moment of the user interaction. For example, one user may be interested in listening to music in the morning, and watching movies in the evening. To deal with various granularities, attributes are organized hierarchically (year, quarter, month, etc.).

The spatial dimension constitutes one of the most important context characteristics. It encompasses all information that characterizes the place from which the users interact. Since users' interaction behaviors change with respect to their geospatial situation, we have enumerated two generic spatial situations: static and on move. Both of these situations may be described with a simple coordinate (e.g., GPS, address, etc.) and a locality label if the place is known (e.g., at home, in train).

The equipment dimension characterizes the media used by the user to interact with the application. Three aspects are described in this dimension: details about the used device (e.g., type, autonomy, memory storage, and computing power), the software that is used (e.g., operating system, mail client, etc.), and characteristics of the connection (e.g., Type: WiFi, 3G; Rate; Services: FTP, MMS).

The environment dimension concerns elementary sensors that may inform applications about external characteristic of users' environments. These sensors may be used to capture temperature, luminosity, and so on. Smart home applications [47],

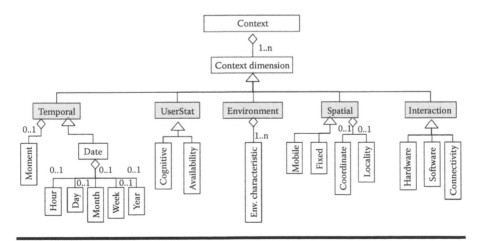

Figure 10.2 Context model.

for example, aim to automate the management of these sensors in order to satisfy users' preferences.

The user-state dimension provides information about the user availability and cognitive emotions. Some applications such as automated hotline agents deal directly with users; therefore, they need to infer the user emotions in order to adapt the dialog. Further, interactions have to be adapted to the user availability, for example, when a user is in a meeting, the personalized application will contact him by email instead of SMS.

Similarly to the profile model, the context model can be customized by the same set of operations (hiding model elements, defining new elements, and renaming elements) to fulfill specific application needs.

10.2.3 Profile and Context Management Platform

Profile and context models constitute a foundation to understand or build various personalized applications. They may also be considered as a mean to interoperate between heterogeneous personalized applications as they make their specific profiles and contexts simple customizations or instances of our generic models. Customization and instantiation are then model management operations that allow users, application designers, or any learning-based agents to adapt these models to application needs and create concrete profiles and contexts. The set of customization operations consists in hiding model elements (dimension, subdimension, or attribute), adding new model elements, and renaming model elements. The instantiation operation allows one to construct concrete profiles and contexts from the customized models (see Figure 10.3). Notice that instantiation operates on customized models only.

Besides customization and instantiation, the model management platform provides other functionalities also, such as import/export of profiles/contexts and matching between profiles or contexts (showing particularly the difference between

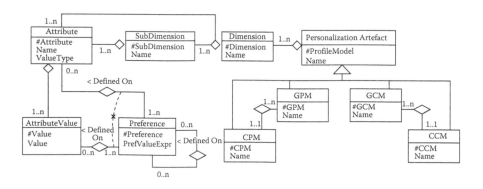

Figure 10.3 Profile and context instantiation.

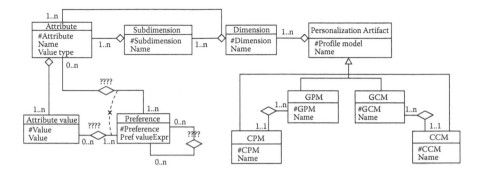

Figure 10.4 Profile and context generic model.

two profiles/contexts). Contextualization operation that links profile elements to contexts (see Section 10.3.1) can also be seen as one of the fundamental operations of this platform.

The profile and context models described above have been implemented and stored in a model management platform which provides developers and users with the necessary functionalities for profiles and contexts management. The global model underlying this platform is given in Figure 10.4.

In this figure, a personalization artifact (PA) is either a GPM (respectively a generic context model (GCM)) or a customized profile model (CPM) (respectively a customized context model (CCM)). Any of these models is defined in a generic way by a set of dimensions, subdimensions, attributes, and preferences. Notice that GPM and GCM are abstract entities that have no instances, while CPM and CCM have, respectively, concrete profiles and concrete contexts as instances. The relationships between GPM and CPM (respectively GCM and CCM) represent customization relationship; the version number discriminates between different customized models.

Profiles and contexts are key elements of any personalization and/or contextualization application. Therefore, these concepts constitute a basis of our PAM for SDPs that provides a set of personalization services using them. In the next section, we introduce the PAM, present the services it provides, and discuss the different deployment architectures it supports.

10.3 A PAM for SDPs

Our definition of a PAM aims to provide a generic set of concepts and techniques that should be deployed over a given system architecture to make the target applications adaptable to users' profiles and contexts. Figure 10.5 gives an overview of the main components of a PAM. The PAM is composed of three layers: (1) a persistency layer, (2) a functional layer, and (3) a communication layer.

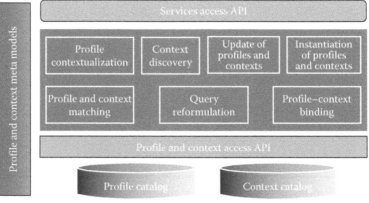

Figure 10.5 Personalized access model architecture.

The persistency layer deals with the storage and the access to the profiles and contexts. It includes the profile and the context catalogs.

The functional layer is composed of the profile management services, context management services, and personalized access services. Among these services, we distinguish offline services and online services. The former set is used at the design time of a personalized application, while the latter is used at the execution time. A more detailed description of the services offered by the platform follows.

Finally, the communication layer provides a communication interface between the PAM and users or applications. The role of this layer is, on one hand, to give access to the profiles and the contexts base and, on the other hand, to enable calling the PAM services.

The components of the three layers of the PAM are built around profile and context meta models that are generic enough to be adapted to a wide range of applications and that are open to integrate specific knowledge not included in the initial modeling. All PAM components are developed in conformance with the meta models. Thus, all messages addressed to the PAM should respect the meta models format.

The rest of this section is devoted to the detailed description of the main services. Section 10.3.1 presents offline services: context discovery and contextualization. Section 10.3.2 describes online services: binding, matching and query reformulation. Finally, Section 10.3.3 introduces various instantiations of the PAM.

10.3.1 Offline Services

Offline services are used to define profiles, contexts, as well as their relationships. Definitions are either done manually by human designers or automatically by software processes, for example, by exploiting log files.

10.3.1.1 Context Discovery

The role of this service is to identify, for each user, regular contexts within which he interacts with the system. For instance, the system can record the fact that each morning, at work, a user reads news about *Java programming* whereas at home, in the evening, the same user consults blogs about *board games*. If such observations are recurrent, the contexts *morning_at_work* and *evening_at_home* can be identified. An overview of this service is given in Figure 10.6.

The context discovery service takes as input a trace of the user activity. This activity is generally available as log files on an application server. By analyzing these logs, the service has to point out relevant contexts for each user with their description. Before performing this task, log files have to be preprocessed and cleaned.

The discovery of contexts from the cleaned logs necessitates regrouping each user action in relevant clusters. The clustering task takes into account log attributes that describe the situation (date, time, location, etc.) in which the user interacts with the system. The resulting clusters represent the contexts of the user. Discovered contexts can then be explicitly labeled (e.g., Holiday context, Office context).

10.3.1.2 Contextualization Service

The contextualization service enables users to express relationships between elements of their profiles and contexts. These relationships can be viewed as mappings between user preferences and contexts in which these preferences are relevant.

For instance, the domain of interest of a user can be centered on *Java programming* at work (i.e., in the context *work*) and on *board games* at home (i.e., in the context *home*). Preferences related to each domain will largely differ depending on the context. Moreover, some attributes have values that depend on the context. The address is such an attribute. The user can specify an address in Versailles in the context *work* and an address in Paris in the context *home*.

The contextualization mappings can be constructed manually or automatically. In the former approach, users are asked to specify explicitly which profile parts depend on a particular context. It is obvious that manual construction is not scalable. The latter approach overcomes this limitation. Indeed, mappings are automatically discovered by analyzing the user behavior contained in application server log files. An overview of an automatic contextualization service is given in Figure 10.7.

The *results partitioning* step consists in identifying which results the user liked and which he disliked. For example, if a user spent a large amount of time in looking at a given result, this result is considered as relevant for him. This step generates

Figure 10.6   The context discovery service.

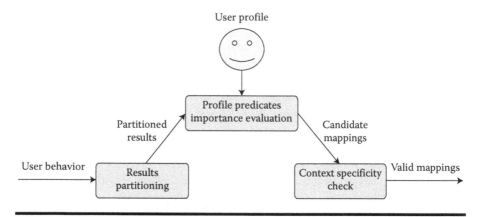

Figure 10.7 The contextualization service.

two sets (called *POS* and *NEG*) containing, respectively, relevant and irrelevant results for each context.

The *mapping initialization* step consists in initializing the possible mappings between the user profile predicates and the application contexts. This task is achieved by evaluating the score of each profile preference in each context. This is done by computing the occurrence frequencies of each profile predicate in the *POS* and *NEG* sets. Then, candidate mappings are created with a weight computed from their frequency.

Finally, the *context specificity check* phase prunes mappings that relate the same profile predicate to all contexts. Indeed, these mappings concern profile predicates that do not have to be contextualized as they have to be considered in all contexts. The result of this step is a set of valid mappings representing relationships between the user profile and the application contexts.

10.3.2 Online Services

Online services are used at application runtime to provide effective personalization to the application. They are obviously based on results of offline services.

10.3.2.1 Binding Service

The *binding service* exploits the mappings generated by the contextualization service to derive the operational profile. It consists in identifying the parts of the user profile that are related to a given context. Basically, the binding filters profile preferences that are not relevant to the active context of the user, and in adapting the remaining ones to this context.

An interesting problem with the binding service is the computation of the final preferences taking into account both the initial value of the preference and the weight computed by the contextualization service.

For instance, if a user is interested in *Java programming at work* but not *at home*, this preference has not to be used in the *home* context. It is thus filtered before applying personalization.

The binding service takes as input a user profile, the set of mappings issued from the contextualization, and the current user context. It returns the contextualized profiles that have to be considered by the application in the current context. Figure 10.8 gives an overview of the process.

The *profile content selection* step consists in selecting profile preferences that have to be taken into account in the actual context. This corresponds to the union of noncontextualized preferences and preferences related to the current context.

The *mappings score considering* step transforms the contextualized user preferences into classic preferences. It consists in merging the initial preferences with the weight computed by the contextualization service.

10.3.2.2 Matching Services

The *matching* service allows capturing the semantic similarity between two concepts. The similarity value can take into account, for example, the structure of concepts, user ratings, or ontologies describing the concepts.

In our context, matching algorithms can be applied to several kinds of concepts. The matching between two user profiles can be used in a collaborative recommender system to find similar users. The matching between a profile and a content has application for content-based recommendation. The matching between two contents can also be used to find similar products. Finally, the matching between two contexts is used in context-aware applications.

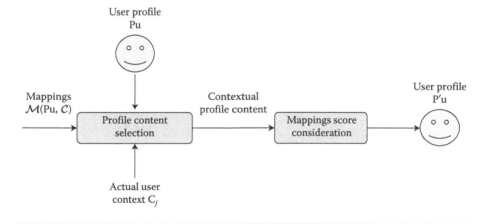

Figure 10.8   The binding service.

10.3.2.3 Query Reformulation

Query reformulation is a high-level service which enables personalizing a query processing with respect to a given user profile [31]. This service conforms to the following assumptions: (1) the users interact with a mediation system where mappings between the virtual schema and the source schemas are defined in a Local As View manner, (2) the user profile is restricted to the domain of interest dimension and is composed of a set of weighted selection predicates, (3) user queries are conjunctive queries of type Select-Project-Join, and (4) terminological problems between terms (attribute and table names) are supposed to be resolved.

The goal of query reformulation is to answer the following questions: (1) How to take into account the multiple data sources? and (2) How to consider the user preferences? To answer these questions, two types of techniques are used: (1) query rewriting which is used to select data sources and to substitute the query variables and (2) query enrichment which exploits the user profile. The query rewriting process consists in transforming the user query expressed on the virtual schema into expressions defined on the data sources [25]. The query enrichment process exploits the user profile to enhance expressiveness of a given query by integrating specific knowledge taken from the user profile [34].

Our reformulation process consists in four steps (Figure 10.9).

The first step (*query expansion*) consists in expanding the initial query with relevant virtual relations. The set of selected virtual relations is chosen according to two criteria: (1) the profile predicates that can be expressed on these relations and (2) the size of the set. The first criterion has to be maximized to allow a good personalization whereas the second has to be minimized to not penalize the rewriting

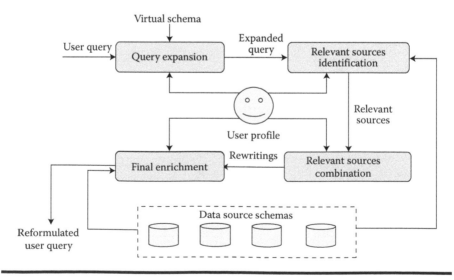

Figure 10.9   Query reformulation process.

process too much. The selected relations are then integrated to the initial query. The result of this step is an expanded query on which a sufficient number of profile predicates can be expressed. The second step (*relevant sources identification*) includes identifying contributive data sources for rewriting the query as well as filtering irrelevant data sources according to the user profile. The result of this step is a set of relevant data sources that can contribute to the query rewriting. The goal of the third step (*relevant sources combination*) is to produce query rewritings by combining sources selected in the previous phase. In this step, knowledge from the profile is used to prune irrelevant combinations. Finally, the fourth step (*final enrichment*) expands the candidate rewritings with profile predicates. This step is quite simple because the concerned profile predicates have been identified during the previous stages of the reformulation process.

This approach has been implemented and evaluated on a significant benchmark [31]. Integrating virtual relations to the user query during query expansion is a difficult task similar to the *Steiner Tree Problem* [29]. Due to its complexity, this task is solved using the *Minimum Cost Path Heuristic* [45]. Relevant data sources are identified using the first part of the MiniCon algorithm [37] with an additional pruning rule. Relevant sources are then combined using an adaptation of the Apriori algorithm [9] which allows pruning irrelevant combination as soon as possible.

It is important to mention that the query reformulation service can be partially used to perform only query rewriting without personalization or to perform only query enrichment according to a user profile.

10.3.3 PAM Deployment

The PAM can be used to build new personalized applications or to provide personalization features to existing applications (*legacy applications*). Even if the PAM is application independent, its deployment depends on the application needs and the architecture. We have identified three use cases of PAM usage: mediation-improvement PAM, application-improvement PAM, and service-providing PAM. This section presents the deployment and the usage of the PAM in each of these architectures.

10.3.3.1 PAM-M: Introducing Personalization in Mediators

The PAM-M is used to enhance a mediation system with personalization services. A mediation system gives a transparent access to a set of applications (sources). Each application communicates with the mediator through a particular wrapper. When a user sends a query to the mediation system, it is reformulated in order to be evaluated on the real applications. The PAMs deployment on a mediation system is shown in Figure 10.10.

The role of the PAM-M is to adapt evaluation of the user query according to the user preferences and to the user context. In this case, applications do not have to be

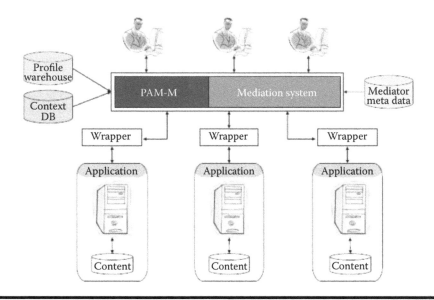

Figure 10.10   Introducing personalization in mediators.

aware of the personalization or of the PAM-M. The knowledge about users (profiles) and contexts is stored at the PAM-M level and personalization is achieved at the mediator level. For example, when a user issues a query, the mediator can call the profile-context binding service in order to get the contextualized user profile. This profile and the initial query can then be sent to the query reformulation service in order to get a set of personalized query rewritings. The enriched rewritings can then be sent to the real applications in order to get relevant results according to the preferences stored in the user profile.

10.3.3.2 *PAM-A: Introducing Personalization in Applications*

The PAM-A provides applications with personalized services and a persistent layer for profile and context management. The PAM-A is considered as being a part of the application (see Figure 10.11). Each application has its own PAM-A and thus, its own container of profile and context instances. However, in order to interoperate between applications, all PAM-A should share the same profile and context meta models.

The PAM-A can be used in different manners according to the implication of the application in the personalization process. In the first scenario, the application assigns all personalization tasks to the PAM-A. This scenario behaves similarly to the PAM-M. When a user submits a query, the application calls the query reformulation service before evaluating the enriched query. Another personalization step can be performed after the query evaluation by using the profile–content matching in order to filter the nonrelevant results.

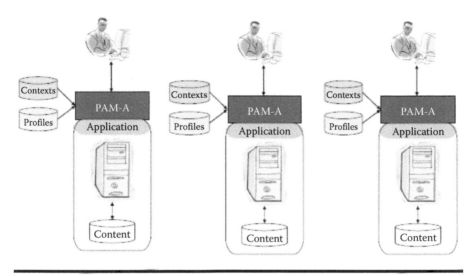

Figure 10.11   Introducing personalization in applications.

In the second scenario, the application aims to have more control on personalization. The PAM-A is thus more strongly coupled with the application and services are used during different steps of the query life cycle. For example, the user query can be enriched with profile preferences during the query compilation step and then the query execution can be done using the profile–content matching service.

In the last scenario, the application performs all personalization tasks and uses only the PAM-A profile and context management components. For instance, the application may query the PAM-A to get a user profile adapted to a given context.

10.3.3.3 PAM-S: Providing Personalization as an Autonomous Service

The PAM-S can be seen as a toolbox of reusable services which can be called by the applications (see Figure 10.12). The PAM-S only receives messages (service invocations), executes the services, and sends the results to the applications. Every application can call the PAM-S services if messages conform to the meta models. Indeed, the PAMs services can only be applied on profiles, on contexts, and on contents which are instances of the profile or context meta model. Consequently, applications have to be aware of the meta models.

In this use case, the PAM-S does not store profile and context instances and is considered to be stateless. Consequently, each application manages its own user profiles and contexts, and supervises the service calls. For example, to personalize a query according to a given user profile in a particular context, the application sends the profile and the context to the PAM-S and calls the profile–context binding service and then sends the bound profile and calls the query reformulation service

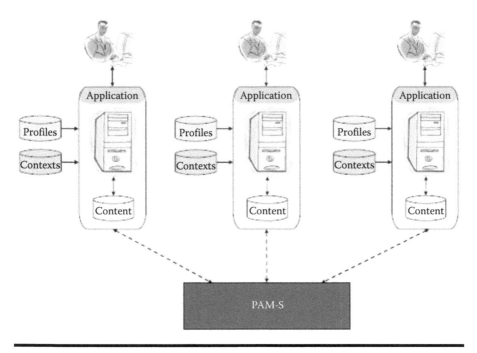

Figure 10.12   Providing personalization as an autonomous service.

to get a personalized query. In this example, the PAM-S will deal with both service calls independently. It is the application responsibility to call the services by sending the appropriate inputs.

The PAM-S can be also deployed for a set of applications that belong to the same information system. In this case, applications will probably share the same user profiles and contexts. Thus, the profile and context management can be performed in the PAM-S layer. However, a protocol has to be developed to deal with concurrent accesses.

In the next section, we show how profiles, contexts, and PAM services can be instantiated in a specific content delivery platform (CDP) through recommender system application.

10.4 Application: Context-Aware Recommender System

Once profiles, contexts, and contextual mappings are modeled and instantiated, they are ready to be used by any personalized application. This section illustrates the use of these concepts within an application scenario borrowed from the recommender system (RS) in content delivery environment. We show through this section how PAM services can be deployed in order to model a context-aware recommender system (CARS).

10.4.1 Why Is Context Important for Recommender Systems?

The importance of an RS is now well established. Netflix organizes a contest* in which one million dollar is offered for any better recommendation engine. This contest shows, on the one hand, the importance that industrials give to an RS, and, on the other, that better recommendations worth a lot. Two important questions arise from the Netflix contest: (1) What is a better RS? and (2) How can a system provide the best recommendations?

In our vision, a better RS is the one that delivers recommendations that best match with users' preferences, needs, and hopes at the right moment, in the right place, and on the right media. This cannot be achieved without designing an RS that takes into account all information and parameters that influence user's ratings. These information may concern demographic data, preferences about user's domain of interest, quality and delivery requirements, as well as the time of the interaction, the location, the media, the cognitive status of the user, and his availability. This knowledge is organized into the two concepts of user profile and context. The user profile groups information that characterizes the user himself, while the context encompasses a set of features that describe the environment within which the user interacts with the system.

We claim that taking into account both profiles and contexts in a recommendation process benefits to any RS for many reasons:

1. Users' preferences/ratings change according to their contexts [8,42].
2. The additive nature of the traditional RS [12] does not consider multiple ratings of the same content.
3. An RS may fail in providing some valuable recommendations as their similarity distance is uniformly applied to user preferences without analyzing the discrepancies introduced by the context.

Based on the PAM, and following the efforts started by Adomavicius et al. [6,7] and Anand et al. [12], we proposed a CARS which is based on both user profiles and contexts. In our approach, the same user who interacts from different contexts is provided with different recommendations.

Our main contributions include the following:

■ We present a general architecture for a CARS based on a set of personalization services.
■ We extend the PAM with a new context learning service that enables concrete construction of users' contexts from their log files.

---

* http://www.netflixprize.com

■ The contextualization service proposed in [4] is improved by combining both *support* and *confidence* of ratings within a given context instead of considering their frequencies only.

10.4.2 CARS Requirements

We focus on an RS that combines content-based techniques and collaborative filtering (CF). The content-based approach permits to learn users' profiles by analyzing content descriptors. Resulting profiles are, then, matched and compared to determine similar users in order to make collaborative recommendations. The CF approach allows exploiting the ratings given by the Top K neighbors of the active user in order to derive the missing rating by aggregation function. Figure 10.13 gives a flavor of the CARS global architecture.

This architecture is composed of the left-upper block that concerns knowledge acquisition and the right-bottom block that concerns personalized recommendation.

10.4.2.1 Knowledge Acquisition Processes

This block is responsible for the acquisition and the management of knowledge a CARS needs to process recommendations. This knowledge is grouped into four entities user profiles, content descriptors, contexts, and contextual mappings (relationships between profiles and contexts). Profiles and contents (items) were subject to extensive research on how they are generated and modeled. Thus, we focus in this chapter on the two remaining entities, or more precisely on the processes that generate them: (1) context acquisition from log files and (2) discovery of relationships between these contexts and user profiles elements (called contextual profiles).

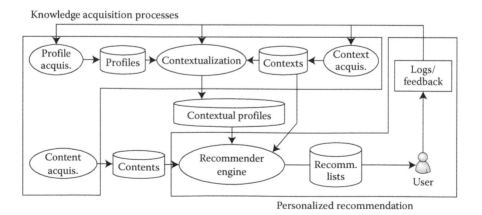

Figure 10.13   Global architecture of CARS.

10.4.2.2 Personalized Recommendation

This block encompasses actual RS operations. The recommender engine (RE) takes as inputs (1) the contextual profiles generated in the knowledge acquisition block and (2) the current context of the target user in order to compute a list of contextual recommendations. The behavior of the target user when consuming these recommendations is stored in log files on which acquisition processes are based to update profiles, contexts, and their mappings; hence, closing the cycle between knowledge acquisition and their exploitation.

The design of such CARSs is driven by the following requirements:

1. Distinguish between user profiles and contexts to guarantee a well-understanding definition and evolution of both entities.
2. Enable the discovery of users' contexts through a concrete approach and widely available data (e.g., standard log files).
3. Take into account both user profile and context in the recommendation process such that a user who interacts from different contexts would be provided with appropriate recommendation lists.
4. Propose an efficient profile matching approach to solve the data sparsity problem which may be increased when considering context in the recommendation process [12].

As a consequence of the separation of profiles and contexts, different privacy rules can be applied. In some applications, user profiles may be more sensitive than user contexts (e.g., anonymity), while in others, contexts may be more sensitive (e.g., user location or itinerary). Besides, this separability allows applications to be only aware of profiles or contexts or of both.

The achievement of such recommender systems needs a concrete instantiation of different PAM services presented in Section 10.3. To simplify the understanding of our approach, we present, hereafter, some definitions of the main concept we manipulate, and then, an example scenario that combines PAM services to provide users with context-aware recommendations. Details on each service are presented later.

10.4.2.2.1 Main Concepts

This section provides an informal definition of the concepts that will be used in the following sections.

User Profile: A user profile provides an extensive definition of the preferences that a user has in a given domain of interest. Concretely, a profile is a set of preferences, each being an (attribute, value) couple rated by the user or the RS on the basis of the user's actions.

Active User: In an RS, an active user is the one for whom recommendations are calculated.

Context: A context is the description of the features characterizing the environment within which users interact with an RS. As for the profile, the context is a set of contextual attributes, acquired from user log files.

Active context: It is the context within which the active user interacts with an RS. It is also called current context.

Contextualized Profile: It is a set of mappings that relate a subset of profile preferences to the context in which they are defined (i.e., the context in which recommended items are rated).

Operational Profile: It is a runtime binding of a contextualized profile to the current context of the user. It corresponds to the subset of user profile that CARS processes to provide contextual recommendations.

10.4.2.2.2 Simple Scenario

The scenario below gives an intuition about the way CARS provides users with contextual recommendations. Figure 10.14 focuses on the RE sketched in Figure 10.13. Dashed arrows represent the inputs of the RE.

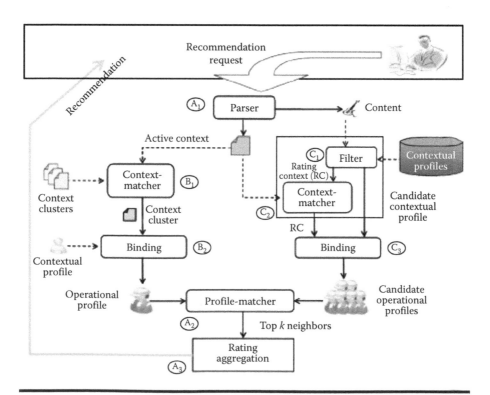

Figure 10.14 Contextual CF-RS.

When recommendations are requested, explicitly by the users (here John) or implicitly by the applications, the RE computes them following three processes A, B, and C shown in Figure 10.14.

First of all, a parser extracts the active context (c_a) of the user, and designates the candidate content to recommendation (step A1). In step B1, the active context is matched with all context clusters in order to determine the one to which c_a belongs. This context cluster is used in step B2 for binding John's contextual profile and producing his operational profile. At the same time, the RE looks for the candidate neighbors of John (process C). A candidate neighbor is a user who rated the content to be recommended (filtering: step C1), in a context c_r similar to the active context c_a (step C2). In step C3, the contextual profile of each candidate user is bound to its corresponding context c_r, resulting in a set of operational profiles. Step A2 synchronizes the two processes B and C. It consists in determining the top k neighbors of John among the candidate users. This is done by capturing the similarity between each candidate profile and John profile; only top k similar profiles are retained. Step A3 aggregates ratings given to the recommended content by the top k returned neighbors. Finally, a decision is made to know whether it is interesting or not to recommend the candidate content (e.g., through a threshold).

In what follows, we focus on PAM services that are used by CARS; we start by detailing offline services (i.e., context discovery and contextualization).

10.4.3 Context Discovery and Profile Contextualization

This section details the two services that acquire the necessary knowledge to efficiently run the CARS. These services are an answer to the first two requirements imposed in Section 10.4.2. For each service, we give its definition, its deployment on CARS, and details on its operational semantics.

10.4.3.1 Context Discovery Service

The approach we propose to discover contexts is based on the analysis of logs to capture regular contexts. Hereafter, a motivating example is given to understand the approach.

Example. Assume that a user, John, interacts frequently with an RS. Then, thanks to information contained in his log file (see Table 10.1), the system can capture that John's interests (activities) change according to the IP address, the date and time, the device, and so on. This can be pointed by a lookup to his log file from where two contexts can be derived: office context during working days and home context during weekends.

Obviously, the content to be recommended will change according to John's context (e.g., scientific papers for office contexts and movies or music for home context). The discovery of contexts from logs assumes the manipulation of a well-structured log file. Hopefully, the analysis of the structure of the most known log

Table 10.1 Log File Example

Id	IP	Date	Time	Agent	Device	Content	Others
John	192.168.53.25	May 4	17 h:00	winXP,En, Firefox5.0	Dell	SDP.pdf	…
John	192.168.53.25	May 4	17 h:10	winXP,En, Firefox5.0	Dell	Personalization.ps	…
John	192.168.53.25	May 5	16 h:30	winXP,En, Firefox5.0	Dell	contexts.pdf	…
John	254.13.70.1	May 9	12 h:00	MacOS X,En,Safari	MacBook	Avatar.avi	…
John	254.70.70.1	May 9	20 h:00	MacOS X,En,Safari	Macbook	Invictus.avi	…

file formats [1,2,36] reveals that most of them already contain some contextual information (fields) that correspond to the context meta model proposed in [5]. Among these fields there are, for example, the IP Address, the DATE and TIME of the request, and the USER-AGENT (browser, operating system, etc).

Log File Format. The log file format we used (see Table 10.2) is based on the W3C log file format [36] which is enough extensible (thanks to 10 W3C General Purpose Fields) to adapt it to various applications.

Notice that this log file format defines two new fields only: *c – action* and *c – device.* The first one is well known in application server log files: it usually contains the *id* of the action (e.g., Buy) that user applies to a given content. The second field informs about the device used (e.g., a laptop, a mobile phone, a remote control, etc.).

Among the log file format fields, five are contextual:

- *Date:* The domain of date is organized into a hierarchy in such a manner that we can know whether it is a working day or a weekend, holiday or not, and so on.
- *Time:* A day is partitioned into periods that are organized into a hierarchy. The time can inform whether it is day, night, morning, afternoon, and so on.

Table 10.2   Log File Format

Date	Date on which the request was received by the server
Time	Time at which the request was received by the server
`cs-uri-stem`	Stem (path) component of the requested URI
`cs-uri-query`	Query component of the requested URI
`c-auth-id`	Name or identifier of the authenticated user
`cs-method`	HTTP method of the request
c – ip	IP address of the client making the request; May captures the location
`c-dns`	Resolved Domain Name System (DNS) hostname of the host (client) making the request
cs(UserAgent)	Information about the user agent (browser, OS, language) originating the request
`sessionId`	Sessionization of user activities
c – Action	User action on the content (buy, add to panel, delete, etc.)
c – Device	Captures material media (laptop, mobile phone, PDA, etc.)

Source: Data from B. Behlendorf and P. M. Hallam-Baker. Extended log file format. Technical report, W3C Working Draf WD-logfile-960323. http://www. w3.org/TR/WD-logfile, 1996.

- *c − ip*: The IP address is used to localize the user; the localities are organized into a hierarchy of continent, region, country, city, town, and so on.
- *cs (User Agent)*: Gives information about the user Browser and operating system (OS) further than their languages.
- *c − Device*: Characterizes the used device for the user interaction; devices can be segmented into a hierarchy.

A conjunction of these attributes represents contexts within which users interact with the RS. We propose to group these contexts into clusters, each representing a regular context or situation such as home, labs, and so on.

Context Discovery Algorithm. Log files are transformed and uploaded into a database relation having the schema of Table 10.1. Based on this table, contexts are discovered by mining the five contextual attributes given above.

Let a context tuple be a conjunction of the five contextual attributes, which corresponds to one row (see Table 10.1). The idea is to cluster the set of all tuples into a finite set of clusters, each cluster representing a particular regular context. Thus, we start by applying the *Agglomerative Hierarchical Clustering (AHC)* [26] in order to estimate the number of clusters k, and then, we apply the *k-means* algorithm with the k center of resulting clusters in *AHC* algorithm instead of choosing arbitrarily the initial k centers. Obviously, both *AHC* and *k-means* use a metric to capture the similarity between contexts. This is given by a matching service that will be presented later in Section 10.4.4.

ALGORITHM 1 AUTOMATIC CONTEXTS DISCOVERY

```
Input: the user log file L, the threshold α.
Output: the set of user contexts C = {c₁, . . ., cₘ},
 1:  C ← Ø, E := ∞
 2:  CI ← CONTEXT(L) {extract context instances}
 3:  C ← AHC (CI, α) {apply AHC with stopping threshold α}
 4:  for all cluster cᵢ ∈ CIdo
 5:  update its mean mᵢ := NewMean(cᵢ)
 6:  repeat
 7:  E_old := E
 8:  for all context instance tⱼ∈ CI do
 9:  assign tⱼ to its closest cluster c* ∈ C such that ∀cᵢ ∈ C,
      MATCH(tⱼ, cᵢ) ≤ MATCH(tⱼ, c*)
10:  for all cluster cᵢ ∈ C do
11:  update its mean mᵢ ← NewMean(cᵢ)
```
12: compute $\mathcal{E} := \sum_{i=1}^{k} \sum_{t \in c_i} \left| MATCH(t, m_i) \right|^2$
```
13:  until E ≥ E_old
14:  return C
```

Algorithm 1 starts by extracting context instances from the user log file (Line 2). In Line 3, *AHC* is applied on these instances in order to compute the initial *k* clusters. Then, from Lines 4 to 13, *k-means* is applied as an iterative relocation algorithm in order to improve the initial clustering obtained with *AHC*. The algorithm iterates until the error ε achieves its minimal value.

10.4.3.2 Contextualization Service

Contextualization process takes as input a user profile P_u and the user history H corresponding to the user feedback in contexts C discovered previously. It returns a set of contextual mappings M representing dependencies between elements in P_u and contexts in C. With regard to CARS, this set of mappings will constitute a contextual profile CP_u defined below.

Contextual Profile: In the traditional RS, a user profile is a set of ratings $P_u = \{R_1, \ldots, R_n\}$, where each rating R_i is composed of a predicate pr_i and a rate (weight) r_i, that is, $R_i = (pr_i, r_i)$. The rate r_i is a real number expressing the importance of the predicate pr_i for the user. pr_i is a triplet *<concept, operator, value>*, for example, *Genre="Drama"*. *Concepts* may be *items* (content) that a user consumed (CF-RS), *features* of these items (CBF-RS) or both of them (Hybrid-RS).

Based on this profile, a contextualized user profile CP_u is defined as a set M of contextual mappings which relates user ratings to a set of contexts C:

$$CP_u = \{m_{ij}(R_i, c_j, s_{ij}) \mid R_i \in P_u, c_j \in C, s_{ij} \in [-1, 1]\}$$

Each contextual mapping m_{ij} is defined by a rating R_i, a context c_j, and a score s_{ij}. The score s_{ij} is a real number expressing the importance of taking into account the Rating R_i when the user interacts from the context c_j. Hereafter, we describe the way a contextual user profile is constructed using log files.

Contextualization Algorithm in CARS: The user history H on which the contextualization service is based is obtained by transforming the user log file L (Table 10.1). A sample of this history is presented in Table 10.3.

User behavior is captured in a relational table where each row is of the form: `<id, Context_id, Content_id, Feature, Action_type>` expressing that a user consumed a content (item) having some features (predicates) in a specific way (`Action_type`) within a given context (`context_id`). `Action_type` specifies whether the action applied by a user when consuming the content is positive (e.g., Buy) or negative (e.g., Ignore). `Content_id` and

Table 10.3 A Sample of User History

Id	Context-Cluster	Content	Feature	C-Action	Others
100	c_1	Avatar.avi	Action	+	…
101	c_2	SDP.pdf	Service	–	…
102	c_3	Avatar.avi	Sci-Fic	+	…

Features are obtained by transforming the URI of the consumed content (which is a log file field) into more significant information (e.g., http://www.imdb.com/title/tt0448157/ becomes Title: *Hancock*, Genre: *Action, Comedy, Crime*). Context_id specifies the context cluster within which the content was consumed; it is computed by the context discovery service.

The algorithm of contextual profile construction captures for each user rating (more precisely, predicate), its importance in each user context. Notice that, in each context, user activities may be of two kinds: positive (belongs to liked contents) and/or negative (relative to disliked contents), according to the type of actions a user applied on contents. Therefore, the importance of each predicate must be captured in both positive and negative activities.

As presented in [4], the importance of a predicate pr_i within a context c_j is captured by computing its frequency in this context. However, we claim that the frequency, alone, does not reflect the real importance of a predicate for one user. Therefore, we propose to model the importance of each predicate pr_i for the positive (resp. negative) activity within a given context as an association rules of the form $\langle pr_i \rightarrow + \rangle$ (resp.$\langle pr_i \rightarrow - \rangle$). Then, the importance of pr_i is obtained by combining both the *support* (*sp*) and the *confidence* (*cf*) of its corresponding rules.

The meaning of these two metrics is given through the following example.

Example. Assume that John is provided by 1000 movie recommendations in a given context, 701 are *action* movies, and 299 are *comedy* movies. Among the 701 *action* movies, John liked 503 and disliked 198, whereas he liked the 299 *comedy* movies. Then, supports (frequencies) of corresponding rules are *support* (<*genre=action*>,+)=(503/1000), *support* (<*genre=action*>,−)=(198/1000), and *support* (<*genre=comedy*>,+) = 299/1000, *support*(<*genre=comedy*>,−)= (0/1000), which clearly shows that the user prefers *action* movies to *comedy* movies, or in other words, that <*genre=action*> predicate is more important than <*genre=comedy*> one.

However, John's behavior can be interpreted differently: among 299 *comedy* movies, John liked all of them, while he liked only 503 *action* movies among 701, that is, *comedy* genre becomes more important than *action* genre. This semantics is captured by confidences: *confidence*(<*genre=action*>,+) = (503/701), *confidence* (<*genre=action*>,−)=(198/701) *confidence*(<*genre=action*>,+)=(198/701) and *confidence*(<*genre=comedy*>,+) = (299/299).

To consider both frequencies (supports) and confidences, many functions may be used to aggregate them. The most natural way to do that is to compute the mean between support and confidence as shown in the following equation:

$$merge(cf^*(pr_i), sp^*(pr_i)) = \frac{cf^*(pr_i) + sp^*(pr_i)}{2} \tag{10.1}$$

where * takes one value among {+ , −} at a time.

Algorithm 2 shows the way contextual profiles are constructed based on user's interaction histories.

The contextual profile constructed in algorithm 2 allows positioning each profile rating in each context through a score. The next section discusses the manner this score is exploited to make contextual recommendations.

10.4.4 Binding and Matching Services

This section details runtime services. They constitute an answer to the last two requirements of CARS.

ALGORITHM 2 AUTOMATIC CONTEXTUAL MAPPINGS DISCOVERY

```
Input: the initial user profile P_u = {p_1, ..., p_n}, the user
       behavior H'
Output: the contextualized user profile CP_u
1:  CP_u ← ∅, cf := 0, sp := 0
2:  C ← CONTEXT (H')
3:  for all R_i ∈ P_u do
4:  for all c_j ∈ C do
5:  compute cf⁺ := confidence(pr_i → +, c_j, H')
6:  compute sp⁺ := support(pr_i → +, c_j, H')
7:  compute s⁺_ij := merge(cf⁺, sp⁺)
8:  if s⁺_ij ≥ γ then
9:  CP_u ← CP_u ∪ (R_i, c_j, s⁺_ij)
10: compute cf⁻ := confidence(pr_i → -, c_j, H')
11: compute sp⁻ := support(pr_i → -, c_j, H')
12: compute s̄_ij = merge(cf⁻, sp⁻)
13: if s̄_ij ≥ γ then
14: CP_u ← CP_u ∪ (R_i, c_j, s̄_ij)
15: return CP_u
```

10.4.4.1 Binding Service

The profile binding takes as input the contextual user profile CP_u and the active context of the user c_i. It returns an operational profile OP_u which contains only profile elements that have to be considered by applications within the context c_i.

The operational profile is produced by filtering user ratings that are not relevant in the active context c_a, and by combining the remaining user ratings with contextual mapping scores.

The operational profile is always related to one context c_a (active context) and can be defined as a set of contextual ratings (and not mappings): $OP_u = \{CR_{1l}, ..., CR_{nk}\}$. Each contextual rating is of the form $CR = <pr_i, w>$, and is derived from a contextual mapping $m_{ia}(R_i, c_a, s_{ia})$ with $R_i = pr_i \times r_i$. Notice that

$w = aggregate(r_i, s_{ia})$ is the weight (contextual rating) of the predicate pr_i (e.g., $item_i$) within the context c_a.

The two reals r_i and s_{ia} have particular semantics. In fact, r_i is a rating representing the absolute importance of a given predicate pr_i (item or feature). In other words, it is the preference that a user has on the predicate independently of contexts. However, a daily analysis of the user behavior can reveal that the importance of the predicate pr_i on which r_i was expressed changes with respect to contexts, leading to the definition of a contextual importance s_{ia}.

Because of these semantics, these two numerics can be aggregated with different manners. One possibility is to multiply r_i by the contextual score s_{ia}. In this way, the higher the importance of the contextual score of a given pr_i is, the more is its related rating r_i preserved. One other possible exploitation of these numerics is to consider only (s_{ia}) in the recommendation process. This means that the initial user rating r_i is ignored.

10.4.4.2 Matching Services

The matching service aims at measuring the semantic similarity between profiles, contexts, and contents. According to CARS need, we focus on two kinds of matching services: the profile/profile matcher that captures the similarity between two profiles and the context/context matcher that measures the similarity between two contexts.

10.4.4.2.1 Profile-to-Profile Matcher

This matcher takes as input two operational user profiles OP_{u1} and OP_{u2}. Each profile contains a set of contextual ratings $CR = <item, w>$. Thus, we will use one of the well-known vector metrics, such as cosine, Pearson's or Spearman's correlation, to measure the similarity between the two profiles. However, before applying these metrics, we must ensure that the two compared profiles OP_{u1} and OP_{u2} are homogeneous.

Two profiles OP_{u1} and OP_{u2} are said homogeneous iff:

- $|OP_{u1}| = |OP_{u2}| = n$
- Under a certain order, O_D, $\forall_i \in [1, n]$: $OP_{u1}(i).item = OP_{u2}(i).item$

Due to the well-known problem of rating sparsity, profiles are rarely homogeneous. Thus, using Generalized Cosine Max metric [11] is one possible way to efficiently overcome the sparsity problem. Similarly, we propose a technique that uses ontologies to homogenize the two compared profiles. The main idea is based on the assumption that a *user likes similarly similar items*.

Example. Consider two PhD students, John: (<*Drama*, 0.2>, <*Action*, 0.8>, <*War*, 0.6>) and Julia: (<*Drama*, 0.9>, <*Action*, 0.3> <*War*, 0.4>, <*Social*, 0.7>), with their respective profiles related to movie genres. To homogenize John and Julia profiles, we have to predict the rating r_1 John would give to *Social* movies. r_1 is estimated as follows: $r_1 = Similarity$ (*Social, c, O*) × *rating(c, John)*, where $c \in$ {Drama, Action, War} such that it is the most similar concept to the concept *Social. Similarity* (*Action, c, O*) is the semantic similarity captured from the ontology *O* (e.g., TvAnytime ontology) between the *Social* concept and its closest neighbor in John's profile. Finally, *rating (c, John)* is the rating that John gave to the closest neighbor of *Social*.

Once profiles are homogenized, vector metrics such as *cosine* can be applied to capture their similarity.

Context-to-Context Matcher This matcher measures the similarity between two contexts. A context is seen as a conjunction of contextual attributes such as time, address, device, and so on. Each contextual attribute takes values from a hierarchy of values as shown in Figure 10.15.

Based on these hierarchies, we propose a hybrid metric to measure the similarity between two different values of the same contextual attribute within a given hierarchy of values. This similarity combines the shortest path length (SPL) and the Depth-relative (DEPTH) distance [44]. SPL specifies that the smaller the number of edges between two values is, the more are these values similar. DEPTH metric specifies that deeper values in the hierarchy are more similar than higher ones because deeper values share more information. The depth of two values is given by the depth of their lowest common ancestor. Hereafter, we propose the formula

$$sim(v_1, v_2) = 1 - \frac{SPL(v_1, v_2)}{DEPTH(v_1, v_2)} \tag{10.2}$$

The similarity between two contexts c_1 and c_2 is then computed as follows:

$$MATCH(c_1, c_2) = \frac{1}{|c_1|} \sum_{i=1}^{|c_1|} sim(c_{1i}, c_{2i}) \tag{10.3}$$

According to the CARS architecture we propose, the context-to-context matcher is used at design time when clustering contexts and at runtime when identifying the cluster to which belongs the active context.

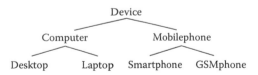

Figure 10.15 Example of contextual attribute hierarchy.

10.4.5 Toward Service-Based CARS

Once contexts are learned and contextual profiles are constructed for each user in the RS, we show how runtime services (matching and Binding) can be deployed in CARS in order to provide users with contextual recommendations.

The key idea behind the CARS we propose is to base the Top k neighbors detection only on the profile parts relevant to a given context (i.e., operational profile) instead of on the whole user profile. Top k neighbors of the active user (u_a) are the k most similar users to him in terms of their profiles. This means that profiles of all users are filtered and adapted to the context of the active user before comparing them. Algorithm 3 details the process of computing the Top k contextual neighbors.

The algorithm starts by finding the context cluster to which the active context c_a belongs (Line 2). Thus, it measures, with the matching service, the similarity between c_a and each context cluster to determine the closest one. In Line 3, the *Binding service* is invoked to produce the operational profile of u_a. Then, in Lines 4 through 7, candidate neighbors of u_a are computed. Every user who rated the item being recommended *It*, in a context similar to c_a, is considered as a candidate neighbor. His operational profile is computed and added to the set *CN*. Line 8 ranks candidate neighbors with respect to their similarity with the operational profile of u_a by applying the profile-to-profile matcher service presented in Section 10.4.4. It extracts, and then, the Top k of them (those which are the most similar to u_a within the context c_a). The algorithm returns this list of Top k neighbors (Line 9).

ALGORITHM 3 CONTEXTUAL TOP K NEIGHBORS

Input: $CP_{ua} = \{m_{11}, \ldots, m_{nm}\}$: Contextual active user profile, $\mathcal{U}$: set of all users, c_a: the active context, $\mathcal{C}$: the set of all contexts, *Item*: the item candidate to recommendation, k: the number of neighbors to consider, γ: the threshold.

Output: the TOP k neighbors

1: $CN \leftarrow \emptyset$ {Set of Candidate Neighbors}
2: $c\star := c \in \mathcal{C} \mid \forall c_i \in \mathcal{C},\ MATCH\ (c_i,\ c_a) \leq MATCH\ (c\star,\ c_a)$
3: $OP_u := BIND(CP_u,\ c\star)$
4: for all $CP_{ui} \in \mathcal{U}$ do
5:   if $m < (Item,\star),\ \star,\star > \in CP_{ui}$ then
6:   $OP_{ui} := BIND(CP_{ui},\ c_a)$
7: $CN \leftarrow CN \cup \{OP_{ui}\}$
8: $TOPK \leftarrow \{OP_{ui} \in CN,\ i = 1,\ k\}$ such that $\forall OP_{uj} \in CN,\ OP_{uj} \notin TOPK,$ $MATCH\ (OP_{uj},\ OP_{ua}) \leq MATCH\ (OP_{ui},\ OP_{ua})$
9: return $TOPK$

10.4.5.1 Aggregating Neighbor Ratings

Once Top k neighbors of the active user u_a are determined, ratings they gave to the item (*It*) to be recommended must be aggregated. The result of this aggregation allows deciding whether it is relevant or not to recommend *It* to the user u_a. There exist several techniques for aggregating these ratings [12,15]. The one we considered is the *Weighted Mean Aggregation* (Equation 10.4) where the contribution of each neighbor rating is weighted with the similarity between this neighbor and the active user u_a.

$$Rating(u, It) = \alpha \sum_{u_i \in TopK} Match(u, u_i) \times Rating(u_i, It), \qquad (10.4)$$

where α is a normalizing factor.

According to the value of the aggregated rating, the RS decides whether *It* could be recommended or not.

10.5 Conclusion

In this chapter, we investigated the personalization paradigm in SDPs through user profiles and contexts. We showed that profile and context are different concepts that cover different knowledge. The concept of profile represents user knowledge regardless to any personalized application, while the concept of context represents specific environments that an application is able to recognize independently of any user. We provided generic models for both concepts. These models can be used independently of each other or in conjunction depending on whether application is only profile based, context based, or profile and context based. We claim that these two concepts are fundamental baselines to build personalized applications or interoperate between existing ones; therefore, a generic model synthesizing both models is abstracted from the previous ones and a set of customization and instantiation operations have been defined to respectively adapt the generic models to specific user/application needs and to generate concrete instances.

We, also, proposed a PAM that encompasses profile and context meta models that serve as a foundation for the interoperability between services and applications. The PAM provides a set of general personalization services such as context discovery, contextualization, matching, binding, and so on. We argued that PAM can be deployed over different architectures to make applications adaptable and sensitive to user profiles and contexts.

The global personalization approach we proposed was instantiated in a CDP environment to build an actual CARS. To this end, profile and context meta models are given a specific semantics and structure. At the same time, PAM services are defined in a sharp way, and detailed through process and algorithms.

References

1. Apache common log format (clf). Available at http://httpd.apache.org/docs/1.3/logs.html, May 06, 2009.

2. Microsoft is log file format. Available at http://msdn.microsoft.com/enus/library/ms525807.aspx, May 06, 2009.

3. S. Abbar, M. Bouzeghoub, D. Kostadinov, and S. Lopes. Toward a meta model for user profile definition and management. Technical report, Alcatel-Lucent, 2007.

4. S. Abbar, M. Bouzeghoub, D. Kostadinov, and S. Lopes. A contextualization service for a personalized access model. *EGC*, RNTI-E-15: 265–270, 2009.

5. S. Abbar, M. Bouzeghoub, D. Kostadinov, S. Lopes, A. Aghasaryan, and S. Betge-Brezetz. A personalized access model: Concepts and services for content delivery platforms. In *Proceedings of the 10th International Conference on Information Integration and Web-based Applications and Services*, Linz, Austria, pp. 41–47, 2008.

6. G. Adomavicius, R. Sankaranarayanan, S. Sen, and A. Tuzhilin. Incorporating contextual information in recommender systems using a multidimensional approach. *ACM Transactions on Information Systems*, 23(1):103–145, 2005.

7. G. Adomavicius and A. Tuzhilin. Toward the next generation of recommender systems: A survey of the state-of-the-art and possible extensions. *IEEE Transactions on Knowledge and Data Engineering*, 17(6):734–749, 2005.

8. R. Agrawal, R. Rantzau, and E. Terzi. Context-sensitive ranking. In *ACM SIGMOD International Conference on Management of Data*, Chicago, USA, pp. 383–394, 2006.

9. R. Agrawal and R. Srikant. Fast algorithms for mining association rules. In *Proceedings of the International Conference on Very Large DataBases (VLDB)*, Santiago, Chile, pp. 487–499, 1994.

10. G. Amato and U. Straccia. User profile modeling and applications to digital libraries. In *Proceedings of the Third European Conference on Research and Advanced Technology for Digital Libraries*, Paris, France, pp. 184–197, 1999.

11. S. S. Anand, P. Kearney, and M. Shapcott. Generating semantically enriched user profiles for web personalization. *ACM Transactions on Internet Technology*, 7(4): 22, 2007.

12. S. S. Anand and B. Mobasher. Contextual recommendation. From web to social web: Discovering and deploying user and content profiles. *Workshop on Web Mining*, Berlin, Germany, pp. 142–160, 2007.

13. R. Belotti, C. Decurtins, M. Grossniklaus, M. C. Norrie, and A. Palinginis. Modelling context for information environments. In *Proceedings on Ubiquitous Mobile Information and Collaboration Systems (UMICS)*, CAiSE, Riga, Latvia, 2004.

14. K. Bradley, R. Rafter, and B. Smyth. Case-based user profiling for content personalisation. In *Proceedings of the International Conference on Adaptive Hypermedia and Adaptive Web-based Systems*, Trento, Italy, August 2000.

15. J. S. Breese, D. Heckerman, and C. Kadie. Empirical analysis of predictive algorithms for collaborative filtering. Technical report, MSR-TR-98-12. Microsoft research, Redmond, Washington 98052, 1998.

16. L. Bright and L. Raschid. Using latency-recency profiles for data delivery on the web. In *Proceedings of the 28th Conference on VLBD*, Hong Kong, People's Republic of China, pp. 550–561, 2002.

17. S. Buchholz, T. Hamann, and G. Hbsch. Comprehensive structured context profiles (cscp): Design and experiences. In *Proceedings of the 2nd IEEE Annual Conference on Pervasive Computing and Communications Workshops*, pp. 43–47, 2004.

18. M. Cherniack, Ed. Galvez, M. Franklin, and St. Zdonik. Profile-driven cache management. In *Proceedings of the 19th ICDE*, Bangalore, India, pp. 645–656, 2003.

19. J. Chomicki. Querying with intrinsic preferences. In *Proceedings of the 8th EDBT*, Prague, Czech Republic, pp. 34–51, 2002.
20. L. Cranor, B. Dobbs, S. Egelman, G. Hogben, J. Humphrey, M. Langheinrich, M. Marchiori, M. Presler-Marshall, J. Reagle, and M. Schunter. The platform for privacy preferences 1.1 (p3p 1.1) specification. Technical report, W3C, 2005.
21. P. Dell'Acqua, L. Pereira, and A. Vitoria. User preference information in query answering. In *Proceedings of the 5th International Conference on Flexible Query Answering Systems*, Copenhagen, Denmark, pp. 163–173, 2002.
22. K. L. Dempski. Real time television content platform: Personalized programming over existing broadcast infrastructures. In *Proceedings of the 2nd International Conference on Adaptive Hypermedia and Adaptive Web-based System*, USA, 2002.
23. A. K. Dey. Understanding and using context. *Personal and Ubiquitous Computing*, 5(1): 4–7, 2001.
24. P. Germanakos and C. Mourlas. Adaptation and personalization of web-based multimedia content. In *Digital Multimedia Perception and Design*, G. Ghinea and S. Y. Chen (Eds.), Idea Group, Inc., Hershey, USA, pp. 284–304.
25. A. Y. Halevy. Answering queries using views: A survey. *The VLDB Journal*, 10:270–294, 2001.
26. J. Han and M. Kamber. *Data Mining: Concepts and Techniques*. Morgan Kaufmann Publishers, New York, USA, 550pp. ISBN 1-55860-489-8, 2000.
27. O. Hasan, M. E. Atwood, J. Waters, and B. W. Char. A context-sensitive search mechanism. In *Proceedings of INMIC. 8th International Volume*, Lahore, Pakistan, pp. 368–374, 2004.
28. S. Holland and W. Kiessling. Situated preferences and preference repositories for personalized database applications. In *ER-Conceptual Modeling*, Shanghai, China, pp. 511–523, 2004.
29. F. K. Hwang, D. S. Richards, and P. Winter. *The Steiner Tree Problem*. Elsevier, North-Holland, Netherlands,1992.
30. W. Kiessling. Foundations of preferences in database systems. In *Proceedings of the 28th Conference on VLDB*, Hong Kong, China, pp. 311–322, 2002.
31. D. Kostadinov, M. Bouzeghoub, and S. Lopes. Query rewriting based on user's profile knowledge. In *Actes des 23emes Journées Bases de Données Avancées*, Marseille, France, 2007.
32. G. Koutrika and Y. E. Ioannidis. Personalization of queries in database systems. In *Proceedings of the 20th ICDE*, Boston, MA, pp. 597–608, 2004.
33. G. Koutrika and Y. E. Ioannidis. Constrained optimalities in query personalization. In *Proceedings of the ACM SIGMOD*, Baltimore, MD, pp. 73–84, 2005.
34. G. Koutrika and Y. E. Ioannidis. Personalized queries under a generalized preference model. In *Proceedings of the 21st International Conference on Data Engineering (ICDE 2005)*, Tokyo, Japan, pp. 841–852, 2005.
35. F. Naumann, J. C. Freytag, and M. Spiliopoulou. Quality driven source selection using data envelope analysis. In *Proceedings of the MIT Conference on Information Quality (IQ'98)*, Cambridge, MA, pp. 137–152, 1998.
36. B. Behlendorf and Phillip, M. Hallam-Baker. Extended log file format. Technical report, W3C Working Draf WD-logfile-960323. http://www.w3.org/TR/WD-logfile, 1996.
37. R. Pottinger and A. Y. Halevy. Minicon: A scalable algorithm for answering queries using views. *VLDB Journal*, 10(2–3):182–198, 2001.

38. D. Rocacher and L. Litard. Prfrences et quantits dans le cadre de l'interrogation flexible: sur la prise en compte d'expressions quantifies. In *Actes des 22e Journes Bases de Donnes Avances (BDA)*, Lille, France, 2006.
39. C. Santos and N. Vieira. Use reformulated profile in information filtering. In *Proceedings of the AAAI Workshop on Semantic Web Personalization*, CA, 2004.
40. B. N. Schilit, N. L. Adams, and R. Want. Context-aware computing applications. In *IEEE Workshop on Mobile Computing Systems and Applications*. Santa Cruz, CA, USA, 1994.
41. W. Siberski, J. Pan, and U. Thaden. Querying the semantic web with preferences. In *Proceedings of the 5th International Semantic Web Conference*, Athens, GA, 2006.
42. K. Stefanidis and E. Pitoura. Fast contextual preference scoring of database tuples. In *11th International Conference on Extending Database Technology*, France, pp. 344–355, 2008.
43. T. Strang and C. Linnhoff-Popien. A context modeling survey. In *Workshop on Advanced Context Modelling, Reasoning and Management*, Nottingham, UK, 2004.
44. M. Sussna. Word sense disambiguation for free-text indexing using a massive semantic network. In *CIKM '93: Proceedings of the Second International Conference on Information and Knowledge Management*, pp. 67–74, New York, NY, 1993. ACM, Washington, DC, USA.
45. H. Takahashi and A. Matsuyama. An approximate solution for the Steiner problem in graphs. *Mathematica Japonica*, 24: 573–577, 1980.
46. V. Vieira, P. Tedesco, A. C. Salgado, and P. Brzillon. Investigating the specifics of contextual elements management: The cemantika approach. In *Proceedings of the 6th International Conference on CONTEXT*, Roskilde, Denmark, 2007.
47. E. Vildjiounaite, O. Kocsis, V. Kyllonen, and B. Kladis. Context-dependent user modelling for smart homes. In *Proceedings of the 11th International Conference User Modeling*, Corfu, Greece, pp. 345–349, 2007.
48. W3C. Composite capabilities/preferences profile. In www.w3.org/Mobile/CCPP.
49. N. Zemirli, L.Tamine-Lechani, and M. Boughanem. Prsentation et valuation d'un modle d'accs personnalis l'information bas sur les diagrammes d'influence. In *XXVme congrs INFORSID*, Perros-Guirec, France, pp. 89–104, 2007.

Chapter 11

On Secure JAVA Mobile Application in SOA-Based e/m-Government Systems

Milan Marković and Goran Đorđević

Contents

11.1 Introduction

This work is related to the consideration of some possible Service Oriented Architecture (SOA)-based e/m-government online systems, that is, about secure mobile communication between citizens and companies with the small and medium governmental organizations (SMGOs), such as municipalities, or other governmental organizations and/or agencies. We have considered a general model of such systems consisting of three main parts:

- Secure mobile or desktop client application
- SOA-based e/m-government platform
- External entities, such as Public Key Infrastructure (PKI), STS, XML Key Management Specification (XKMS), Time-stamping Authority (TSA), and Universal Description Discovery and Integration (UDDI)

Although the generic e/m-government model has been proposed and considered, a main emphasis and contribution of the study is consideration of the Secure JAVA Mobile Web Service application communicating with the Web Service of the proposed platform. In other words, although the clients of the proposed platform could be desktop or mobile, as well as that mobile clients could be JAVA or .NET-based applications, we mainly considered the Secure JAVA Mobile Web Service application and all security aspects of its mobile communication with the Web Service of the proposed platform.

The work presented and examples described are included in the general framework of the EU IST FP6 SWEB project (Secure, interoperable cross-border m-services contributing toward a trustful European cooperation with the non-EU member Western Balkan countries, SWEB) [1].

In a process of development the secure JAVA Mobile Application we have used the J2ME development environment. The J2ME is a runtime environment for resource-constrained environments and development. J2ME includes specific virtual machines, configurations, and profiles for various environments and needs. With an appropriate configuration and profile, J2ME applications could be executed within pagers, mobile phones, Personal Digital Assistants (PDAs), set-top boxes, and automobile navigation systems, just to mention some [2].

This chapter mainly identifies the need for security in mobile communications, such as mentioned in [3]. It mainly presents a secure mobile framework that is based on widely used XML-based standards and technologies such as XML-Signature, XML-Encryption (XML-security), and Web Services Security (WS-security). It describes the main entities that participate in the communication, and illustrates its operation with a fully deployed mobile scenario.

Besides security aspects of the XML communication, a possible Federation ID system based on Security Token Service (STS) is considered. The security token represents a collection of claims about an entity. Typically, this token is used to authenticate the sender or responder in a message transaction. In this work, Security Assertion Markup Language (SAML) tokens/assertions have a role of security tokens. Communication between JAVA mobile application, or the SOA-Based platform itself, and STS server is realized by using WS-Secured Simple Object Access Protocol (SOAP) communication. The communication flows in the following way. The JAVA mobile application (or the platform itself) sends the Request for Security Token (RST) to the STS server and, if everything is ok (if the user is successfully authenticated and authorized for using the particular Web Service), receives back the RST Response (RSTR) which consists of the SAML token with the user's role on the proposed e/m-government platform.

We also investigated the possibility of using XML Key Management Specification (XKMS) protocol [4] in the proposed e/m-government system. The XKMS protocol constitutes a lightweight front end for accessing PKI services, possibly used in the wireless communication which should be of primary benefits. It enables the integration of keys and certificates into mobile applications as well as the implementation of PKI X.509v3 digital certificate registration, revocation, and update mechanisms.

Besides STS and XKMS, the client applications and the platform used also the time-stamping functionalities in order to create timely valid electronic documents with digital signatures of long-term validities. In this sense, a suitable TSA also represents an important part of the proposed model.

Keeping in mind the above-mentioned text, the main contributions of the study are

- Proposal of a possible secure e/m-government model based on JAVA mobile/ desktop application and an SOA-based e/m-government platform
- Usage of secure JAVA mobile application in which all modern security techniques are implemented (XML-security, WS-security, SAML, time stamping, PKI, and XKMS) which are used in an optimum way in order to cope with majority of security issues of the mobile Web Service communication

- Usage of SOA-based request-response e/m-government platform (Web Services) instead of session-based Web application platform which is far more suitable in the mobile communication systems
- Usage of XKMS service which is more suitable for mobile PKI system since it outsources complex operations such as PKI validation services to the external entity—the XKMS server, compared to usages of other techniques [4]

The chapter is organized as follows: A description of the possible m-Governmental architecture is given in Section 11.2 while the possible m-Governmental scenario is considered in the Section 11.3. Section 11.4 is dedicated to the consideration about secure JAVA mobile Web Service application while conclusions are given in Section 11.5.

11.2 Possible m-Government Architecture

The proposed m-government model is presented in Figure 11.1 [1,3]. This model consists of

- Mobile users (citizen, companies) who send some Web Services requests to e/m-government platform based on Web Services for the purpose of receiving some governmental documents (e.g., residence certificate, birth, or marriage certificates, etc.). These users use secure JAVA mobile Web Service application for such a purpose.

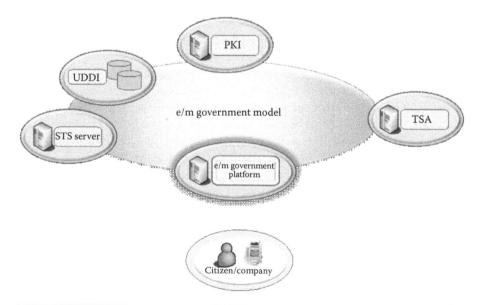

Figure 11.1   Proposed m-government model.

- Fixed/Desktop users connecting to the proposed Web Service governmental platform through some desktop secure Web Service application (could be JAVA based too). Since everything regarding functionality is similar and almost the same between the two client options, in the remaining part of the text we will consider only the secure JAVA Web Service mobile application and m-government solution model.
- Web Service endpoint implementation on the platform's side that implements a complete set of security features—the same set as implemented in the secure JAVA mobile and the secure desktop applications. Well-processed requests with all security features positively verified, the Web Service platform's application proceeds to other application parts of the proposed SOA-based platform, including the governmental Legacy system for issuing actual governmental certificates requested. In fact, the proposed platform could completely change the application platform of some governmental organization or could serve as the Web Service "add-on" to the existing Legacy system implementation. In the latter case, the Legacy system will not be touched and only a corresponding Web Service interface should be developed in order to connect the proposed SOA-based platform and the Legacy governmental system.
- External entities, such as a PKI server with an XKMS server as a front end, UDDI, and TSA.

11.2.1 External Entities

The following are the functions of the proposed external entities:

- *STS server* is responsible for strong user authentication and authorization based on PKI X.509v3 electronic certificate issued to users. Communication between the STS server and user's JAVA mobile application is SOAP based and secured by using WS-security features. After the succesful user authentication and authorization, the STS server issues to the user a SAML token which will be subsequently used for the user authentication and authorization to the Web Service of the proposed m-government platform, or any other Web Service of other governmental organizations—acting as the Federation ID or Single-Sign-On instrument. The SAML token is signed by the STS server and could consist of the user role for platform's user authentication and authorization and eventually of URL of the requested governmental Web Service, found by contacting the suitable UDDI server.
- *PKI server* is responsible for issuing PKI X.509v3 electronic certificates for all users/actors in the proposed m-governmental platform (users, civil servants, administrators, servers, platforms, etc.). Since some certificate processing functions could be too heavy for mobile users, the PKI services are exposed by the XKMS (XML Key Management Specification) server which could

register users, as well as locate or validate certificates on behalf of the mobile user. This is of particular interest in all processes that request signature verification on the mobile user side.

■ *TSA server* is responsible for issuing time stamps for users requests as well as for platform responses (signed m-documents).

The secure JAVA mobile application communicates with all mentioned external entities, that is, it has all security functions mentioned implemented, that is,

■ Secure JAVA mobile application sends RSTs to the STS server by using WS-Secured (WS-Signature and WS-Encryption) SOAP communication.
■ Secure JAVA mobile applications sends digitally signed (XML signature) request for mobile governmental document to the Web Service of the proposed platform by using WS-Encrypted SOAP communication.
■ The sent request includes the SAML token issued and signed by the STS server.
■ The request is time stamped by sending a time-stamp request and obtaining the corresponding time-stamp response (digitally signed by the TSA).
■ The secure JAVA mobile application also receives the signed and time-stamped m-governmental document from the platform through WS-Encrypted communication and performs all necessary signature verifications and certificate validations (with the help of the XKMS server) actions.

In the sequel, we will describe in more detail a possible m-governmental scenario.

11.2.2 Possible m-Government Community

We considered a possible m-government community, consisting of [5–7]:

■ Citizens
■ Civil servants
■ Administrators

Depending on the scenario, it might be necessary to introduce some other roles, like delegates of either citizens or civil servants and several levels of administration here, but it is assumed that for the functional purpose of the proposed system those roles do not matter as they usually do not influence the platform processes directly. Instead, the proposed system may use the delegates themselves that are actually belonging to one of the mentioned groups.

Citizens are the primary user targets of the proposed system. Using a mobile device, they access the system, initiate requests, or receive notifications pushed by the platform. Citizen delegates are handled like Civil servants, as they cannot access the system for someone else, due to the nature of the authentication mechanism. Instead they are forced to get help from a Civil servant, actually initiating the request.

Civil servants are the right hand of the proposed platform. Where the platform itself and automatically is only able to check user requests for security constraints (WS-security, XML-security, time stamping, SAML, and XKMS features), Civil servants may approve or decline requests on a semantic legal level that is elusive by computer systems. They are also necessary when it comes to delegate requests by other citizens or Civil servants from other municipalities/governmental bodies/agencies.

Administrators are responsible for administration of the community as a whole or the platform and the involved community members in detail. There are administrators who are actually handling the technical maintenance and administrators who are able to hand out, for example, PKI electronic certificates to the users and/or Civil servants.

Those three roles are actually directly mapped to system roles, when it comes to the technical realization. While administrators are mainly used for PKI administration and security certificate handling, the roles of Civil servants and citizens/users need to be integrated into the platform logic directly to distinguish between them, when it comes to access control, authorization, and business logic decisions. Therefore, the decision to use SAML assertions with integrated roles for the authentication and authorization purposes came naturally.

By using SAML together with WS-security, it was a small step to extend the server–server communications to use the same technology as well. For that reason, internally a fourth role was defined. The role of each server is important as it is necessary to be defined for intercommunication between the various proposed m-government platforms. However, although the communication is established between two servers, the documents delivered are meant to be assured and signed by Civil servants, to assure responsibility by a human being.

11.3 Possible m-Residence Certificate Service Scenario

One example of the mobile government online services is particularly emphasized: sending m-residence certificate request and obtaining the m-residence electronic document as a municipality's response (mRCertificate—m-Residence Certificate).

A citizen of city A needs a certification for his principal residence in city A. He will contact the municipality of city A for that.

In this process, he sends a request to this municipality first. The municipality creates his mRCertificate. He gets a final notification message and can pick up his mRCertificate afterwards. See Figure 11.2 for the scenario overview.

In a more detailed view, there are three system objects belonging to the municipality. It is the platform, the local IT Infrastructure (Legacy system), and the Civil

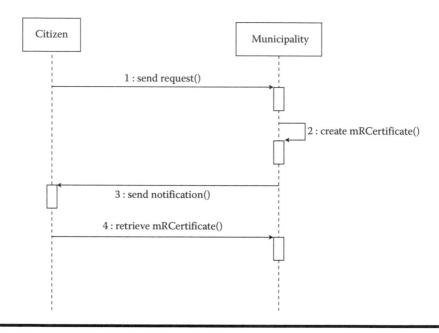

Figure 11.2   Municipality's service overview.

servant as the human actor. The citizen sends his request to the platform, which in return first sends a notification back about the incoming request and then forwards the request to the Civil servant for approval. After this, the request is sent to the Legacy system, where the mRCertificate is created.

After that, the Civil servant has to approve this mRCertificate. Furthermore, there is a final notification sent to the mobile to inform the citizen that he can pick up his mRCertificate. Finally, the mRCertificate needs to be retrieved by the citizen using the document retrieval service described before (see Figure 11.3).

11.3.1 Secure Service Request Creation and Dispatch

This section describes all the steps required from the mobile user who uses the JAVA mobile application in order to securely create a request for m-government online service and dispatch it to the platform.

Time-stamping authority is an entity which provides valid and digitally signed time-stamp tokens.

In order for a mobile user to request a specific m-government service, following all the steps described in this paragraph, he/she must successfully complete the following:

■ To successfully install the secure stand-alone JAVA application on the mobile device.

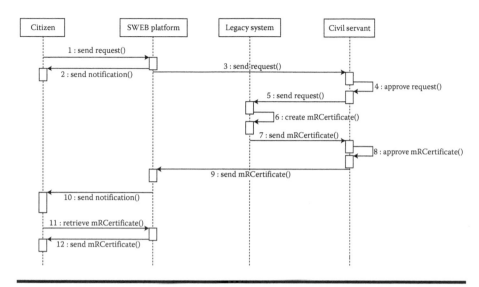

Figure 11.3 Municipality's scenario details.

■ To be successfully authenticated to the application, using the cryptographic credentials associated to the application.

The request creation and dispatching phase is depicted in the sequence diagram (see Figure 11.4). This phase starts with the creation of the m-Document that contains the service request. Specifically, the mobile user activates the mobile application and fills in the required fields in the predefined sequence of electronic forms (Step 1).

When ending this process, the mobile application automatically creates an m-Document (Step 2). The secure mobile application digitally signs the m-Document (Step 3), using the cryptographic credentials associated with the operation of the mobile application and the specific user.

If not, the process ends and has to be restarted from the first step (Step 1).

The next step involves the creation of a request to the time-stamping authority, in order to obtain a valid time-stamp token. The mobile application automatically creates this request, while previously creates a hash value of the final m-Document. It attaches the hash value of the m-Document in this message (Step 4) and dispatches it to the time-stamping authority (Step 5). The TSA processes the received request, issues a valid time-stamp token (Step 6), and responds to the mobile device with the signed and time-stamped hash value of the m-Document (Step 7).

Once the m-Document has been digitally signed, the mobile application creates a SOAP message (Step 8) and embodies the signed document into it.

In addition to what was last mentioned, the secure application embeds the required SAML authentication and authorization token for accessing the selected m-government service. The procedure that describes the way of obtaining a valid

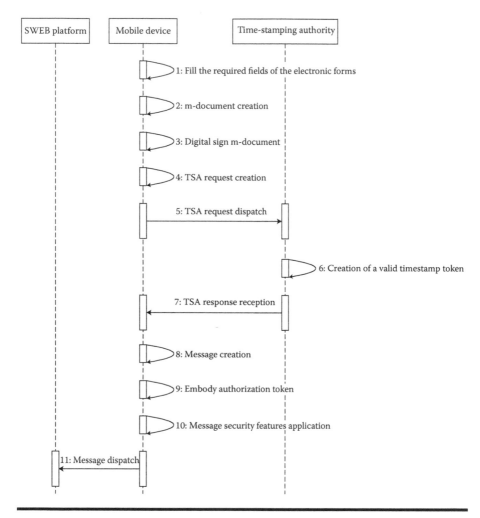

Figure 11.4 Secure service request creation and dispatch.

authorization token is described in detail in the next section. This SOAP message is the one that will be sent to the platform.

The next step for the mobile application is to automatically apply strong security features on the message (Step 10) and to dispatch it to the platform (Step 11).

11.3.2 Certificate Management Protocol

In order to register and retrieve a PKI certificate from the corresponding Certification Authority (CA), a mobile user with an appropriate secure mobile application could use the XKMS protocol (XKMS 2.0), and particularly its XML Key Registration Service Specification (X-KRSS) part.

A possible scenario of the generation of the user asymmetric key pair and a corresponding PKI certificate by using the secure mobile application and the XKMS protocol is described below.

This scenario is shown in Figure 11.5.

1. The user will be registered in a registration office for this service. The user will obtain an "empty" secure mobile application at the registration office along with a PIN letter with an initial username/password for accessing the application and a XKMS authentication code. The password will be enforced

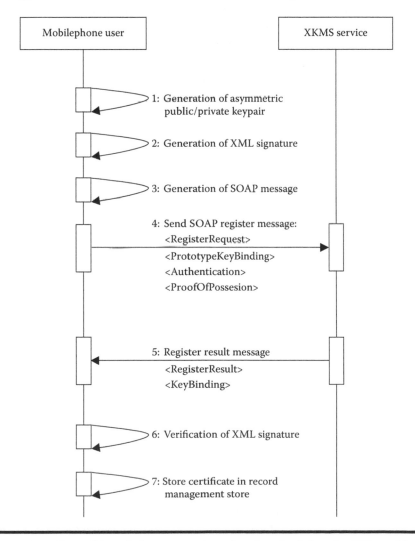

Figure 11.5 Procedure of the user registration for obtaining the PKI certificate from the XKMS server.

to change upon initial starting of the application. The XKMS authentication code will be valid for a specific period of time.

2. In a specific configuration file of his mobile application, the user will have URL and a name of XKMS server defined during the registration process.

3. The user will start the secure mobile application and start a procedure for generating asymmetric public/private key pair. The generated asymmetric private key will be protected by a separate PIN code defined by the user. The key pair could be generated inside the secure mobile application or could be generated on the SIM PKI smart card of the mobile phone.

4. The user will generate a Register XKMS request to assert a binding of information to the generated public key and send it to the registration service of the XKMS server. The registration of a key binding will result in the issue of a PKI certificate. The registration service will require additional information to authenticate this request which should be provided by the client–the XKMS authentication code and a Proof of Possession of the private key. The <Prototype-KeyBinding> element contains the public key that has to be registered. The <ProofOfPossession> element contains an XML Signature element. The user's private key—a cryptographic pair with the public key contained within the <PrototypeKeyBinding> is used to create the signature.

5. On receiving the registration request, the registration service of the XKMS server will verify the XKMS authentication code and the Proof of Possession information provided.

6. If the registration service accepts the request, a corresponding PKI certificate will be generated and key binding will be registered.

7. On receiving a registration result message from the XKMS server, the secure mobile application will verify the XML signature of the received message and validate the received PKI certificate.

8. If the mentioned verifications are successful, the secure mobile application will store the received PKI certificate in a Record Management System (RMS) of the mobile application.

We considered a Web Service scenario where the mobile phone user produces a cryptographic signature in the JAVA application using the private key stored either in JAVA keystore or in PKI SIM smart card.

Data are encrypted using a crypto MIDlet or Xlet JAVA application installed on mobile phone with CLDC or Connected Device Configuration (CDC) configurations. The user uses standard WS-Signature mechanism (WS-security) to wrap a cryptographic signature into the SOAP request and sends the request over to the remote Web Service endpoint implementation. Web Service performs request processing and sends SOAP response back to the JAVA-based mobile phone application. The mobile application processes the SOAP response and displays the status to the mobile user.

All servers mentioned in this chapter are deployed as SOAP-based Web Service using the Sun Glassfish application server.

The user registration process through XMKS service was implemented in J2ME CDC 1.1 environment. We partly modified and used an open-source Apache *Trust Services Integration Kit* (TSIK) package (http://svn.apache.org/viewvc/incubator/tsik/) for the implementation of digital signature of XML documents. In the XML signature procedure, we used the 1024-bits RSA key.

This document was then sent to an XKMS server using HTTP protocol. For the implementation of HTTP communication, we used mobile phone J2ME-embedded methods that implemented the javax.microedition.io.HttpConnection interface.

11.3.3 Certification Validation Schema

XML Key Information Service Specification (X-KISS) is a second part of the XKMS protocol which supports a delegation of the processing of PKI certificates for a purpose of XML-security or WS-security from an application (in this case, it is the secure mobile application) to the XKMS server.

In this sense, the secure mobile application will include the following functionalities: Locate or Validate requests as specified by the X-KISS.

The Validate request of the X-KISS protocol will be used by the secure mobile application in order to be able to complete the verification of an obtained digitally signed document (or message) onto the mobile terminal. This verification consists of two parts:

- Verification of the document (or message) signature
- Validation of the associated PKI certificate of the signer

The first step, that is, the actual verification of message signature, is done by corresponding modules (e.g., WS-security or XML-security) of the secure mobile application.

After that, the signer's PKI certificate validation (or a complete certificate chain validation) is done by sending a Validate request message to the XKMS server. In this way, complex functionalities of certificate validation will be removed from the secure mobile application of the mobile terminal and delegated to the XKMS server.

One of the main benefits in this case is that the secure mobile application does not need to implement a certificate revocation status check (either by downloading a corresponding Certificate Revocation List (CRL) or by implementing the Online Certificate Validation Protocol (OCSP)). In this way, this complex functionality is completely outsourced to the XKMS server.

A possible scenario of validation of the received X.509v3 electronic certificate by using the secure mobile application and the XKMS protocol is described below (see Figure 11.6):

1. The mobile user receives a message with an XML Signature. The secure JAVA mobile phone application verifies the XML Signature by using a public

key obtained from X.509v3 certificate. The next step is a validation of the user's certificate by using the ValidateRequest message of the XKMS protocol. In this sense, the mobile application creates an XML ValidateRequest message.

2. The mobile application further creates a SOAP message in which a body part is the already created XML ValidateRequest message in the previous step.

3. The mobile user application sends the SOAP message to the XKMS server.

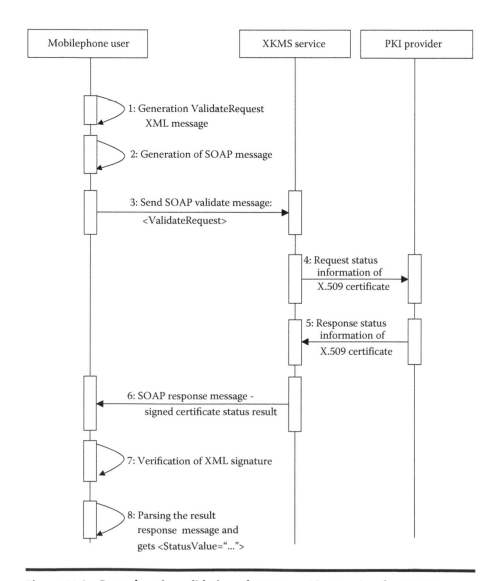

Figure 11.6   Procedure for validation of X.509 certificate using the XKMS server.

4. In order to validate the received X.509v3 certificate, the XKMS server communicates with the PKI server beyond that publishes CRL (or uses the OCSP).
5. PKI server responds with status information about received X.509v3 certificate.
6. The XMKS server creates the XML response message and generates a corresponding XML signature and sends it to the mobile user.
7. The mobile application receives the ValidateResult message and verifies the XML signature.
8. If the signature verification is successful, the secure JAVA mobile application parses the received result message and gets information about a StatusValue of the requested X.509v3 certificate.

If the received StatusValue of the requested X.509v3 certificate is correct, then the secure JAVA application finalized successfully the verification of the signature of the received signed document.

If the signer's certificate is not included in the signed message, the Locate request function should be done (based on some appropriate information about a location where this certificate could be found (e.g., some URL)) before using the Validate request function in order to get the appropriate PKI certificate from the XKMS server.

Also, the Locate request could be used to request a corresponding PKI certificate from the XKMS server in order to be able to perform encryption functionality (encryption part of the WS-security or XML-security) of some XML or SOAP messages.

All testing and simulations of the mentioned functionalities of the XKMS protocol are implemented by using a simulator for the development of JAVA application supporting also the mobile phones: Sun Toolkit NetBeans 5.5 Mobile Edition (http://netbeans.org/).

11.3.4 Component Orchestration

The functionality of the above components is orchestrated by the main application (see Figure 11.7). As soon as the user selects which scenario is to be run, a specific set of actions are performed. Figure 11.7 depicts a running scenario, communication between internal components of the application, and communication with external entities.

11.4 Secure JAVA Mobile Web Service Application

This is a functional description of the developed secure JAVA mobile Web Service application for a purpose of secure communication with the described m-government

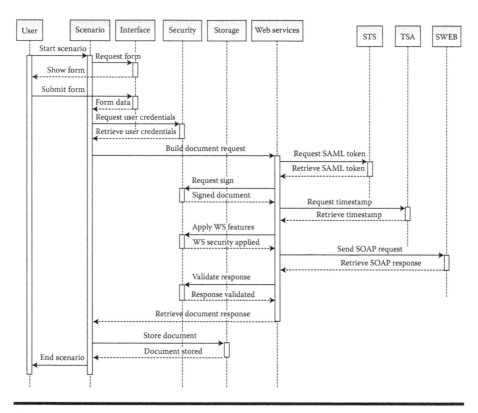

Figure 11.7   m-government scenario sequence diagram.

SOA-based platform. This JAVA application was developed in the framework of the FP6 SWEB Project.

The assumption is that the user already has the JAVA application on his mobile phone/terminal and thus a procedure of downloading and activating the application is beyond a scope of this document. Possible usages are described in [5].

This client application comprises the following functionalities:

■ Graphical User Interface (GUI) for presenting business functionalities to the end user
■ Business (core) functionalities of the application–m-government functionality, for example, m-residence certificate
■ Security functionalities
■ Communication

The proposed m-government secure JAVA mobile application objects and community are represented in Figure 11.8.

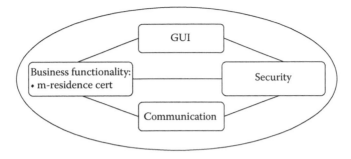

Figure 11.8 Possible m-government JAVA mobile WS application community.

11.4.1 GUI Object

The GUI object of the proposed JAVA mobile Web Service application is responsible to show user interface that enable calling of authentication function for authentication of the end user and presenting the core functionalities to the end user.

According to this, the GUI object communicates with the following modules:

- User Authentication module for JAVA application of the Security object.
- User PKI Registration module (XKMS module) of the Security object.
- User Authentication and Authorization module for the m-government platform (SAML module) of the Security object.
- m-residence module of the Business object.

User Authentication module will be described in Section 11.4.3 (Security object) and other modules will be described in the next section.

11.4.2 Business Functionalities Object

This object of the SWEB JAVA mobile application is responsible for realization of the core SWEB client-based functionalities:

- Secure requesting and receiving the m-residence certificate from the corresponding municipality SOA-based platform, receiving a notification and delivering the obtained certificate to some interested party.
- Secure sending of other kind of predefined message (e.g., m-invoice) to the corresponding municipality platform and receiving the notification from the platform.

Both business functionalities have links to Security and Communication objects.

11.4.3 Security Object

The Security object of the considered JAVA mobile application is responsible for overall application-level security functionalities. It consists of the following modules:

- Authentication module to the JAVA application
- XKMS module
- STS module
- XML-security module
- WS-security module
- Time-stamping module

The above-mentioned modules will be briefly described in the following text.

11.4.3.1 User Authentication Module to the JAVA Mobile Application

An end user could have the proposed JAVA application installed on his mobile phone/terminal following one of the two possible ways:

- Installing the JAVA mobile application by using the desktop computer via wired (corresponding cable) or wireless (e.g., Bluetooth, Infrared, WiFi) transmission path to the mobile terminal—offline approach.
- Installing the JAVA mobile application through some Over-The-Air (OTA) services—online approach.

User authentication for using the JAVA mobile application should be a two-step process:

- The first step should be a combination of username/password for accessing the application (the password should be changeable by the user). This should be done immediately after starting the application. These credentials will be generated during the user registration process. During the initial phase of the application installation, the user will obtain the username and some default password. The application has to force the user to change the password on the first application start.
- The second step should be a corresponding PIN code for accessing the asymmetric private key just before signing the m-residence certificate request.

Generation of the user asymmetric public/private key pair and the corresponding digital certificate should be done through the user registration function of the XKMS protocol which is described in Section 11.3.2. The User Authentication module is called from the GUI object.

11.4.3.2 XKMS Module

The XKMS protocol comprises two parts:

- The X-KISS—a protocol to support the delegation by an application to a service of the processing of key information associated with an XML signature, XML encryption, or other usage of the XML Signature <ds:KeyInfo> element.
- The X-KRSS—a protocol to support the registration of a key pair by a key pair holder, with the intent that the key pair subsequently be usable in conjunction with the X-KISS or a PKI.

XKMS enables one to simplify the use of PKI by client systems. It provides them the actual information about keys issued by CAs of the PKI. We used partly modified OpenXKMS—an open-source implementation of XKMS. OpenXKMS is JAVA-based and platform-independent (http://sourceforge.net/projects/xkms/). It consists of two main parts: the OpenXKMS Web Service and the OpenXKMS Client API which is used by a client to prepare requests to the OpenXKMS Web Service.

11.4.3.3 XML-Security Module

XML-security module is responsible for implementation of standard XML signature and XML encryption components. The XML Signature Module provides message integrity, signature assurance, and nonrepudiation over Web data (XML). This also provides means procedures for computing and verifying such signature. It is a method associating a key with referenced data. Data objects are digested, the resulting value is placed in an element (with other information) and that element is then digested and cryptographically signed.

XML Encryption is the part of XML-security package that enables the confidentiality service. XML encryption module provides that XML content can be transformed such that it is discernable only to the intended recipients, and opaque to all others. XML Encryption has the key feature of being able to encrypt part of an XML document, like a single element or a subtree of elements, but it is also possible to encrypt several elements using several different encryption keys, just like XML signature—this enables for different entities to access different parts of the document exclusively.

XML Signature Module consists of

- Implementation of the RSA private key operation for creating digital signature, as well as a function for signature verification
- Implementation of hash functions (SHA-1, SHA-256, MD5)
- Implementation of different symmetrical cryptographic algorithms (AES, IDEA, Serpent, TDES)

■ Implementation of the RSA private key operation for decryption of encrypted session symmetric key in digital envelope, as well as a function for the creation of an encrypted session symmetric key.

11.4.3.4 WS-Security Module

WS-security provides a mechanism that allows one to digitally sign (using XML Signature) all or part of a SOAP message and passes the signature in the SOAP header. It provides a mechanism that allows one to encrypt (using XML Encryption) all or part of a SOAP message. It also provides a way to pass information in a SOAP header about the encryption keys needed to decrypt the message or verify the digital signature. And it allows trust assertions about the SOAP message to be passed in the SOAP header as well.

WS-security module is implemented as standard security mechanisms for the protection of SOAP messages. WS-security module is a very important module of the security object, because it is used for protection:

■ Communication with STS server—STS server
■ Communication with the proposed municipality's SOA-based platform
■ Protection of corresponding SOAP message with m-residence certificate

In this way, the WS-security module communicates with STS module of the security object as well as with Business functionalities object.

11.4.3.5 STS Module

The STS module is responsible for communication with the STS server in order to receive a SAML assertion (token) that will be used afterwards to enable access to the proposed online m-residence certificate services by the client.

The client first sends a RST message to the STS server by using a SAML protocol. Security is done by using WS-security. After successful authentication of the user based on the client's X.509v3 digital certificate, the STS server issues a SAML token to the user which is digitally signed by the STS server. This token is securely communicated to the end user by using the WS-security mechanisms.

After successful checking of WS-security mechanisms (decryption and verification of digital signature on the level of SOAP message), the end user prepares a message with the request of m-residence certificate service. This message consists of the SAML token and is additionally secured by WS-security mechanisms. The actual URL of the municipality's online services is found in the SAML token. After successful decryption and verification of digital signature on the level of SOAP message (WS-security), as well as after successful verification of the SAML token's signature from the municipality's platform, the end user is granted access to the desired platform services. After that, the user could send business requests (m-residence certificate) to the platform.

The SAML module communicates with WS-security module of the security object as well as with the Communication object.

11.4.3.6 Time-Stamping Module

The time-stamping module is responsible for communication with the time-stamping server. A time-stamping service supports assertions of proof that a datum existed before a particular time. In order to associate a datum with a particular point in time, a TSA may need to be used as a trusted third-party (TTP) service.

As the first message of this mechanism, client application requests a time-stamp token by sending a request (TimeStampReq) to the TSA. As the second message, the TSA responds by sending a response (TimeStampResp) to the requesting entity.

11.4.4 Implementation Aspects

The Java Specification Request 172 (JSR 172) specifies standardized client-side technology to enable J2ME applications to consume remote services on typical Web Services architectures [7]. JSR 172 defines a standardized API that J2ME clients can use to invoke SOAP and XML-based Web Services. This API is in the form of an optional package for J2ME, and is referred to as Web Services APIs (WSA) for J2ME. WSA is actually a subset of the Java API for XML-based Remote Procedure Call (JAX-RPC) defined by JSR 101 [8].

With the ever-present concern over security, software applications must consider how to secure confidential data. A mobile application is not immune from privacy concerns. In fact, mobile devices and their software application have special considerations given that most people carry these devices wherever they go.

11.4.4.1 Bouncy Castle APIs

In order to encrypt sensitive data, we used Bouncy Castle Cryptography APIs. Bouncy Castle is an open-source Java API for encrypting and decrypting data. There is a lightweight package that is suitable for MIDP applications where only a fraction of the API will be used at any one time.

11.4.4.2 Obfuscation Process

One problem inherent in most mobile devices is the limited amount of memory. As with most libraries you use, only a small portion of the code is typically needed by your application. One common way to eliminate an unused code, and at the same time make it more challenging to reverse engineer an application, is to use a Java obfuscator. Reverse engineering of Java programs is not too difficult. As a matter of fact, there are free decompilers that will do the work for you. To make it a little more challenging to reverse engineering applications, many Java developers use an obfuscator to rename classes, methods, and fields. The intention of this renaming process is to make the source more unreadable. A side effect of the obfuscation process is the reduction of class file size. This is accomplished in two steps. First, a lot of bytes can be saved by replacing names of classes, methods, and field names that are one or two characters in length. In addition, obfuscators will remove unused classes, methods, and fields. The combination of these two steps can significantly reduce the size of the final application. We used an open-source obfuscator named ProGuard.

11.4.4.3 Security and Trust Services API

Security and Trust Services API (SATSA) is a new API that provides additional security capabilities to the J2ME CLDC platform. It specifies a collection of APIs that provide security and trust services for J2ME CLDC by integrating a Security Element (SE) [9].

The SE is a hardware or software component in a J2ME device. It provides the following features:

- Secure storage to protect sensitive data
- Cryptographic operations

With these features, J2ME applications would be able to have secure key stores as well as encryption and decryption capabilities. These features could be used to provide security services for applications such as e-payments, mobile commerce, and so on. An SE can be (1) deployed as a smart card in wireless phones (e.g., SIM PKI cards) or, (2) can be implemented by a handset itself (e.g., embedded chips or special security features of the hardware) or (3) may be entirely implemented in software [10].

The support for cryptographic smart cards is of particular interest to developers writing J2ME applications for smart phones. Keys and certificates can be stored on the smart card and data can be signed without the private key ever leaving the card. High-end smart cards are tamper resistant and provide authentication schemes, such as requiring a PIN or a password before access to the smart card is granted. In this way, security is dependent on the smart card not being compromised. Private keys do not have to be stored on diverse insecure clients, enabling vendors to focus on keeping the smart card secure from physical tampering and, just as important, smart card API exploitation [11,12].

11.4.5 Java Mobile Client

The Java mobile client used for communication with an SWEB platform is developed by using a J2ME CDC1.1 platform. According to this, a lot of technologies that exist in J2SE have been employed in the J2ME application.

Figure 11.9 describes the forms (screens) on mobile phone application used to perform communication with an SWEB platform. The order of form appearance according to the button pressed is presented only for the most important functions.

The first form is "Logon form." The user should enter the username and password after which verification will be passed to the next form. Also, the language that will be used in whole application can be chosen on this form.

After successful verification of username and password, "Functions form" will be passed to the user. This is the main form for the user, where a task that needs to be done can be chosen.

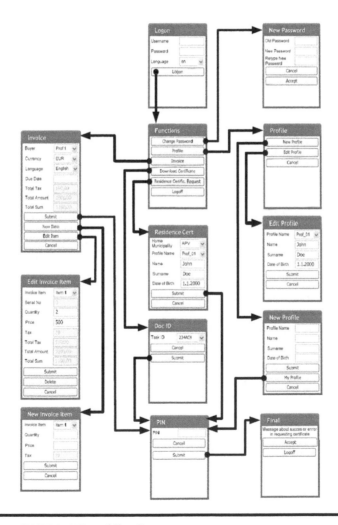

Figure 11.9　SWEB J2ME mobile client.

The available tasks are

1. Change Password—used for changing the login password in order to access the mobile application.
2. mResidence Certificate Request—used for sending request for mResidence certificate to the municipality.
3. Download mRCertificate—used for downloading prepared mResidence certificate from the municipality.
4. Send m-Invoice—used for preparing and sending m-invoices to the municipality.

Changing the login password can be done via "New Password" simple form.

By pressing button for sending of mResidence Certificate Request, user will jump to "Residence Cert" form where the Municipality should be chosen from the list that appears on the form.

The next step is entering a PIN used for reading the user private key that is stored in KeyStore on the file system on the user mobile device. This should be done on the "PIN" form. The result of request processing (error or success) is displayed on the "Final" form. All communications between client and servers are synchronous here. It means that each request produces a response.

After successful processing of the user's mResidence Certificate Request, the SWEB platform prepares the required mRCertificate and sends the SMS message to the user mobile device. This part of communication between the user and the SWEB platform is asynchronous. Arrived message is a signal for the user to perform downloading of mRCertificate via the option of Download certificate. On the "Doc ID" form, the Task ID should be chosen from the list. The Task ID uniquely identifies the mRCertificate that should be downloaded. After that, on the next form, the PIN, for obtaining the user's private key, should be entered. The result of download will also be displayed on the "Final" form.

The first step for sending m-Invoice is populating the necessary form fields on the "Invoice" form. The Municipality should also be chosen from the list. The next form is the form for entering a PIN used for reading the user private key. The result of m-Invoice sending (error or success) will also be displayed on the "Final form."

In order to realize the above-mentioned functionalities, the mobile JAVA application communicates with the following external entities:

■ STS server
■ XKMS (XML Key Management Specification) server
■ TSA server
■ SWEB platform—Interaction Tier Manager of the municipality

Communication between the JAVA mobile application and the STS server is realized by using WS-secured SOAP communication. According to the scenarios, the JAVA mobile application sends the RST to the STS server and, if everything is ok, it receives back the RSTR which consists of URL of the municipality and the SAML token with the user's role on the SWEB platform.

Communication between the JAVA mobile application and the SWEB platform of the municipality is realized as WS-Encrypted SOAP communication with the Interaction Tier Manager of the municipality. According to the scenarios, the JAVA mobile application sends the signed mRCertificate request or m-invoice (signing is done by using XML signature mechanisms) to the Interaction Tier Manager of the municipality.

Before sending it to the municipality, the signed mRCertificate request or m-invoice must be time stamped. In order to accomplish this, the JAVA mobile application communicates with time-stamping server via HTTP communication. In this sense, the JAVA mobile application sends a hash of the signature of the mRCertificate or m-invoice to the time-stamping server and receives back a time stamp (signed hash with added time information) which is signed by the private key of the time-stamping server.

Only in the mRCertificate scenarios, when the mRCertificate is ready for delivery at the platform, the SWEB platform sends a SMS to the mobile user informing him that the mRCertificate with the given TaskID is ready for download. After that, the JAVA mobile application will send a request for mRCertificate download also as a signed and time-stamped request in a body of the WS-Encrypted SOAP message to the Interaction Tier Manager of the municipality.

During the above-mentioned communication, in order to verify signatures and validate different X.509v3 certificates, the JAVA mobile application needs to communicate with the XKMS server which outsources a part of the time- and resource-consuming PKI functionalities to the JAVA mobile application. Namely, the JAVA mobile application could obtain a suitable certificate from the XKMS server (by using LocateRequest XKMS function) and, more importantly, could validate certificate of some party (by using ValidateRequest XKMS function). In this way, the most time-consuming PKI operations, like certificate validation, will be excluded from the mobile phone. The communication with the XKMS server is a SOAP communication without applying security features. Only, the XKMS server's response is always digitally signed by using the XML signature mechanism.

11.4.5.1 Interface Description

The interfaces of the SWEB mobile client application can be divided into four groups:

1. Graphical User Interfaces—responsible for visual representation of the application, and for event handling.
2. Security interfaces—responsible for encryption, signing, checking password, and other security issues.

3. Communication interfaces—used for making and exchanging messages with the SWEB server platform (SOAP messages, time-stamp requests, etc.).
4. Store interfaces—used to load resources to and from stores. Store can be file system, RMS, KeyStore, or something else.

According to the specific design of GUI which relays on XML file definitions, there is no direct call between GUI implementations and other interface groups. Indirectly, all three other interface implementations can be called in GUI via developed service mechanism defined in XML files. It means that GUI can be developed separately from three other implementations.

The core functionality of mobile application includes APIs for sending and receiving protected XML messages. The security functionality is implemented in the following modules:

- XML encryption module—provides an interface to the XML Encryption procedure.
- XML digital signature module—provides an interface to the XML Digital Signature procedure.
- WS-Trust (STS) module—the interface is used to establish the communication with the STS Server in order to get the SAML token.
- WS-security module—provides an interface to WS-Signature and WS-Encryption procedures.
- Time-stamping module—enables the communication with a TTP (Time-Stamping Authority, TSA) in order to obtain certified time data that will be cryptographically bound to a document.
- XKMS module—the interface is used to establish the communication with the XKMS Server in order to locate and validate certificates.
- Certificate module—the interface which is used to access and parse the X.509v3 electronic certificate stored in JAVA store of the mobile phone.
- WebService module—provides an interface to the SOAP parser module. The SOAP parser is built on a generic XML parser with special type-mapping and text data-marshalling mechanisms.

Software and tools used for the development of Java mobile application are

- NetBeans IDE 5.5 development tool—vendor Sun Microsystems Inc. The Netbeans is a free, open-source Integrated Development Environment for software developers. The NetBeans IDE Bundle for Mobility is a tool for developing applications that run on mobile devices; generally mobile phones, but this also includes entry-level PDAs, among others (http://www.netbeans.org/community/releases/55/).
- Sun Java™ Toolkit 1.0 for CDC based mobile devices. The Sun Java Toolkit for CDC is a set of tools for developing applications that run on a range of

network-connected consumer and embedded devices that support the Java ME CDC application framework, including high-end wireless devices (http://www.java.sun.com/products/cdctoolkit/).

■ Open-source BouncyCastle Crypto Package—mobile edition (J2ME). The package contains a lightweight J2ME API suitable for use in any mobile environment with the additional infrastructure to conform the algorithms to the JCE framework. During the development of the java mobile application it used the mobile edition package—version 1.37. (http://www.bouncycastle.org).

■ Apache TSIK open-source library with security features for Web services, such as XML Signature, XML Encryption. Apache TSIK is a client and server toolkit for creating secure XML applications (https://svn.apache.org/repos/asf/incubator/tsik/).

■ Open XKMS 2.0 library. The Open XKMS is an open-source implementation of the W3C Recommendation of the XKMS 2.0. It is a compound of an XKMS Server and a Client API to access to the Server via Web Service (www.sourceforge.net/projects/xkms/).

11.5 Conclusions

In this chapter, we present a possible model of secure SOA-based e/m-government system. In fact, this work is related to the consideration of some possible SOA-based e/m-government online systems, that is, about secure mobile communication between citizens and companies with the SMGOs, such as municipalities. We elaborated an m-government framework which is based on secure JAVA mobile application, PKI certificates, SOA-based platform, XML-security, WS-security, SAML, time stamping and XKMS. In this sense, main contributions of this study are (1) proposal of a possible secure m-government model based on JAVA mobile application and SOA-based m-government platform, (2) usage of secure JAVA mobile application in which all modern security techniques are implemented (XML security, WS-security, SAML, time stamping, PKI, XKMS), (3) usage of SOA-based request-response m-government platform Web Service instead of session-based Web application platform which is more suitable for usage of secure mobile application, and (4) usage of XKMS service which is more suitable for mobile PKI system since it outsources complex operations such as PKI validation services to the external entity—the XKMS server, compared to usages of other techniques. The work presented and examples described are included in the general framework of the EU IST FP6 SWEB project.

Acknowledgments

This work was carried out in the context of the IST international cooperation project SWEB (044979). This study is based on the work performed within the context of this project and the authors would like to acknowledge all SWEB partners.

Disclaimer

This research outlined in this chapter has been undertaken with the financial assistance of the European Community. The views expressed herein are those of SWEB Consortium and therefore can no way be taken to reflect the official opinion of the European Commission. The information in this document is provided as is and no guarantee or warranty is given to state that the information is fit for any particular purpose. The user therefore uses the information at their sole risk and liability.

References

1. SWEB Project Homepage. http://www.sweb-project.org
2. Kolsi, O. and Virtanen, T. MIDP 2.0 security enhancements, *Proceedings of the 37th Hawaii International Conference on System Sciences*, Big Island, HI, USA, 2004.
3. Papastergiou, S., Karantjias, A., Polemi, D., and Marković, M. A secure mobile framework for m-services, *The Third International Conference on Internet and Web Applications and Services, ICIW 2008*. Athens, Greece, June 8–13, 2008.
4. Lee, Y., Lee, J., and Song, J. Design and implementation of wireless PKI technology suitable for mobile phone in mobile-commerce. *Computer Communications* 30(4): 893–903, 2007.
5. Cuno, S., Glickman, Y., Hoepner, P., Karantjias, T., Marković, M., and Schmidt, M. *The Architecture of an Interoperable and Secure eGovernment Platform which Provides Mobile Services, Collaboration and the Knowledge Economy: Issues, Applications, Case Studies*. P. Cunningham and M. Cunningham (Eds), IOS Press, Amsterdam ISBN 978-1-58603-924-0, pp. 278–256, 2008.
6. Đorđević, G. and Marković, M. Simulation and JAVA programming of secure mobile web services, Balcor 2007, *Balkan Conference on Operational Research*, Zlatibor, October 2007.
7. Marković, M. and Đorđević, G. Java-based secure mobile Web Service scenario, *Proceedings of INFOTECH—JAHORINA 2009*, Jahorina, Republic Srpska, Bosnia and Herzegovina, March 18–20, 2009.
8. Ortiz, C.E. Introduction to J2ME Web Services, http://developers.sun.com/techtopics/mobility/apis/articles/wsa/, April 2004.
9. *IBM Workplace Client Technology Micro Edition Version 5.7.1: Application Development and Case Study, Redbook*, sg246496.pdf, www.ibm.com/redbooks, June 2005.
10. *Building a Secure SOAP Client for J2ME, Part 1: Exploring Web Services APIs (WSA) for J2ME*. Bilal Siddiqui, http://www-128.ibm.com/developerworks/edu/, June 16, 2006.
11. Ortiz, C. E. Understanding the Web Services Subset API for Java ME. http://developers.sun.com/techtopics/mobility/midp/articles/webservices, March 2006.
12. MIDP *2.0: SATSA-APDU API Developer's Guide, version 1.0*, February 2nd, 2007. Forum Nokia, *Handbook*. University Science, Mill Valley, CA, 2007.

Chapter 12

Tele-Measurement Services for m-Learning

E. Kayafas, F. Sandu, A. V. Nedelcu, C. Costache, M. Demeter, and D. N. Robu

Contents

12.1 Technical Aspects of Mobile Data Streaming

Wireless cellular networks are nowadays omnipresent, being available now to more than 3 billion subscribers worldwide and creating a great impact on their personal and professional life.

The permanent and almost ubiquitous world cellular networks can be used for data transmission in civil/general purpose. The use of data acquisition and data transmission over PLMN (Public Land Mobile Networks) can save time and enhance accessibility (from multiple locations) with minimum footprint from the technology and equipment point of view. Due to the fact that the infrastructure is already in place, the added value of mobile remote measurement is cost efficient. The infrastructure is constantly extended and new wireless transmission technologies are developed (like WiMAX as a telecommunication backbone for broadband Internet access). Hence, the quality and availability are constantly improved and maximized for data transmission, thus, making the wireless solutions available for small business and/or remote areas. Availability and mobility (handover and roaming) provided applicability and increased usability to further domains of activity such as remote measurement and measurement on the move that are presented hereby.

The technical aspects are introduced based on a real case study [1] based on a workbench with high-speed data acquisition by a digital oscilloscope and with real-time transmission [2] via mobile networks—GSM/UMTS-GPRS [3,4] that provided the transport. The "real-time" concept in mobile networks is characterized by the fact that allowed transfer times are limited (lower) by the conversational nature of the architecture. It must also be considered that the time relationship between information entities ("time-stamps" of "events"—in the interrupts' treatment perspective) should not be affected by streaming. In multimedia, delays, or jitter have rather perceptual-physiologic (upper) limits—as much as human perception of video and audio allows. In this case study, time-relationship is carefully controlled by the data acquisition subsystem—before packing at transmitter side, acquired data can be sent without strict timing, provided the transfer is fast-enough, and reconstruction at the receiving side can be done precisely, inside the same perceptual delay allowed: for instance, for an usual oscilloscope, this must be lower than "retina persistence" which must be lower than "phosphor persistence"—so, again, a human perceptual limit like for normal video streaming (e.g., mobile video-tele-phony, mobile "video on demand," mobile IP-TV, etc.).

Characteristic to this approach of data streaming for remote-measurement is the use of a real, existing PLMN. The testbed [5] had the architecture of Figure 12.1, with a Radio Access Network (BTS-Base Transceiver Station/Node B-UMTS Base station, BSC-Base Station Controller/RNC—Radio Network Controller), Switching Center (MSC-Mobile Switching Center and TRAU-Transcoding and Rate Adaption Unit), and a Data Center (SGSN-Serving GPRS Support Node and a GGSN-Gateway GPRS Support Node from CISCO, [6]).

Both the probe and the client can be fixed or mobile, local or remote (see the configurations of Figures 12.2 and 12.3). Different scenarios can be imagined, with mobile units under test (e.g., vehicles, patients) having attached data loggers and transmitters and/or mobile surveillance-diagnose-intervention teams, and so on.

From the measurement point of view, the authors considered the state-of-the-art in the field of remote measurement and distributed measurement systems relying on wireless telecommunications networks.

Previous literature presented different remote measurement platforms and distributed configurations (networked devices as well as distributed databases and even distributed processing). Specific transfer protocols can be implemented on Microsoft .NET [7] or other frameworks. General purpose data sockets can be used

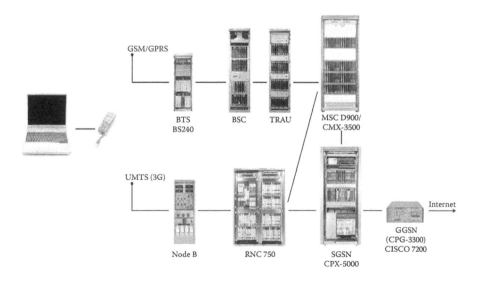

Figure 12.1   The testing environment with Siemens PLMN equipment (including a GGSN from CISCO).

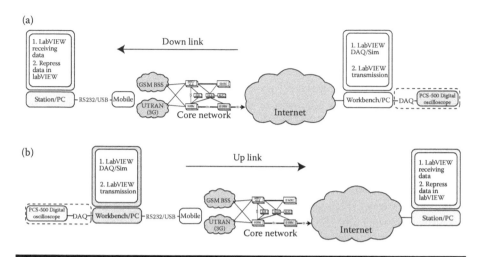

Figure 12.2   Scenarios for remote measurement over PLMN: (a) with DL (down link) transmission; (b) with UL (up link) transmission.

or proprietary solutions, for example, DataSocket Transfer Protocol "DSTP" [8], an application-layer protocol (implemented on top of TCP) for high-speed real-time mass data sharing [8] via the dedicated port 3015.

As DSTP support was not extended for all "handhelds," they were considered other solutions for wireless measurement systems [9], starting from different wireless modems and continuing with Bluetooth (IEEE 802.1a), WiFi (IEEE 802.11 b/g), WiMAX (IEEE 802.16) and, mainly, with the mobile GSM/UMTS—GPRS. The

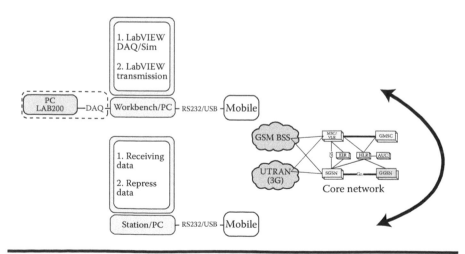

Figure 12.3   Remote measurement over PLMN with both probe and client mobile.

specific advantage of GPRS—it does not require establishing a connection, it is always connected—is considered crucial for applications where time and quick reaction to events is important. The trade-off for mobility, carefully addressed hereby, is the lower rate (compared with fixed networks), and dependency on radio environment and number of users (hereby addressed by the QoS). Measurement of data streaming QoS was also documented [10]; considering mobility, they were interesting contributions in IPv6 test-beds—our work aims to extend such research in cellular communications data streaming.

Different applications were considered, like meteorological distributed processing platforms or telemedicine [11], industrial communications (closest to this case study).

12.1.1 Objectives

The goal was to *establish a remote-measurement connection* which is reliable and *to perform several tests* in order to determine the best method to transmit electrical experimental data aiming for a real-time effect in manipulation and operation of remote test equipment like being local, on-site. TCP/UDP transmission [12,13] was used. The network elements in this case are divided between the network *transport medium* which enables the transmission and the *data acquisition* equipment. Identifying the impact of their technical limitations on the remote measurement was one of the purposes of our study.

12.1.2 Implementation

Data Acquisition (DAQ) equipment (together with sensors/transducers and local signal conditioning) applied to the unit-under-test, attached to the portable ("palm-top" or "lap-top") or fixed ("desk-top") PC, is identified as *"the probe."* The portable or fixed PC at the receiving side is *the client* that collects the remotely measured values. The probe and the client can both have mobile connections (e.g., wireless access using a mobile modem in order to connect to a PC)—see Figures 12.2 and 12.3.

12.1.2.1 Mobile Network

The PLMN was formed of real equipment starting with core network components (see Figure 12.1): SGSN, MSC (HLR—Home Location Register and VLR—Visitor Location Register), GGSN and continuing with the GSM-UMTS/GPRS Enhanced Radio Access Network. All these components are interconnected but transmission errors can occur by interferences, deficient network processing, and so on. The interfaces used for data transmission [14–16] cover the physical range from air (radio waves), to electrical and optical: Uu/Um air interfaces for GSM/GPRS/UMTS, optical UMTS interfaces Iub, Iu, Iur, Ethernet electrical/optical Interfaces Gn and Gi. Most of the tests were performed on a GPRS/EDGE platform.

From the radio access point of view, the transmission is Up Link (UL) if it is from the mobile terminal towards the network and Down Link (DL) if it is vice versa. Generally, the network improvements are applied first to the DL, in order to enlarge the download bandwidth.

Enhanced Data Rates for GSM Evolution (EDGE) is a bolt-on enhancement of the 2G/2.5 GSM. Data Rates possible for DL are up to 236.8 kbit/s for 4 timeslots (theoretical maximum is 473.6 kbit/s for 8 timeslots) in the packet mode. After Context Activation is accepted by the network at the request of the mobile terminal (for a valid SIM/subscriber) the mobile/subscriber and the PC attached to it via USB/Bluetooth are able to communicate on the Internet or with other subscribers in the network based on the IP (assigned by the GGSN). At the other end, the station or client (depending on the situation, see Figure 12.2) are connected on the Gi Interface/Internet.

12.1.2.2 Data Acquisition: The Digital Oscilloscope

Data were acquired using a Digital Oscilloscope [17] (see Figure 12.4) either on simulated data (in the "demo mode") or with real measurement. The oscilloscope

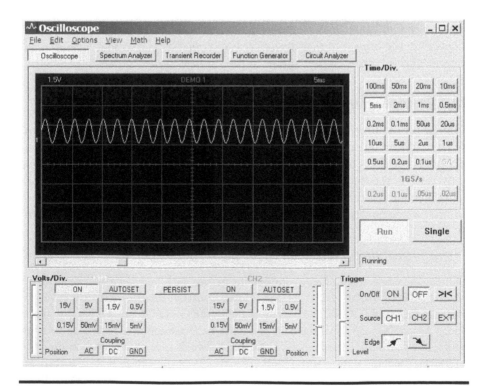

Figure 12.4 PC graphical user interface (GUI) of the Velleman PCS-500 digital oscilloscope.

acquires an adjustable number of samples per transmitted frame (from 200 to 5000, the capacity of the buffer; default value is 1000) of the measured signal—real or internally simulated.

The two main rates applied to data transmission are: the sampling rate (an adjustable value from 1250 to 50,000,000 samples/s) applied to the measured signal, and the rate at which the buffer is read. These two values must be correlated; the reading of the buffer should not be done earlier than the next storage in the buffer—see formula (Equation 12.1) below.

The actual signal measurement with the oscilloscope produces a 5000 samples cluster ("frame" or "window") at once, that is, the effective buffer capacity of the I/O memory—1st location stores the *Sampling Rate* (SR) [samples/second], 2nd location the Full Scale (mV), 3rd location the Offset (mV), and the others, the 5000 acquired values (with 8 bits per scaled and centered sample). The *Time to Wait* (TTW) until the buffer is full and can be read again is given by the formula

$$\text{TTW} = \frac{5000}{\text{SR}} \ (s) = \frac{5000 \times 1000}{\text{SR}} \ (\text{ms}) \qquad (12.1)$$

12.1.2.3 TCP/UDP Transport

The difference between TCP and UDP consists in the acknowledgment of the received packets. TCP works on the basis of sending data as soon as the last sent data is positively acknowledged [18] by the receiving party (establishment of end-to-end operations). In UDP, the Datagrams are sent regardless of the successful receiving [19].

In case of TCP *Round-Trip Time* (RTT) is observed and studied. This is the time from the moment one IP packet is sent until the corresponding acknowledgment is received. This is the delay that could be perceived by an operator.

In case of UDP the jitter is one parameter observed and studied. This is the variance of the time interval between the samples. This has a similar effect on the human operator as the RTT in case of TCP. Before each measurement, the data link is investigated with a bandwidth and reliability test (loss of packets, delays) using a dedicated tool, Iperf (with the Jperf graphical interface) for measuring maximum TCP/UDP bandwidth [20]. For bandwidth measurement (and for other QoS parameters) Ethereal/WireShark [21] was also used. Jperf Bandwidth and the Ethereal I/O Graphs are expressed in kbps (Y-kbps, X-s).

The samples are read from the memory buffer and then transmitted to the client via IP—TCP with its own error detection mechanism and UDP with simple error detection (the connection is ended when samples are lost). In this process, the samples are not altered in *any way* and so at the client we will have the same metrological characteristics as those of the digital oscilloscope used.

12.1.2.4 Probe and Client Programs

In order to program the data transmission and reception at the probe and client level it was employed LabVIEW [22], which can also run on smartphones (many times throughout the implementations of this chapter, the HTC P3600—GSM-GPRS-EGDE/3G-UMTS-HSDPA/WiFi/Bluetooth/IrDA/GPS/Windows Mobile 5.0—running LabVIEW for PDA-Pocket PC was used). Data streaming was done via TCP or UDP by the already established PDP (Packet Data Protocol) between the mobile terminal and the PLMN (GGSN) using the defined LabVIEW functions for TCP and UDP Transport.

Using these functions, four peered virtual instruments (VIs) were developed for client DL and probe UL transmission on TCP or UDP. From the "Probe" front panel (see Figure 12.5), the number of points (samples) to be sent can be chosen. The maximum value is 5000. All tests were done with 200 samples. For the VI-s at client side, an example is presented in Figure 12.6 (the front panel) and Figure 12.7 (the diagram—including contextual details).

At the "probe" (the server side), the VI that controls the digital oscilloscope (using the drivers provided by the producer) as a "digitizer" reads the data from the memory buffer (where it was acquired), with a frequency which depends on the value of the time base (this is the first one to be read from the memory). The bytes are then sent to the TCP connection, preceded by the number of samples to be sent. Thus, the length of the IP packets is proportional to the number of samples that are sent.

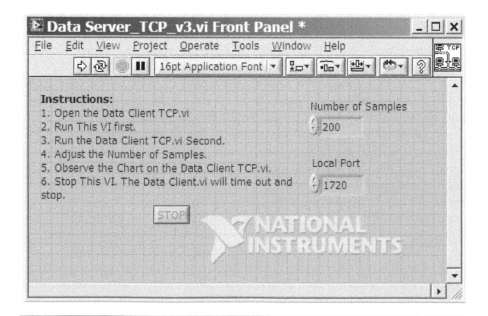

Figure 12.5   TCP LabVIEW probe front panel.

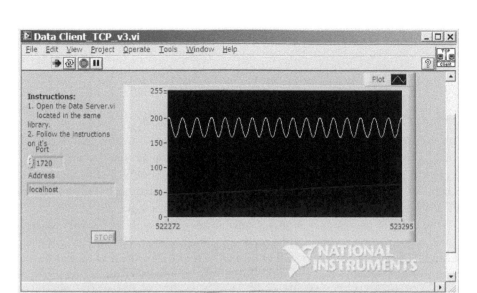

Figure 12.6   TCP LabVIEW client front panel.

The programs can recognize and treat transmission errors. The Digital Oscilloscope screen was emulated at the client (receiving party) processing the data stream sent from the probe. The good QoS allowed the remote client to perceive the measured data and to interact with the emulated oscilloscopes' interface ("skin") immediately, in real-time (without a perceptible delay, as it is defined in 3GPP QoS

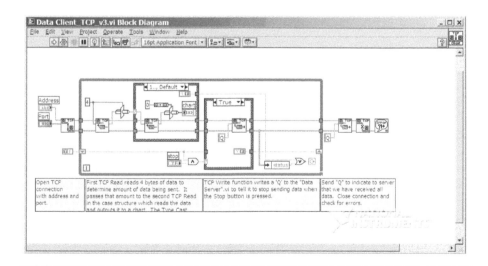

Figure 12.7   TCP LabVIEW client diagram.

document [2]), like operating it on-site. In case a transmission error occurred on the transmission media, this gets translated into delays at the receiving party's end and, as a consequence, the displayed waveform ("oscillogram") is stumped.

12.1.3 Methodology for Performance Testing

This case study evolves from previous solutions in remote measurement developed at the "Transylvania" University of Braşov [23,24] in cooperation with the National Technical University of Athens [25], based on client/server (Web server and workbench servers—the "probe" level with National Instrument PCI-6024E or AT-MIO-16E10 DAQ) involving Ethernet transmission media.

The focus of this study was on key capabilities of the cellular network for real-time data streaming, assessed in DL and UL: in DL, the mobile "client" consisted of the GSM/GPRS/UMTS/HSDPA terminal connected to the data representation hardware and software (see Figure 12.2a); in UL, the mobile "probe" consisted of the GSM/GPRS/UMTS/HSDPA terminal connected to the acquisition hardware and software (see Figure 12.2b). The mobile terminal is connected to the network with the data packet mode activated. Simulated or real measurement data is transmitted between the client and probe VI-s (see Figures 12.5 through 12.7 and the examples for *different scenarios*). The client is signaling the data receipt by an elementary "synchronization" (acknowledge by a simple character sent to the server—see the client VI diagram in Figure 12.7). After every 50 s, the Time/Div setting on the oscilloscope was decreased from the starting rate of 100 ms to lower ones as 1 μs. This was causing more frequent buffer filling and to a higher rate of transmission. The higher is the rate, the higher is the probability of longer RTT (TCP) or higher jitter (UDP) [26].

12.1.3.1 EDGE DL Access

12.1.3.1.1 EDGE DL Access on TCP

The first scenario is represented in Figure 12.2a. The available bandwidth is measured with Jperf (see Figure 12.8) and Ethereal/Wireshark (see Figures 12.9 and 12.10), both at the client and at the probe.

As illustrated by Figure 12.8, at the start of the transmission, for about 2 s, there is no data transfer on the "probe" (server) side. This is caused by the setup of the radio channels. The peak on the client side is caused by the filling of the transmission buffers. After the radio channels are set up, the buffers are emptied which is causing a peak in the data transmission. The small (but steady) increase of the available bandwidth is caused by the adjustment of the TCP window size.

Interpretation of Figure 12.9: The test case considers a step-by-step increase of Time/Div at the time base of the oscilloscope. It can be seen that the bit rate is increasing proportionally. At the moment should the time base be switched, there

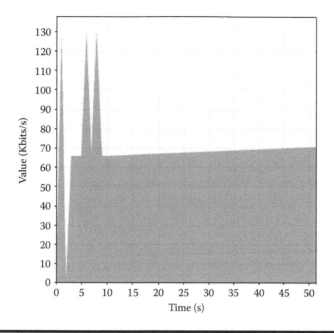

Figure 12.8   Jperf/Iperf measurement of the bandwidth (kbps) on DL TCP for 70 kbps at the client.

will be some spikes caused by the fact that in those moments the oscilloscope is not providing the samples at the same speed (a small time is needed for the stabilization of this speed).

Out of the measurement, it can be noticed that most of the RTT values are situated around 700 ms. There are also packets with a RTT of a few seconds. These delays are caused by loss of packets on the radio interface. Also, one can notice the almost constant throughput when maintaining the first Time/Div values. In the second part, reaching channels' capacity ("congestion"), when the needed bit rate approaches the available bandwidth, more and more packets are lost causing big variations in the throughput (which is dramatically dropping during negotiation of the retransmission).

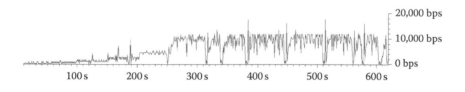

Figure 12.9   Ethereal/Wireshark measurement of the EDGE TCP transfer (bps) on "probe" side (sending) for Time/Div. from 100 ms to 5 μs.

12.1.3.1.2 EDGE DL Access on UDP

The same scenario of Figure 12.2a was used. Similar charts were plotted (see Figure 12.10) for the same benchmarks like in the previous DL case (Section 12.1.3.1.1). The following interpretation of results is pointing also the main differences from TCP DL. No effect of the buffering is visible at the probe (server) side, because UDP is a connectionless protocol. Buffering has visible effects on the receiving (client) side. Because of the initial buffering and because there is no TCP windowing mechanism, the available bandwidth is slightly decreasing.

Interpretation of Figure 12.10: As emphasized previously, buffering is not visible on the probe (server) side because of the connectionless nature of the UDP protocol so the bit rate is not influenced by the retransmission or loss of packets. Buffering is visible on the receiving (client) side, although not as much as for TCP (no retransmission implemented in the UDP). When the bit rate approaches the available bandwidth the number of lost packets is increasing, making the transmission unreliable. The jitter is slightly increasing during the transmission because the packets are stored in the buffers where they spend more and more time.

12.1.3.2 EDGE UL Access

12.1.3.2.1 EDGE UL Access on TCP

It used the scenario of Figure 12.2b. First the bandwidth was investigated using Jperf/Iperf and Ethereal/WireShark, both at probe and client side. In case the probe and the client are both mobile, the UL limitations apply in both cases. This means a series of delays, which actually limit the transfer rate by UL access. The UL bandwidth is very similar to the DL bandwidth but proportionally lower (due to the reduced number of GPRS/EDGE channels available for UL)—this observation is valid for all the performance measurements related to the TCP UL transfer.

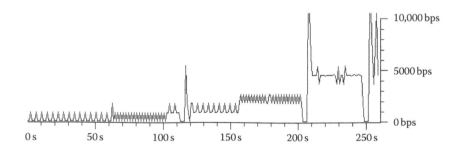

Figure 12.10 Ethereal/Wireshark measurement of the EDGE UDP transfer (bps) on client side (receiving) for Time/Div. from 100 to 2 ms.

12.1.3.2.2 EDGE UL Access on UDP

It uses the scenarios as in Figure 12.2b. The bandwidth is investigated first, using Jperf/Iperf both at the probe and at the client (Figure 12.11).

The same phenomenon can be observed for UDP as for TCP, the buffers at the server first fill up, then they are emptied causing a peak in the traffic. It can be seen that the network accepts very short peaks of data. The available bandwidth is smaller in UL than in DL (because usually fewer radio channels are available for DL than for UL) but the overall behavior is similar to the DL case. One can notice that jitter is not constant at the beginning of the transmission due to the radio channels setup. The connection is reliable as soon as all the radio channels are setup, after a measured time of approximately 10 s.

Concluding these technical considerations, the *performance* of the transmission was assessed at the upper OSI layers. The study was not done on the physical radio-wave characteristics, but to the *service level* provided by the radio capabilities. Nevertheless, we observed that, for all test scenarios, the setup period (~10 s) involved a large variation of the transmission rate due to the establishment of radio channels. Similar fluctuations were determined also by the interruptions caused by buffer filling (and unfilling).

The performance of the streaming and the quality of measurement was given, in our case study, directly by the accuracy of the received data, as no modifications

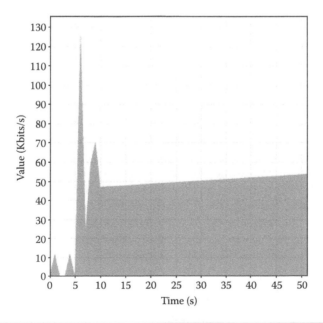

Figure 12.11   Jperf/Iperf measurement of the bandwidth (kbps) on UL UDP for 61 kbps at the probe.

were done on the communication channel. Data itself were completely discarded if transmission failed, with the advantage that *no erroneous information was on display at the receiving side* at any time.

The disturbance in the packet-oriented network resulted in the loss of packets; in case of TCP they are retransmitted (affecting the Round Trip Time, etc.) and, in case of UDP, transmission is discontinued—which is essential from the metrological point of view, as *no altered* waveforms could be transmitted. Moreover, in order to avoid such distortion, QoS configuration can forbid *delivery of erroneous PDU-s.*

The subjective effect of good QoS was the "online" feeling due to real-time reception and remote operation by the graphic user interface ("skin") of the oscilloscope emulated at the client side. For TCP DL transmission the radio interface allowed 40–50 kbps EDGE and the Time/Div. of the oscilloscope could be decreased to 1 μs keeping the overall real-time transmission performance. However, starting with 5 ms (Time/Div.), some retransmissions were needed. For TCP UL the radio interface allowed lower rates and limitations of the real-time transmission effect was reached faster. For UDP DL, the real-time transmission effect was blocked faster than expected and the Time/Div. could only be lowered to 2 ms. For UDP UL the probe could transmit at lower rates and the "online feeling" was disturbed even more (the Time/Div. could be lowered only to 20 ms). This was caused by the fact that in GPRS/EDGE/UMTS there are less traffic channels or radio resources allocated for UL and their priority is lower compared to the down link channels.

For noncritical data streaming applications, UDP packet loss could be ignored and transmission could simply continue (with newly received packets)—the effect could be an overall faster "noninterlocked" link (compared with TCP).

The case study allocated the same attention to UL as for DL, because the telecommunication market—that was primarily targeting the download of content—is gradually shifting to upload access, allowing customers to provision content. More and more 3G operators will give importance to data streaming (mostly multimedia—as the forecast for the near future, but also industrial—as we illustrated here) activating the *QoS traffic class "streaming"* which provides *guaranteed bit-rate and constant delay* (which lowers the requirement for buffering at receiving part). For both sides (providers and their customers), the performance assessment techniques presented hereby, could be very useful. Last, but not the least, these techniques can be easily adapted to other data acquisition systems—generally to any other equipment that can be driven by any other general-purpose instrumentation software (e.g., Keithley "TestPoint" or Agilent "VEE")—in this case LabVIEW (as drivers are also compliant to VISA—Virtual Instrument Software Architecture [27] and IVI—Interchangeable Virtual Instruments [28]).

The techniques presented will be further extended in the next paragraphs considering newest protocols and controls for the programming of distributed applications in remote monitoring—National Instruments developed Data-logging and Supervisory Control (DSC) modules [29] for remote monitoring with the networking of devices for the transfer of acquisition (output)/or (input) control data to/from distributed databases [30].

12.2 Remote Access to Real and Emulated Experiments

Mobile phones are the most popular technological devices among the younger population [31–34]. There are approximately 10 million handsets sold monthly worldwide. They increasingly include palm-like functionality: the "smartphones" are handheld devices that integrate personal information management (PIM) and mobile phone capabilities (either adding phone functions to already capable personal digital assistants—PDAs—or putting "smart" capabilities, such as PDA functions, into mobile phones): MMS/e-mail/news/chat clients, scheduling software and contact management, built-in MP3/MP4 players, portable gaming, photo and video cameras, digital radio, TV or GPS receivers, the ability for streaming and to read files in a variety of formats (e.g., MS Office, Adobe PDF or including Macromedia Flash) stored in larger amounts in high-capacity internal memory, add-on memory cards or even micro hard-disks. Many of the smartphones have Symbian, Palm, Android, Linux or mobile-Windows operating system and extended with built-in Web (and WAP) browsers for Internet access, supporting TCP/IP, HTTP, WAP, PPP, POP3&SMTP, and Java J2ME (or BREW) applets.

Mobile Internet (accessed by cell phones, handhelds, and notebook computers) has as main attribute *ubiquity* (for "anywhere and anytime" connections—besides accessibility, richness, efficiency-usability, flexibility, security, reliability, interactivity. Mobile wireless access reached very high data transmission speed (10 Mbits/s and up, like the actual LAN-s). It is now *pervasive*—multimodal (users can be simultaneously connected to several wireless access technologies—GSM/GPRS/ UMTS or WiFi/ZigBee and can seamlessly move between them). "Mobile landscapes" are connected frameworks of wireless networks.

Two recent, opposite, trends are more and more acute:

- More and more young adults, particularly those unemployed, those in low-skilled jobs or even homeless, have poor literacy and numeracy, and do not take part in learning after leaving school; to struggle against these evidences, modern learning no longer needs to be classroom—or course—bound but more and more in short flexible programs.
- On the other side, technology is endorsing e-Learning as a deeply personal act, offering *on-demand* and *just-in-time* support, allowing personal relationships with information, strengthening motivation, focusing attention, increasing performance; technology can engage learners by structuring and organizing information, by displaying and demonstrating procedures and operations, by simulating a range of conditions, immersing people in virtual environments, and providing safe practice opportunities.

Mobile learning ("m-Learning") represents the next step in a long tradition of technology-mediated learning in ODL (open and distance learning), with computer-based-training (CBT), Web-based training (WBT), and so on. It features new strategies, practices, tools, applications, and resources to implement *ubiquitous, pervasive,* personal, and connected learning. As mobile phones are *two-way* communication devices, collaborative and peer-to-peer learning is also supported (including interactive network game-based learning). m-Learning infrastructures can include learning management systems (for wireless services as rights management, content management, search management) that facilitate access to m-Learning materials and services from a variety of mobile-(besides Web- or TV-)access ways. Interfacing is available also for devices with minimum multimedia functionality, for the benefit of learners with sensory difficulties that can use speech ↔ text- and special SMS/MMS facilities.

12.2.1 Automated Test and Measurement Laboratories with Internet Access

The infrastructure of the automated Test and Measurement laboratories developed at "Transylvania" University—Braşov—Romania and at the National Technical University of Athens was "published" in the Internet as the result of the "Leonardo da Vinci" programs RO/01/B/F/PP141024 "Virtual Electro-Lab," continued with RO/06/B/F/NT175014 "VET-TREND."

One of the main goals of these programs was to enable Internet access not only to information, but also to experimental resources (*real* and emulated) [35]. Students, teachers, and researchers from any location (e.g., even from home) and even on the move [36] are granted access to high-tech laboratory instruments for real test and measurement combined with analytical simulation-emulation. After this remote-accessed measurement process and data tele-transmission, mathematical postprocessing of experimental data-bases, multicriteria comparison with theoretic calculus and results of complex simulation on behavioral (and extendable) models complete the capabilities of the *e*-University of tomorrow.

The electronic core of this implementation can be multidisciplinary extended/adapted by appropriate sensors and transducers at one edge and by drive elements that interface with external equipment, at the other edge.

12.2.1.1 Client–Server Architecture

Implementation for Internet (Figure 12.12) was oriented on Microsoft systems (Internet Explorer browsers and Microsoft Mobile Explorer, Front-Page for Web site development, etc.). For WAP pages accomplishment, one can use, for instance, Nokia Mobile Internet Toolkit or any other similar package.

At *Client level* Web (or WAP) forms were used to post stimuli and configurations toward the work-bench (see Figure 12.13 for an example related to the Device-Under-Test, "DUT," Junction Field-Effect Transistor—N-JFET).

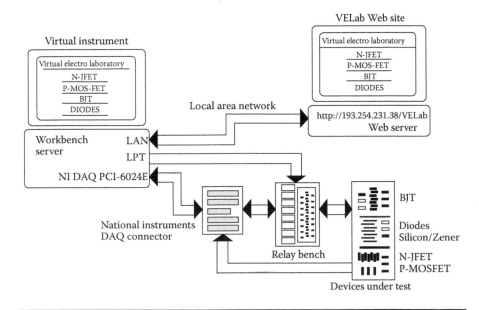

Figure 12.12 DAQ-based architecture for remote measurement.

At *Work-Bench Server level*, the LabVIEW program runs (in "Continuous" mode), in an instrumentation-oriented implementation that processes the stimuli, controls a relay bench and the automated measurement (whose results are incorporated in a .html file published by the virtual instrument directly on the *WebServer*, by a simple local link (in "Network Neighbourhood")—avoiding in this way expensive solutions such as "Internet Toolkit" for LabVIEW (with extra problems of specific methods—for example, "Run-Time Engine"—incorporation at Client level).

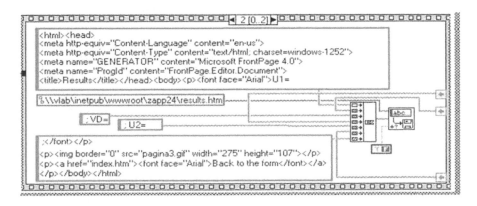

Figure 12.13 VI diagram frame that assembles the results.html file by simple string concatenation.

The experimental configuration controlled by the "workbench server" PC running MS Windows is built around National Instruments (NI) AT MIO 16E10 or PCI6024E analog and digital signals acquisition and generation board, integrated with NI "LabVIEW" in a virtual instrumentation (VI) system with Internet (including WAP) access.

For remote-controlled switching (in order to redirect signals, to reconfigure the subsystems and to adjust the loads), it was accomplished with relays block, controlled (by LabVIEW) via the parallel port (that has the specific advantage of being "latched," keeping this way stable configurations).

These DAQ&D cards given as example have an excellent resolution of 12 bits/sample and peak outputs of ±10 V (then, the LSB corresponds with 10 V/(2^{12}) = 0.00244140625 V which is appropriate also for small-signal studies).

At client level, stimuli are posted with a mouse click on the "Submit" button in experiment's control page (e.g., http://vlab.unitbv.ro/VELab/bjt/bjt.stimuli.php), arriving at Web server into a .TXT file associated to the Web form (e.g., bjt.stimuli.txt for the above particular case of Bipolar Junction Transistor, BJT). The VI runs, in "Continuous" mode, a wait-loop (based on the measurement of input.txt file's length—equal to zero if no new forms were posted from remote; the file input.txt is purged after reading). If stimuli were posted, the step-out from the loop continues with a frame where they are extracted out from the labels (e.g., U1_start: U1_stop: etc.,—by LabVIEW string processing) and directed to the two analog outputs (AO) of DAQ&D card (in this case, for U1 and U2 that supply the base and collector's circuit respectively) in steps that are internally computed (e.g., (U1_stop—U1_start)/U1_steps#) for *multipoint, automated,* measurement.

"*Soft-protections*" are programmed (they can be also programmed as "compatibility formulas"—e.g., stimuli products or ratios that have to be within certain limits—or "exclusion lists"—e.g., to avoid remote configuring that can damage UUT by unwanted short-circuits, etc.) to limit the AO (e.g., preventing dangerous FET grid potentials). The Web pages were developed also to provide remote user's *assistance* by *AO recommended values,* by devices' *data sheets* and *SPICE* models, by *sample test-reports,* Adobe-Macromedia "Flash" animated "*Visual-Help,*" and authors' *published references.*

As soon as acquisitions from the (AI) analog inputs (for base and collector potentials, in the case of BJT) are accomplished, they are either directly incorporated in output files which are assembled progressively with the measurement (by writing in APPEND mode): the .HTML file (for intuitive publication of stimuli and results that are graphically published just on the schematic), and also the .TXT file (including both experimental results and SPICE-emulated results) and the .XLS file (useful for different post-calculations and plots) that can be, both, downloaded by the user. At the end (as already justified), the AO of the card are automatically reset to zero.

The way these files are assembled is the "Columbus egg" for most of these solutions: instead of using expensive "Internet Toolkit" add-ons for NI LabVIEW, direct and "intimate" (nonprotocolled) publication is done by workbench-server, directly on Web server (see, e.g., for FET T&M, in Figure 12.13), as simple string that concatenates any preformatted (e.g., by Front-Page) "results" page (that can encapsulate workbench and UUT images—e.g., .GIF in the example of Figure 12.13—with input and consistent output data—to avoid confusing it with the results requested by other user) with stimuli (inputs sent explicitly or computed by the workbench-server) and measured values; the assembled string becomes a file (e.g., results.htm) that is saved directly in the inetpub/wwwroot directory of the Web server.

12.2.1.2 Emulation Synchronized with Real Remote Experiments

It was implemented, the *backup* and *validation* of real remote measurement with emulation/simulation of systems behavior, based on advanced mathematical models (e.g., PSpice computer-aided solving of electron-devices equations with complex stimuli vectors [37,38]). The user can access on the experiments' server the hyper-schematic of the UUT, for example, developed with Adobe-Macromedia "Flash" and incorporated in courseware book-pages (it was used Click-to-Learn/SumTotal "Tool-Book Editor") where the reader can modify the parameters of these "living figures," *working directly on the e-Book pages* where he/she can post data (stimuli and different parameters—usually more than what can be adjusted in the real remote experiments) for preliminary or postmeasurement computation, in order to compare the measured results with the theoretical ones.

An important step forward is parsing into simulation data (stimuli and/or configuration) of *the same* values/parameters that were sent to the real experiment (see Figure 12.14 for NI LabVIEW implementation in the particular case of BJT)! A great educational impact was foreseen (besides the above-mentioned validation, virtual reality completes the perception and trustful computation and comprehension of theoretical models is encouraged [11]).

12.2.2 Wireless and Mobile Access to Remote Experiments

12.2.2.1 Content to Mobile Terminal Adaptation

The principle of this implementation is the optimization of the processing load-share: central—on the server—and local—on (mobile) terminals—featuring embedded intelligence. Adaptation of normal Web content for mobile terminals should not need rewriting (of initial Web) pages into WAP-pages anymore. It should then be based on the *innovative principle* of (pseudo-)mini-browsers (actually simple J2ME applications, very compact) that cooperate with *middleware* (intermediate servers that perform "*small-screen rendering*" (SSR) to prepare the content exactly for the terminal that requested it—taking over, this way, an important processing load from the terminal). Cheap terminals, provided they are Java

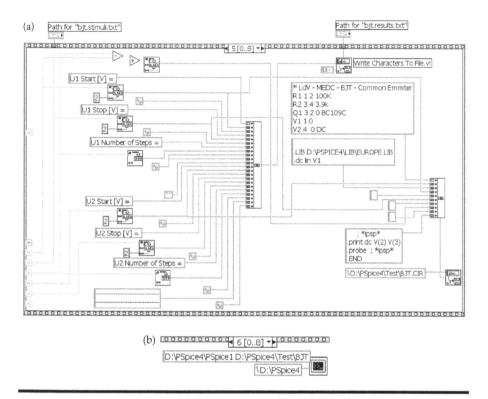

Figure 12.14 **(a) NI LabVIEW frame for the assembly of topological, structural and DC sweep (*stimuli parsing*) information in the PSpice .CIR file; (b) detail on "System-Exec" LabVIEW sub-VI.**

enabled, can receive customized pages as soon as, based on their ID that accompanies any query, the pixels' layout of the screen and its color depth is known to be used in the preparation of interpolated images and optimized fonts with sharp letters (see Figure 12.17)—the pages fit exactly to the screen, without the need for horizontal scrolling.

The most popular mobile browsers are actually: Pocket Internet Explorer (the default browser on Windows CE used on the Pocket PC or Windows Mobile edition on modern SmartPhones), Safari, and Opera Mobile (with the freeware version Opera Mini that can run on devices with extremely low resources). NetFront, Nokia Series 90 Browser, Konqueror Embedded, Minimo ("Mini Mozilla"—using the same engine as Mozilla and Firefox). OperaMini, provided by Norway's Opera Team, illustrates in a very spectacular way, the principle of SSR implemented by AJAX ("Asynchronous JavaScript and XML") for the Opera Platform and enabling the integration of the phone's local applications with online content. Out of the three approaches to create Web sites accessible for handheld devices (creating "cascading style sheets"/creating a mobile portal (Web pages dedicated to mobile terminals)/

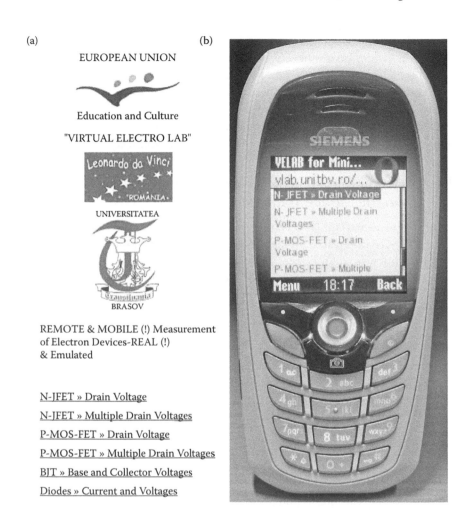

Figure 12.15 (a) Main page of the mini-version of "Virtual Electro-Lab" Web site; (b) Spectacular small-screen rendering on Siemens C65.

improving the HTML site) we choose the last one, which is the most practical: we rearranged the content in a single column; we limited the graphical content (to improve page loading time and reduce the scrolling necessary to view the entire content)—on the other side, there is no more effort needed to prepare all figures in *WBMP* (WAP bit-map) format—that would create restrictions to "backward" access via normal browsers; we used hyperlinks instead of great images or icons; we limited text and icons' size (to be consistent with terminals' standard font sizes) and so on. It was accomplished [39] "mini" version of the Web site—see Figure 12.15.

The "mini" version of the "Virtual Electro-Lab" Web site: The following figures illustrate the Web site we adapted for real and virtual experiments that can be performed not only remotely but also mobile (for *m*-Learning in technical domain) published at http://vlab.unitbv.ro/velab/mini-index.html.

Figure 12.16 presents the results' page of a complete N-JFET measurement. To check saturation, one could easily increase U2 (e.g., a step of 1 V, from 9 to 10 V) and notice, in the new results' screen that drain potential, V_D, increased by the same step (up to approx. 7.6 V). This confirms that voltage on the serial resistor $R = 5.6 \ k\Omega$ remained unchanged then the transistor is *saturated*.

12.2.2.2 Accomplishment of Dedicated Processing on the Mobile Terminal

A pilot implementation (at "Transylvania" University of Braşov, Romania—developer A. Gavrila) placed an important part of the processing, by J2ME—Java 2 Mobile Edition (e.g., for devices' bias-point computation), on the mobile phone. Similar approach can involve software development for MS Windows Mobile (with .NET Mobile Web SDK) or for Symbian OS.

The local computation of bias-point for the simple circuit with semiconductor diode was implemented. In Figure 12.18 one can see the schematic and, aside, the device's equation 12.2 (see also Figure 12.19) and the circuit's equation 12.3.

$$I = I_0(e^{U/(KT/q)} - 1) \tag{12.2}$$

$$E = U + IR \tag{12.3}$$

Figure 12.20 presents (also on mobile phone's screen) the graphical solution of this transcendent system.

The "square snail" iteration for these equations starts from an initial estimation of the bias point at the intersection of circuit's line with the vertical axis, then, iteratively, current and voltage are fixed for the step from exponential/from line to line/ exponential, respectively. The iteration stops when decimals at preimposed position do not change anymore.

12.3 Service-Based Mobile Tele-Measurement

12.3.1 NI Data Logging and Supervisory Control

The Data logging and Supervisory Control module of National Instruments LabVIEW is dedicated to remote control, by networking of devices for the transfer of acquisition (output)/or (input) control data to/from distributed databases [http://www.ni.com/swf/presentation/us/labview/newdsc/]

The LabVIEW *Shared Variables* enable efficient programming of distributed applications [30]—http://zone.ni.com/devzone/cda/tut/p/id/4679. Data can be

N-JFET « Drain Voltage

The N-JFET is open for:
−0,7 V <= U1 <= −0,3 V

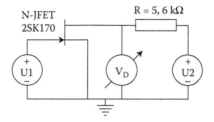

The N-JFET DUT accepts voltages inside this limits:
−10 V <= U1 <= 0 V
0 V <= U2 <= 10 V

Results

N-JFET results file >> Generated by labVIEW.
VI (Virtual Instrument) running on the work bench computer.

N-JFET measurement #290
Date: 1/28/06
Local time (Braşov: GMT+2): 6:11 PM

U1 Start [V] = −4.00E−1
U1 Stop [V] = −4.00E−1
U1 Number of steps = 1

U2 Start [V] = 9.00E+0
U2 Stop [V] = 9.00E+0
U2 Number of steps = 1

IP: 81.12.219.217

Total steps (U1_steps × U2_steps) = 1

U1 (DATA IN) || VD (DATA OUT) || U2 (DATA IN)

−4.00000000E−1 || 6.5673828125E+0 ||
9.00000000E+0

End of measurement.

Figure 12.16   Web page with results of a complete N-JFET measurement.

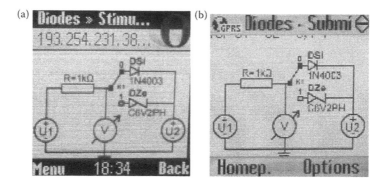

Figure 12.17 Obvious improvement of readability on the screen of Siemens C65, by small screen rendering (a); compared with own browser (b).

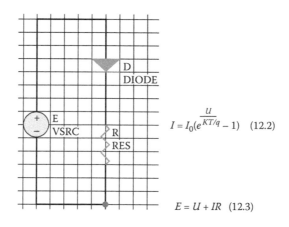

$$I = I_0(e^{\frac{u}{KT/q}} - 1) \quad (12.2)$$

$$E = U + IR \quad (12.3)$$

Figure 12.18 Experiment assisted by J2ME programming of the mobile terminal.

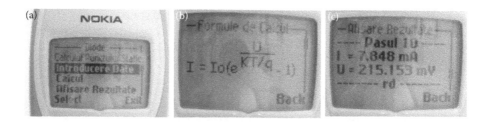

Figure 12.19 (a) Emulated experiment's menu; (b) Device's equation; (c) Iteration step.

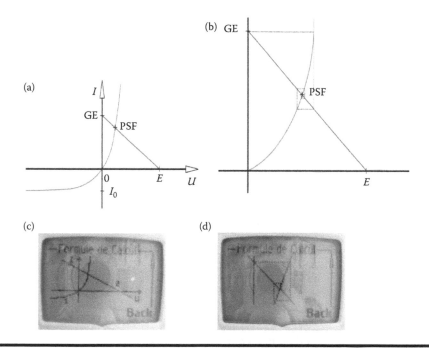

Figure 12.20 (a) Graphical solution of (12.2) and (12.3); (b) Iterative graphical computation; (c) and (d) Plots of (a) and (b) on mobile phone's screen.

shared between VIs running in different nodes of an Ethernet network in an optimized way compared with other data-sharing methods usable in LabVIEW (e.g., UDP/TCP, like shown above, queues, and real-time FIFOs).

Shared variables are configured at project's edit time (using property dialogs) and do not need configuration codes included in the applications. There are three types of shared variables: single-process, time-triggered network-published (the latter ones are used in our appliance).

Creation of shared variables in our project tree was done by a right click on the main computing node of our configuration (hosting the server) hereby called "My Computer" (see Figure 12.21); by selecting *New > Variable*, the shared variable properties' dialog is displayed where the new variable can be configured.

The target to which the shared variable belongs is the node from which LabVIEW deploys and hosts the shared variable (in our case, the above-mentioned "My Computer").

After shared variables are added to the LabVIEW project, they can be simply dragged into the block diagram of the VI (that reads or writes these shared variables or part of them), thus, creating the "Shared Variable" read or write nodes (see the following Server VI and Client VI diagrams).

The NI Publish and Subscribe Protocol (NI-PSP) is a simplified networking protocol optimized to be the transport for network-published shared variables.

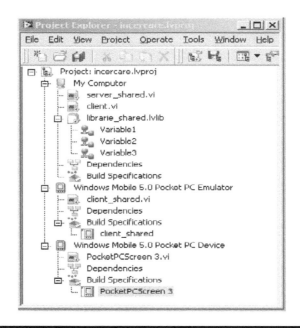

Figure 12.21   Remote and mobile control project tree.

The underlying layer of NI-PSP has been written to be more efficient by optimized architecture of TCP/IP use. This requires LabVIEW to run on all the nodes involved in the communication. The Shared Variable Network Stack is presented in Figure 12.22. LogosXT is the layer that is optimizing the throughput for the shared variable by an 8 KB transmit buffer (*unique* for all the connections between two distinct endpoints) and a 10 ms timer thread (so every network operation has a fixed overhead both in packet size and in latency time, for target stream of data).

Deployment of network-published shared variables is done toward a shared variable engine (SVE) that *hosts* the shared variable values on the network (see

LabVIEW
Shared variable engine
NI-PSP
LogosXT
TCP/IP
Ethernet

Figure 12.22   The NI Shared Variable Network Stack.

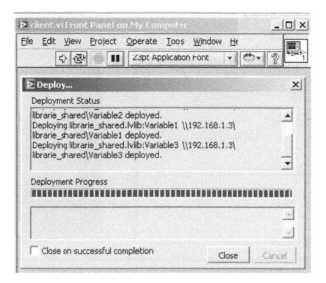

Figure 12.23 Deployment of network-published shared variables.

Figure 12.23). When writing to a shared variable node, LabVIEW sends the new value to the SVE that deploys and hosts the variable (in our case the "My Computer" server with the Intranet address 192.168.1.3). The SVE processing loop *publishes* the value so that any *subscriber* (e.g., a mobile Client SmartPhone Pocket PC) gets the updated value.

The Server VI: The Server VI takes over data from the serial interface—see the VISA ("VXI (plug & play) Systems Alliance") read (R) control in the diagram of Figure 12.24; these data (e.g., temperatures) can be acquired by a laboratory work

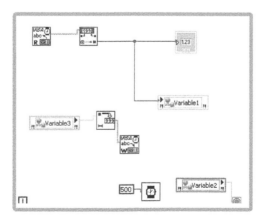

Figure 12.24 The Server VI LabVIEW diagram.

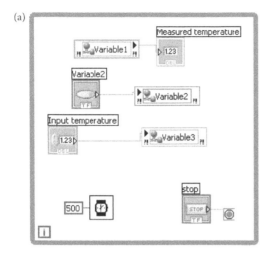

Figure 12.25 (a) The Client VI LabVIEW diagram. (b) Client VI LabVIEW panel, running on HTC P3600–PDA.

bench. The VI converts these data in decimal format and will publish the measured temperature values in the shared *Variable*1. As for the input temperature (that can be sent from the client terminal to an automated temperature control system) their values are read from the shared *Variable*3, converted from decimal and sent (via the VISA write (W) control) to the serial interface.

The Client VI: The VI (with the diagram of Figure 12.25a and the panel of Figure 12.25b reads the shared *Variable*1 (representing the Measured temperature) and displays it. It writes into the shared *Variable*3 the value of the input temperature (taken over by the remote HVAC-automated control system).

12.3.1.1 Remote and Mobile Control Implementation

A pilot test configuration (see Figure 12.26) was built around a PC server, connected in the Intranet of a "Linksys Wireless-G" router.

In order to check its status and configuration, the HTC P3600–PDA runs a "vxUtil" [http://cam.com/vxutil_pers.html] tool from Cambridge Computer Corp. (see Figure 12.27) that enables DNS Audit/DNS Lookup/Finger/Get HTML/Info/IP Subnet Calculator/Password Generator/Ping/Ping Sweep/Port Scanner/Quote/Time Service/Trace Route/Wake On LAN/ Whois.

In the PocketPC PDA Build Specification, Deploy aliases file must be selected (see Figure 12.28).

As shown above, normal Run-Time Engines that are plugged in PC Internet Explorer (allowing granting remote control of LabVIEW panels) are not a solution for PDA browsers.

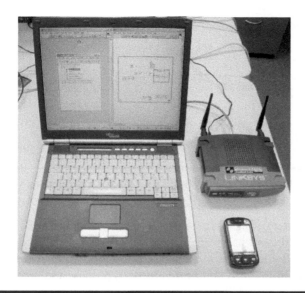

Figure 12.26 **Pilot test configuration for remote and mobile control based on NI shared variables.**

12.3.1.2 Case Study: Mobile Data Acquisition and Tele-Transmission by PDA

The problem addressed is the integration of a system for DAQ + logging and/or tele-transmission that can be personal and portable [41].

Figure 12.27 **"vxUtil" status screen on HTC P3600–PDA run.**

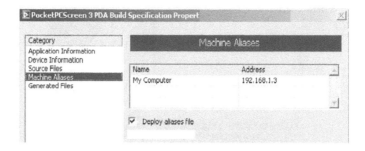

Figure 12.28 The PocketPC PDA build specification properties screen.

Most of the preexisting comparable solutions are highly specialized and propri-etary (e.g., by dedicated hardware–firmware–software of microcontrollers), being not always very affordable. The accomplished configuration is using common PDAs that have a plugged-in micro DAQ card—general purpose and with common inter-face (CF—"Compact Flash"), a high-capacity common memory card (SD—"Secure Digital") and run an universal instrumentation software (e.g., the above-mentioned NI LabVIEW). LabVIEW can perform not only data acquisition and logging but also processing (e.g., digital filtering, identification–classification, compression, etc.) and communication.

If the PDA has also "SmartPhone" capabilities, it can also transfer the data via mobile communications (e.g., toward a central server). If not, a cheap alternative (that was integrated by the authors) is the connection to the owner's mobile phone that can be controlled, for instance, via AT commands (for dial-up, GPRS and/or SMS transmissions). Local data-logging can be used for data acquisition and detailed diagnosis (human and/or automated) while mobile tele-transmission can be used mainly for updates and alarms.

As demonstrated, tele-monitoring is the basis of remote management technol-ogy and can be enhanced with two-way communications that allow better person-alization of support. Personal means of communication, like mobile phones and SmartPhones or PDA-based "communicators" are more and more involved in dif-ferent ways of monitoring (direct physiologic feedback), logging, tracing but also messaging (including alarms when threshold values are reached) and notifications of decisions—all these in a way that should be tailored to the patient's capability and needs.

For remote monitoring, many portable devices were developed for acquisition, logging, processing, and transmission of acquired data. Digital signal processing (DSP) enable complex analysis and allow real-time decisions.

The authors developed different tele-monitoring systems in the last years. A "twin-microcontroller" solution had one microcontroller allocated to the control of data acquisition and processing, and another one to the management of communica-tions (interfacing and control of a GSM modem). Such dual-processor solutions are common even to SmartPhones (only recently DSP and general purpose processing were brought to a common core).

12.3.1.2.1 System's Architecture

The most recent system implemented by the authors, with the architecture depicted in Figure 12.29, consists of 4 channels, 200 kS/s DAQ card, namely NIs CF 6004, connected in the Compact Flash slot of a HP iPAQ 2210 PDA. The authors have used in their test configuration a portable probe consisting of sensors/transducers and signal conditioning circuitry.

Because most Compact Flash PDAs currently available on the market do not embed mobile communications technologies such as GSM/UMTS or WiFi, which would allow the system to transfer the acquired data directly to a remote server, the authors decided to use a common mobile phone as a GSM/GPRS modem. The PDA is connected to the mobile phone through Bluetooth and is using standard AT commands to transfer a message (that incorporates the acquired data and/or specifically parameters computed out of them) to the mobile phone that should transmit these data to a remote server using GPRS (and/or at least SMS).

12.3.1.2.2 Technical Solution and Results

The software for both data acquisition and transmission was implemented using LabVIEW's PDA Module, a subset of NI LabVIEW functions specifically designed for PDAs [40]. Data acquisition is performed using NI DAQmx Base, a set of drivers and functions for the DAQ boards made by NI. A DAQ task has to be defined, specifying attributes such as sampling rate, used channels, number of samples, and so on. DAQ task creation is accomplished using the DAQ Configuration Utility included in LabVIEW.

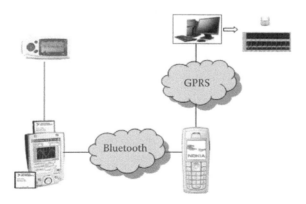

Figure 12.29   Test configuration for mobile DAQ.

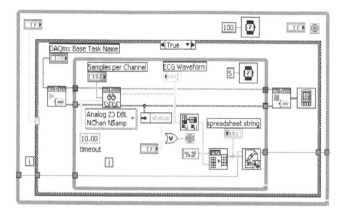

Figure 12.30 Data acquisition virtual instrument for the example of a bio-medical tele-monitoring (ECG_Graph.VI).

A measurement cycle (Figure 12.30) consists of starting a specific DAQ task (using DAQmxBase Start Task.VI), reading data from the DAQ card (using DAQmxBase Read.VI) in a loop until the user stops the system or until an error occurs and then stopping the DAQ task (using DAQmxBase Stop Task.VI).

The acquired data can be viewed both online, on the PDA screen using a Waveform Chart (see the VI Panel on the screen, in Figure 12.31), or offline by logging the data into spreadsheet files (XLS format). A sampling rate of 250 S/s has been chosen (for many industrial or physiologic appliances). A relevant measurement cycle implies acquiring 1000 samples which can be stored in 10 kB spreadsheet files. The log file obtained after 24 h of continuous data acquisition (required for monitoring cardiac activity over a longer period of time) would then be 216 MB

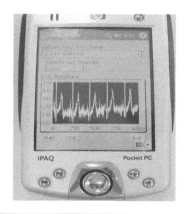

Figure 12.31 Acquired ECG (unfiltered)—in the LabVIEW Panel on PDA.

large. Generated files which are as big as 10 kB, can be directly transferred over GPRS to a remote server. If the acquired signal needs to be monitored over a longer period of time, thus, resulting in larger files which are not fit for transfer over GPRS, these files can be stored on the SD card of the PDA and then "downloaded" to a PC in order to be analyzed offline.

The transfer of the acquired data over Bluetooth to the mobile phone (which acts as a GPRS modem) is handled by a subVI (Figure 12.32) which opens a Bluetooth connection to that phone (using the LabVIEW function Bluetooth Open Connection. VI). The Bluetooth Open Connection function receives two main parameters, namely the MAC address of the mobile phone to which one wants to connect and the channel number (which in the case of the Bluetooth dial-up service is 0). The subVI then sends a sequence of standard AT commands for initiating and performing a GPRS transfer.

These commands are sent over Bluetooth by calling LabVIEW's Bluetooth Write.VI function. They are as follows:

AT + CGATT = 1—attaches the mobile unit to the GPRS network
AT + CGDCONT = 1, "IP," "internet"—defines a mobile operator specific PDP context
AT + CGACT = 1,1—activates that context
ATD *99***1#—connects to the GPRS network

The present implementation proved the possibility to use powerful instrumentation hardware and software (NI CF6004 and NI LabVIEW) in a much miniaturized format (for PDA) for mobile tele-monitoring.

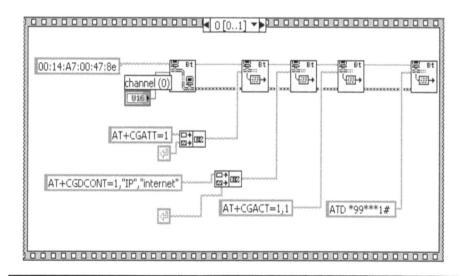

Figure 12.32 SubVI for Bluetooth transmission using AT Commands.

12.3.2 VISA Approach to Instrumentation Accessed via Web Services

Rationale: This section will focus on presenting an approach that allows remote access to a VISA compliant instrument by using Web services [41].

Web services provide a standard means of interoperating between different software applications, running on a variety of platforms and/or frameworks. Since they are based on open standards such as HTTP- and XML-based protocols including SOAP and WSDL, Web services are hardware, programming language, and operating system independent.

By using Web services we have the possibility to decouple clients from the underlaying hardware and platform and also in the same time to provide a standardized way of accessing the hardware resources, namely the VISA instrumentation. The client application is completely decoupled from the instrumentation hardware, and has no dependency on drivers or any other hardware access routines.

With this approach we eliminate the need to install and configure runtime environments (like the LabVIEW Runtime, required to run remote VIs) on the client system. Because Web services are accessed using standard SOAP protocol using the HTTP as transport protocol, which is allowed in most firewall configurations, any computer connected to internet can be used to access the provided Web services.

The communication with the instruments is based on VISA standard, and for this implementation we will use the VISA libraries provided by National Instruments.

The VISA standard is endorsed by over 35 of the largest instrumentation companies in the industry including Tektronix, Hewlett-Packard, and National Instruments, and due to this wide acceptance, VISA unifies the industry to make instrumentation software interoperable and reusable.

12.3.2.1 Remote Instruments' Aggregation with VISA

In 1993, National Instruments joined with several companies involved in the instrument industry like GenRad, Racal Instruments, Tektronix, and Wavetek to form the VXI plug&play Systems Alliance. One of the goals of this alliance was to ensure multivendor interoperability for VXI [VME (Versa-Module Eurocard) eXtended for Instrumentation] systems and to reduce the software development time for an operational system. This was meant to develop a new standard for instrument drivers, soft front panels, and I/O interface software. The term VXI plug&play has come to indicate the conformity of hardware and software to these standards. VISA (Virtual Instrument Standard Architecture) is the VXI plug&play I/O software language that is the basis for the software standardization efforts of the VXI plug&play Systems Alliance. VISA is a standard defined by Hewlett-Packard/Agilent Technologies and National Instruments for communicating with instruments regardless of the interface.

Session-orientation of VISA (based on interrupts' servicing) encouraged us to use the virtual instrumentation software—like NI LabVIEW—together with NI Measurement and Automation Explorer (MAX) in order to integrate remote and local resources and to build test sequences.

Using NI Remote VISA Servers, we have developed a method of sharing instruments to remote users who can integrate them into their experiments.

If they are only didactical, even Units-Under-Test, UUT, can be far from the user, if not, at least UUT and intelligent transducers should be local and other equipment can be remote. The configuration implemented consists of an instrumentation (workbench) PC, running NI Remote VISA Server, to which all the instruments are connected using different interfaces (such as GPIB, USB, serial, and parallel). As shown before, the workbench server is also connected in the intranet of the Web server which is "visible" in the Internet.

A client who wants to have access to the remote instruments needs to know the IP address of the Web server (e.g., 82.78.149.213). The user can view the resources offered by a certain Web server in a tree-like manner using National Instruments' MAX (Measurement and Automation Explorer) (Figure 12.33). In order to add a new remote system in MAX, the user needs to right-click on the *Remote Systems* and to select *Create new*. The user needs to click on *Remote VISA System* and to configure the properties window that appears, as depicted in Figure 12.34.

Port forwarding has also been configured in order to route Remote VISA data (coming through the dedicated port 3537) from the Web server to the workbench server. Any security requirements can thus be fulfilled by configuring access lists for port 3537. A client who has access to the system will use only the IP address of the Web server for remote access to the "published" experimental resources.

Figure 12.33   Viewing the remote instruments using MAX.

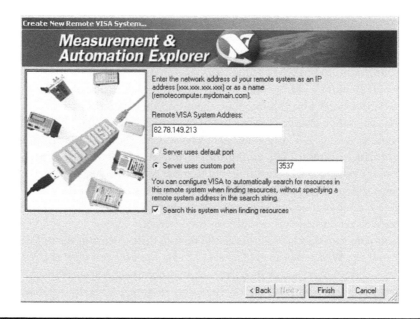

Figure 12.34   Configuring the remote system.

The availability of the remote resources can also be tested in MAX using a graphical user interface (Figure 12.35). In order to do this the user needs to select the desired instrument (e.g., GPIB::4::INSTR) and then click on the "Open VISA Test Panel" button. In the window that opens, the user can enter commands for the instruments (Write function) and read their answer (Read function).

In order to show the manner in which Remote VISA calls can be used to develop remote experiments we have implemented a proof-of-concept application controlling:

- An experiment board UUT (unit-under-test)
- A stimuli generator (programmable power supply—Hameg HM8142 with the GPIB address 4)
- A digital multimeter (HP34401A DMM with the GPIB address 22)
- A relay board connected on the parallel port (for its advantage of being "latched") of the instrumentation work-bench server; the relays are used for (reconfiguration) switching but also for test point multiplexing or load-adjustment and so on; for instance, the Centronics port of the work-bench server, having the address 378 h = 888 (10) can simply be controlled in LabVIEW by

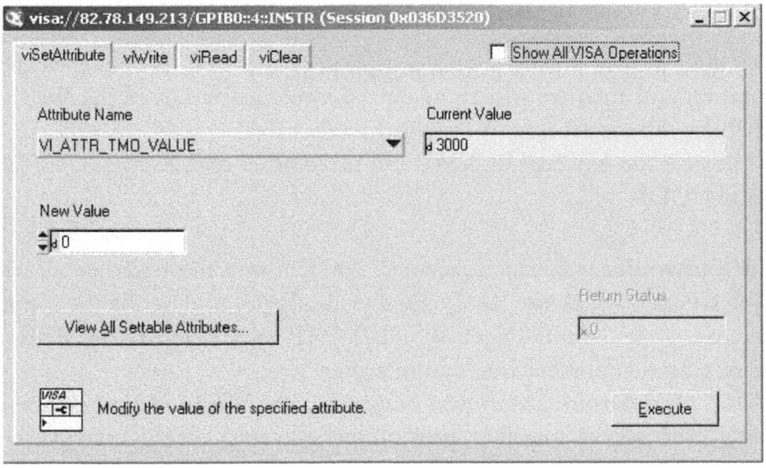

visa://82.78.149.213/GPIB0::4::INSTR (Session 0x036D3520)

viSetAttribute | viWrite | viRead | viClear | □ Show All VISA Operations

Attribute Name

VI_ATTR_TMO_VALUE ▼

Current Value

3000

New Value

0

View All Settable Attributes...

Return Status

0

Modify the value of the specified attribute.

Execute

Figure 12.35   Testing remote resources using VISA Test Panel.

A measurement cycle involves opening a VISA session to each of the instruments used in the laboratory.

The syntax for identifying a remote instrument is as follows: *visa://ip_address_of_the_webserver/instrument_ID*.

After opening a VISA session, an instrument can be controlled using standard SCPI—Standard Commands for Programmable Instruments commands (e.g., the syntax for measuring a DC voltage is: *MEAS:Voltage:DC:?*). Such commands are sent to the instrument using the VISAWrite.VI LabVIEW function. The measured value can be viewed by calling the VISARead.VI function, whose output is converted to a double format and placed in an array.

After being *discovered*—as described before—the VISA entities can be used for a LabVIEW program (e.g., the SCPI command to a remote DMM, for the measurement of a DC voltage—Figure 12.36).

If LAbVIEW "full-development system" is not available, one can download and install the free NI LabVIEW Run-Time Engine, a kind of "LabVIEW player" (http://sine.ni.com/apps/utf8/niup.ni?ap = GB_NIDU&ip = 209&loc = en-US& du = http://joule.ni.com/nidu/cds/view/p/id/679/lang/en) and open the VI application in .EXE format.

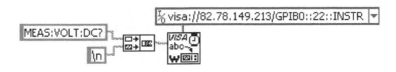

Figure 12.36   VISA measurement on a remote DMM.

12.3.2.1.1 Software Technologies Overview

In this section, we will make a short description of the used technologies and software libraries, and then we will focus on the implementation of the Web service that will allow remote access to VISA instrumentation.

Web services are powered by XML and three other core technologies: WSDL, SOAP, and UDDI.

- *XML (eXtensible Markup Language):* XML forms the basis for all modern Web services, which use XML-based technologies to describe their interfaces and to encode their messages. WSDL, SOAP, and UDDI all use XML-based messaging that any machine can interpret
- *WSDL (Web Services Description Language):* The WSDL definition acts as the initial Web service interface, providing clients with all the information they need to interact with the service in a standard-based way. The WSDL document describes the service's location on the Web and the functionality the service provides.

Through the WSDL, a Web services client learns where a service can be accessed, what operations the service performs, the communication protocols the service supports, and the correct format for sending messages to the service.

WSDL describes four critical pieces of data:

- Interface information describing all publicly available functions
- Data-type information for all message requests and message responses
- Binding information about the transport protocol to be used
- Address information for locating the specified service

Using WSDL, a client can locate a Web service and invoke any of its publicly available functions. With WSDL-aware tools, you can also automate this process, enabling applications to easily integrate new services with little or no manual code.

- *SOAP (Simple Object Access Protocol):* SOAP is an XML-based protocol from the W3C for exchanging data over HTTP. It provides a simple, standards-based method for sending XML messages between applications. Web services use SOAP to send messages between a service and its client(s). Because HTTP is supported by all Web servers and browsers, SOAP messages can be sent between applications regardless of their platform or programming language. This quality gives Web services their characteristic interoperability.

To be able to understand in an easy and elegant way how the SOAP protocol works, we will present a communications example (using SOAP) between a client and

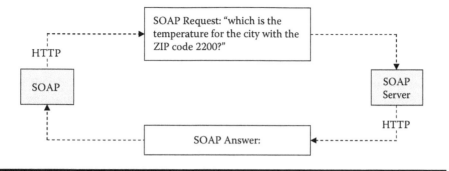

Figure 12.37   SOAP example.

a server. This example (shown in Figure 12.37) presents a very simple weather service, which shows the current temperature for a specified location using the ZIP code.

SOAP messages are XML documents that contain some or all of the following elements:

- Envelope specifies that the XML document is a SOAP message; encloses the message itself.
- Header (optional) contains information relevant to the message, for example, the date the message was sent, authentication data, and so on.
- Body includes the message payload.
- Fault (optional) carries information about a client or server error within a SOAP message.

Example of a SOAP request message (Figure 12.38):

- *VISA:* represents a high-level API that calls in low-level drivers. It can control VXI, GPIB, serial, or computer-based instruments and makes the appropriate driver calls depending on the type of instrument used.

```
<soapenv:Envelope
xmlns:soapenv="http://schemas.xmlsoap.org/soap/envelope/"
xmlns:vel="http://velabservice/">
    <soapenv:Header>
    <soapenv:Body>
        <vel:executeCommand>
            <vel:instrumentDescriptor>visa//192.168.67.101/GPIB0::22::INSTR</
vel:instrumenDescriptor>
            <vel:command>*IDN?</vel:command>
        </vel:executeCommand>
    </soapenv:Body>
</soapenv:Envelope>
```

Figure 12.38   Example of SOAP request message.

The VISA library has standardized the presentation of its operations through a C API exposed from Windows DLL, visa32.dll and over the Microsoft COM technology. Although there are several VISA vendors and implementations, applications written against VISA are (nominally) vendor interchangeable; thanks to the standardization of VISA's presentation and operations/capabilities. Implementations from specific vendors are also available for less common programming languages and software reuse technologies.

On the work-bench server, any *free* VISA controls can be downloaded, for example, NI VISA Run-Time: [http://sine.ni.com/apps/utf8/niup.ni?ap=GB_NIDU&ip=209&loc=en-US&du=http://joule.ni.com/nidu/cds/view/p/id/1071/lang/en]. In this way, the specific instrumentation controls are made free from any "service logic" dedicated software packages such as NI "LabVIEW," Keithley "TestPoint," Agilent "VEE," and so on (as shown before) and can be accessed by simple Web services even from mobile terminals (as a counter-example, LabVIEW Run-Time engine running from a Windows terminal, is installed starting from a ... 24 MB package).

VISA keywords: # Resource—Any instrument in the system, including serial, and parallel ports

Session—You must open a VISA session to a resource to communicate with it, similar to a communication channel.

Instrument Descriptor—Exact name of a resource. The descriptor specifies the interface type (GPIB, VXI, ASRL), the address of the device (logical address or primary address), and the VISA session type (INSTR or Event).

The VISA server: VISA Server runs on computers, controlling the instrumentation work-benches (called by us "work-bench servers" in the generic multitier architecture introduced before). It is needed in order to communicate with VISA-controlled instruments connected on a different machine in the same network or in case of communication with remote resources using Internet connections (usually via HTTP protocol). VISA server is useful when devices need to be shared or when aggregating devices connected to different remote machines (see the case-studies below).

By "VISA resources" they are meant, any kind of instruments or devices connected with the computer by GPIB, VXI, PXI, Serial (including USB), Parallel, or Ethernet interfaces.

Building the Web service for remote VISA instrument access: The communication model proposed by this approach (Figure 12.39) contains the following elements:

- ◼ The client (can be any SOAP client)
- ◼ Web server (handles the HTTP communication)
- ◼ VISA server (handles the communication with the VISA compliant instruments)
- ◼ VISA instruments

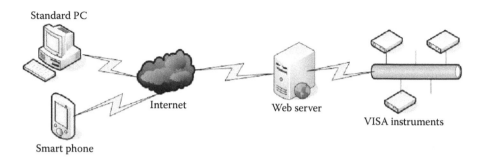

Figure 12.39 **Communication model proposed for the Web service for remote VISA instrument access.**

On the Web server, we have an architectural solution (Figure 12.40) composed of three layers:

Web Server Layer: This is the Web server responsible for the HTTP communication with the client. The Web service SOAP requests and responses are sent using the HTTP protocol.

Taking into consideration the portability, the ease of configuration, and the performance, we have chosen Apache Web Server.

The version of the Apache Web Server used in this implementation is 2.0.63.

Web Service Layer: The Web service layer is responsible for processing the SOAP requests received from the client, executing the required Web service operation, and sending back the response to the client. All the business logic for the Web service operation is implemented at this level.

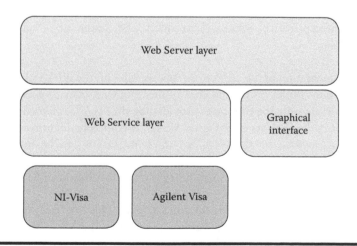

Figure 12.40 **Web server architectural solution.**

Because we have an Apache Web Server and we decided to build the software in C ++, the best solution for the Web service layer is Axis C ++. One major advantage of Axis C ++ is the possibility to install it as an Apache Web Server Module.

VISA Instrument Access Layer: The instrument access and communication will be made using a VISA implementation library and the VISA Server.

There are two major implementations of the VISA standard:

■ NI-VISA the implementation provided by National Instruments
■ Agilent VISA an implementation provided by Agilent

For this implementation, we will use the NI-VISA. This implementation is distributed by National Instruments for both Windows and Linux platforms.

12.3.2.1.2 Implementation of the Web Service

Next, we will present the operations that are implemented by the VeLab Web service. Each operation will be presented focusing on the response and request messages and implementation details.

The VeLab Web service offers the following operations:

■ executeCommand: operation to execute a VISA command on a specified resource
■ getInstrumentList: operation to obtain the list of all available VISA resources
■ getInstrumentStatus: operation to query the status of a specified Visa instrument

The implementation of the service operations relies on the NI-VISA library that provides functions for accessing and communicating with instruments.

We will present only the functions provided by the NI-VISA libraries that were used in building the Web service (Figure 12.41). A complete documentation of all the functions provided by the NI-VISA implementation can be found on the National Instruments Web site.

A central concept in the NI-VISA library is the VISA Resource Manager.

A powerful feature of VISA is the concept of a single interface for finding and accessing devices on various platforms. The VISA Resource Manager does this by exporting services for controlling and managing resources.

CNI-488 Device Function	*C VISA INSTR Operation*	*LabVIEW NI-488 Device Function*	*LabVIEW VISA INSTR Operation*
ibdev	viOpen	<no equivalent>	 VISA Open
ibonl	viClose	<no equivalent>	 VISA close
ibwrt	viWrite	 GPIB Write	 VISA Write
ibrd	viRead	 GPIB Read	 VISA Read

Figure 12.41 Functions provided by the NI-VISA library.

These services include, but are not limited to, assigning unique resource addresses and unique resource IDs, locating resources, and creating sessions.

Each session contains all the information necessary to configure the communication channel with a device, as well as information about the device itself.

[Quote from NI-VISA User Manual]

When trying to access any of the VISA resources, the first step is to get a reference to the default Resource Manager by calling viOpenDefaultRM(). The application can then use the session returned from this call to open sessions to resources controlled by that Resource Manager, as shown in the following examples.

1. GetInstrumentList operation: GetInstrumentList operation will return a list of all available instruments. The returned list contains the VISA address of the instrument, formatted using the VISA standard notation for resource descriptor. The implementation of this service operation uses the functions provided

by the NI-VISA library, to discover the available resources. As mentioned above, we should open the default resource manager:

```
status = viOpenDefaultRM (&defaultRM);
```

Find all the VISA resources in our system and store the number of resources in the system in the "numInstrs" variable. Notice the different query descriptions that are available:

Interface	Expression
GPIB	"GPIB[0–9]*::?*INSTR"
VXI	"VXI?*INSTR"
GPIB-VXI	"GPIB-VXI?*INSTR"
Any VXI	"?*VXI[0–9]*::?*INSTR"
Serial	"ASRL[0–9]*::?*INSTR"
PXI	"PXI?*INSTR"
All instruments	"?*INSTR"
All resources	"?*"

```
ViFindList findList;
ViUInt32 numInstrs;
char instrDescriptor[VI_FIND_BUFLEN];
status = viFindRsrc (defaultRM, "?*INSTR", &findList,
&numInstrs, instrDescriptor);
```

The function viFindRsrc will return a list with all the available instruments. The parameters for this function (Figure 12.42) are:

The viFindRsrc() function matches the value specified in the "expr" parameter with the resources available for a particular interface. You use a regular expression to specify patterns to match in a given string; in other words, it is a search criterion. The viFindRsrc() function uses a case-insensitive compare feature when matching resource names against the regular expression specified in "expr."

On successful completion, this function returns the first resource found in the list and returns a count (retcnt) to indicate if there were more resources found for the designated interface. This function also returns, in the findList parameter, a handle to a find list. This handle points to the list of resources and it must be used as an input to viFindNext().

After obtaining the list of available instrument resources, we loop through this list and by calling the viFindNext function we get all the resource address

Name	Direction	Description
sesn	IN	Resource Manager session (should always be the session returned from `viOpenDefaultRM()`).
expr	IN	This is a regular expression followed by an optional logical expression. Refer to the discussion of the Description String in the *Description* section of this operation.
findList	OUT	Returns a handle identifying this search session. This handle will be used as an input in `viFindNext()`.
retcnt	OUT	Number of matches.
instrDesc	OUT	Returns as string identifying the location of a device. Strings can then be passed to `viOpen()` to establish a session to the given device.

Figure 12.42 Parameters of viFindRsrc function.

descriptors. With this resource descriptor strings we build the list that will be returned by the service operation.

```
xsd__string_Array * returnList=new xsd__string_Array;
if(numInstrs > 0)
{
        xsd__string * instrList = new xsd__string[numInstrs];
        instrList[0] = instrDescriptor;

        for(int i = 1; i < numInstrs; i++)
        {
                char buffer[VI_FIND_BUFLEN];
                status = viFindNext (findList, buffer); /* find
next descriptor */
                if (status >= VI_SUCCESS)
                {
                        instrList[i] = new char[VI_FIND_BUFLEN];
                        strcpy(instrList[i],buffer);
                }
        }
        returnList->set(instrList, numInstrs);
}
```

2. ExecuteCommand operation: ExecuteCommand operation takes as parameters the instrumentDescriptor and the command to be executed. The response from the executeCommand operation contains the following fields:

```
< complexType name = "CommandResult" >
    < sequence >
        < element name = "instrumentDescriptor" type = "string" nillable = "true"/ >
        < element name = "executionStatus" type = "string" nillable = "true"/ >
        < element name = "errorMessage" type = "string" nillable = "true"/ >
        < element name = "result" type = "string" nillable = "true"/ >
        < element name = "executedCommand" type = "string" nillable = "true"/ >
        < element name = "instrumentStatus" type = "string" nillable = "true"/ >
    < /sequence >
< /complexType >
```

- InstrumentDescriptor: the visa address of the instrument which executed the command.
- InstrumentStatus: the status of the instrument after the command execution.
- Result: the result of the command execution returned by the instrument.
- ExecutionStatus: the status of the operation. The value is OK in case of success of ERROR in case something went wrong during the service execution (resource not available, communication errors, various errors).
- ErrorMessage: in case of error, this field will contain the error cause and some text describing the error.

Based on the instrumentDescriptor parameter which contains the instrument address descriptor, we try to open a communication session with the instrument located at that address. This is done using the viOpen() function with the parameters listed below (Figure 12.43).

Name	Direction	Description
sesn	IN	Resource Manager session (should always be the session returned from `viOpenDefaultRM()`).
rsrcName	IN	Unique symbolic name of a resource. See the *Description* section for more information.
accessMode	IN	Specifies the mode by which the resource is to be accessed. See the *Description* section for valid values. If the parameter value is `VI_NULL`, the session uses VISA-supplied default values. Multiple access modes can be used simultaneously by specifying a *bit-wise OR* of the values other than `VI_NULL`.
timeout	IN	Specifies the maximum time period (in milliseconds) that this operation waits before returning an error.
vi	OUT	Unique logical identifier reference to a session.

Figure 12.43   Parameters of viOpen() function.

Name	Direction	Description
vi	IN	Unique logical identifier to a session.
buf	IN	Location of a data block to be sent to a device.
count	IN	Number of bytes to be written.
retCount	OUT	Number of bytes actually transferred.

Figure 12.44   Parameters of viWrite() function.

```
status = viOpen (defaultRM, instrumentDescriptor, VI_NULL, VI_NULL, &instr);
```

The viOpen() function opens a session to the specified resource. It returns a session identifier that can be used to call any other operation of that resource.

After successfully opening a session to the instrument resource, we will try to write the command on this resource. This is done using the viWrite() function provided by NI-VISA with the following parameters (Figure 12.44):

```
ViUInt32 writeCount;
status = viWrite (instr, (ViBuf)command, (ViUInt32)strlen(command),
&writeCount);
if (status < VI_SUCCESS)
{
        status = viClose(instr);
        status = viClose(defaultRM);
        cmdRes->executionStatus = "ERROR";
        cmdRes->errorMessage = "Error writing to the device";
        return cmdRes;
}
```

The viWrite() operation synchronously transfers data. The data to be written are in the buffer represented by variable "buf." This operation returns only when the transfer terminates.

If the writing of the command on the instrument was successful, we will attempt to read the result of this command from the instrument. For this purpose, we use the viRead() function.

```
unsigned char buffer[100];
ViUInt32 retCount;
status = viRead (instr, buffer, 100, &retCount);
```

The parameters required by the viRead function are given in Figure 12.45.

The viRead() operation synchronously transfers data. The data read are to be stored in the buffer represented by variable "buf." This operation returns only when the transfer terminates.

Name	Direction	Description
vi	IN	Unique logical identifier to a session.
buf	OUT	Location of a buffer to receive data from device.
count	IN	Number of bytes to be read.
retCount	OUT	Number of bytes actually transferred.

Figure 12.45   Parameters of viRead() function.

If the read operation was successful, we copy the value from the buffer and set it on the object that will be returned by the service operation:

```
cmdRes- > executionStatus = "OK";
cmdRes- > result = new char[512];
sprintf(cmdRes- > result,"%*s", retCount, buffer);
```

3. GetInstrumentStatus operation: GetInstrumentStatus operation returns the status of an instrument. This operation takes as parameter the visa instrument address and returns the value of the status bit obtained from the instrument.

The instrument descriptor must have one of the formats described above in the getInstrumentList operation. Usually the instrument address (or descriptor) is obtained by invoking the getInstrumentList operation (opening the default resource manager, as shown before).

Based on the instrumentDescriptor parameter which contains the instrument address descriptor, we try to open a communication session with the instrument located at that address.

This is done using the viOpen() function.

```
status = viOpen (defaultRM, instrumentDescriptor, VI_NULL, VI_NULL, &instr);
```

The status of the instrument is obtained by reading the status bit. This is done by using the viReadSTB provided by NI-VISA library which has the following parameters (Figure 12.46):

```
ViPUInt16 instrStatus;
status = viReadSTB(instr, instrStatus);
```

If the status information is only one byte long, the most significant byte is returned with the zero value.

Name	Direction	Description
vi	IN	Unique logical identifier to a session.
status	OUT	Service request status byte.

Figure 12.46   Parameters of viReadSTB() function.

4. "VeLab" Web Service Demo

GetInstrumentList operation: the request of this Web service operation has no parameters. SOAP request has the following format:

```
< soapenv:Envelope xmlns:soapenv = http://schemas.xmlsoap.org/soap/envelope/
xmlns:vel = "http://velabservice/" >
    < soapenv:Header/ >
    < soapenv:Body >
            < vel:getInstrumentList/ >
    < /soapenv:Body >
< /soapenv:Envelope >
```

This Web service operation returns a list of available instrument resources. The response contains an array of strings representing the resource address in VISA syntax.

Example of SOAP response:

```
< SOAP-ENV:Envelope xmlns:SOAP-ENV = "http://schemas.xmlsoap.org/soap/envelope/"
xmlns:xsd = "http://www.w3.org/2001/XMLSchema"
xmlns:xsi = "http://www.w3.org/2001/XMLSchema-instance" >
    < SOAP-ENV:Body >
        < ns1:getInstrumentListResponse xmlns:ns1 = "http://velabservice/" >
            < ns1:return > visa://192.168.67.101/PXI3::2::INSTR < /_return >
            < ns1:return > visa://192.168.67.101/ASRL1::INSTR < /_return >
            < ns1:return > visa://192.168.67.101/ASRL2::INSTR < /_return >
            < ns1:return > visa://192.168.67.101/ASRL10::INSTR < /_return >
            < ns1:return > visa://192.168.67.101/GPIB0::4::INSTR < /_return >
            < ns1:return > visa://192.168.67.101/GPIB0::22::INSTR < /_return >
            < ns1:return > PXI5::9::INSTR < /_return >
            < ns1:return > PXI5::10::INSTR < /_return >
            < ns1:return > PXI5::13::INSTR < /_return >
            < ns1:return > PXI5::15::INSTR < /_return >
            < ns1:return > PXI4::6::1::INSTR < /_return >
            < ns1:return > ASRL1::INSTR < /_return >
        < /ns1:getInstrumentListResponse >
    < /SOAP-ENV:Body >
< /SOAP-ENV:Envelope >
```

The invocation of the Web service operation was done using the SOAP UI testing tool. The following screenshot (Figures 12.47 and 12.48) shows the SOAP request and response for this invocation.

Figure 12.47 SOAP request and response.

On the left side of the image is the request and on the right side the response from the "VeLab" Web service.

ExecuteCommand operation: As presented in the previous section, this Web service operation takes two parameters:

- InstrumentDescriptor: the instrument resource address descriptor formatted using VISA syntax.
- Command: the command to be executed on the specified instrument. This should have a standard VISA command.

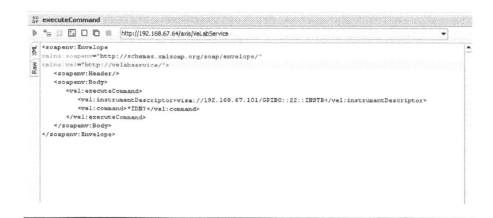

Figure 12.48 SOAP response.

Example of SOAP request:

```
< soapenv : Envelope
xmlns : soapenv = http : //schemas.xmlsoap.org/soap/envelope/
xmlns : vel = "http : //velabservice/" >
    < soapenv : Header/ >
    < soapenv : Body >
< vel : executeCommand >
< vel : instrumentDescriptor >
visa : //192.168.67.101/GPIB0 : :22 : : INSTR
< /vel : instrumentDescriptor >
        < vel : command > *IDN? < /vel : command >
        < /vel : executeCommand >
    < /soapenv : Body >
< /soapenv : Envelope >
```

The response has the following fields:

- Instrumentdescriptor: the instrument where the command was executed
- InstrumentStatus: the status of the instrument after the command execution
- Result: the result returned by the instrument
- executionStatus: the status of the service execution
- executedCommand: the command that was executed
- errorMessage: if an error occurs during the service operation execution, a short message explaining the cause of the error will be returned.

For example: "Error reading a response from the device" or "Could not open a session to the VISA Resource Manager"

Example of SOAP response from executeCommand operation (Figure 12.49):

```
< SOAP-ENV : Envelope
xmlns : SOAP-ENV = "http : //schemas.xmlsoap.org/soap/envelope/"
```

```
<SOAP-ENV:Envelope
xmlns:SOAP-ENV="http://schemas.xmlsoap.org/soap/envelope/"
xmlns:xsd="http://www.w3.org/2001/XMLSchema"
xmlns:xsi="http://www.w3.org/2001/XMLSchema-instance">
    <SOAP-ENV:Body>
        <ns1:executeCommandResponse xmlns:ns1="http://velabservice/">
            <_return>
                <ns1:instrumentDescriptor>visa://192.168.67.101/GPIB0::22::INSTR</ns1:instrumentDescriptor>
                <ns1:executionStatus>OK</ns1:executionStatus>
                <ns1:errorMessage xsi:nil="true"/>
                <ns1:result>HEWLETT-PACKARD,34401A,0,7-5-2</ns1:result>
                <ns1:executedCommand>*IDN?</ns1:executedCommand>
                <ns1:instrumentStatus xsi:nil="true"/>
            </_return>
        </ns1:executeCommandResponse>
    </SOAP-ENV:Body>
</SOAP-ENV:Envelope>
```

Figure 12.49 **SOAP response from the executeCommand operation.**

```
xmlns:xsd = "http://www.w3.org/2001/XMLSchema"
xmlns:xsi = "http://www.w3.org/2001/XMLSchema-instance" >
    < SOAP-ENV:Body >
    < ns1:executeCommandResponse xmlns:ns1 = "http://velabservice/" >
    < ns1:return >
< ns1:instrumentDescriptor >
visa://192.168.67.101/GPIB0::22::INSTR
< /ns1:instrumentDescriptor >
            < ns1:executionStatus > OK < /ns1:executionStatus >
            < ns1:errorMessage xsi:nil = "true" / >
            < ns1:result >
HEWLETT-PACKARD,34401A,0,7-5-2
 < /ns1:result >
 < ns1:executedCommand > *IDN? < /ns1:executedCommand >
            < ns1:instrumentStatus xsi:nil = "true" / >
        < /ns1:return >
        < /ns1:executeCommandResponse >
    < /SOAP-ENV:Body >
< /SOAP-ENV:Envelope >
```

5. Further development: As further development, there are several improvements possible for both the "VeLab" service (responsible for remote instrument access) and to the BPEL (Business Process Execution Language) process.

Improvements for the "VeLab" Web service:

■ Implementation of a better error handling.

At this point, only the basic error scenarios are handled, but there are also several special error cases that need handling.

■ Adding the possibility to refine the search (discovery) of instruments.

This can be achieved by adding parameter to the getInstrumentList Web service operation that will specify the type of instruments to be searched.

For example, passing the string "GPIB" as a filter parameter to the getInstrumentList Web operation, will limit the search only to GPIB instruments.

■ Error messages returned by the Web service in case of failure should resemble the error codes provided by the NI-VISA implementation.

12.3.2.2 Case Study: Remote Experiment: "Bartlett Study of a Bridged-T 2-Port"

They were implemented two approaches of this remote experiment, in order to test the efficiency of the BPEL-based version compared to the "resource-intensive" LabVIEW approach (its panel, with the UUT layout, can be seen in Figure 12.50)

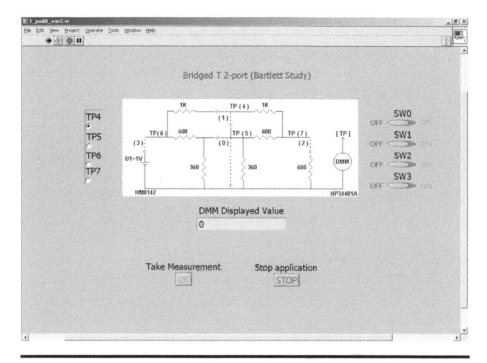

Figure 12.50   VI panel of the "Bartlett study of a Bridget-T 2-port."

The possible remote experiments are of two types:

■ Resistance measurement (when SW3 = OFF <=> voltage source NOT in the circuit); one can measure the resistance with short-circuited semi-section (SW1 = ON, SW0 = ON), and with open-circuit semi-section (SW1 = SW0 = OFF).

■ Voltage measurement (SW3 = ON, voltage source in the circuit it), at the input, at the output and in the intermediary nodes.

BPEL version of the remote experiment on "Bartlett study of a Bridget-T 2-port"
As shown before, the possible remote experiments offered by this scenario are of two types:

■ Resistance measurement (when SW3 = OFF ⇔ voltage source NOT in the circuit); one can measure the resistance with short-circuited semi-section (SW1 = ON, SW0 = ON), and with open-circuit semi-section (SW1 = SW0 = OFF).

■ Voltage measurement (SW3 = ON, voltage source in the circuit it), at the input, at the output and in intermediary nodes.

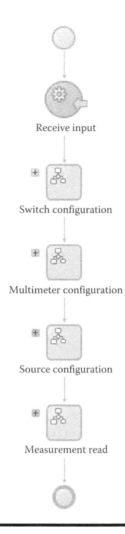

Figure 12.51   BPEL process flow.

The BPEL process flow is shown in Figure 12.51. The process implies several invocations of the VELabService, each enclosed in its own block (scope) as seen in the diagram above:

- Switch configuration (Figure 12.52)
- Multimeter configuration
- Source configuration
- Reading the measurement from the multimeter

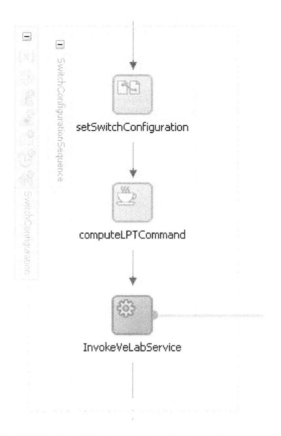

Figure 12.52 Switch configuration block.

The embedded Java activity called "computeLPTCommand" is used to calcu-late the value of the character that will be written on the LPT port. This calculation is based on the input parameter values as seen in the following snippet:

```
< bpelx:exec name = "computeLPTCommand" language = "java"
version = "1.5" >
        < ! [CDATA[short value = 0;
// switch configuration
String sw0 = ((Element)getVariableData("inputVariable","payloa
d","/client:TPoditProcessRequest/client:SW0")).
getFirstChild().getNodeValue();
value += sw0.equalsIgnoreCase("on") ? 1 : 0;
String sw1 = ((Element)getVariableData("inputVariable","payload",
"/client:TPoditProcessRequest/client:SW1")).getFirstChild().
getNodeValue();
value += sw1.equalsIgnoreCase("on") ? 2 : 0;
```

```
☐ <scope name="SwitchConfiguration">
    ☐ <sequence>
        ☐ setSwitchConfiguration
            [2009/02/03 18:43:13] Updated variable "switchConfigurationCommand"less
              - <switchConfigurationCommand>
                - <part xmlns:xsi="http://www.w3.org/2001/XMLSchema-instance" name="parameters">
                  - <executeCommand xmlns="http://velab.service/">
                      <instrumentDescriptor>visa://192.168.67.101/LPT1</instrumentDescriptor>
                      <command/>
                    </executeCommand>
                  </part>
                </switchConfigurationCommand>
        ☕ computeLPTCommand
            [2009/02/03 18:43:13] bpelx:exec executed
        ⊶ InvokeVeLabService
            [2009/02/03 18:43:18] Invoked 2-way operation "executeCommand" on partner "VeLabService".less
              - <messages>
                - <switchConfigurationCommand>
                  - <part xmlns:xsi="http://www.w3.org/2001/XMLSchema-instance" name="parameters">
                    - <executeCommand xmlns="http://velab.service/">
                        <instrumentDescriptor>visa://192.168.67.101/LPT1</instrumentDescriptor>
                        <command>@</command>
                      </executeCommand>
                    </part>
                  </switchConfigurationCommand>
                - <switchConfigurationResponse>
                  - <part xmlns:xsi="http://www.w3.org/2001/XMLSchema-instance" name="parameters">
                    - <ns1:executeCommandResponse xmlns:ns1="http://velabservice/">
                      - <ns1:return>
                          <ns1:instrumentDescriptor>visa://192.168.67.101/LPT1</ns1:instrumentDescriptor>
                          <ns1:executionStatus xsi:nil="true" xmlns:xsi="http://www.w3.org/2001/XMLSchema-instance"/>
                          <ns1:errorMessage xsi:nil="true" xmlns:xsi="http://www.w3.org/2001/XMLSchema-instance"/>
                          <ns1:result xsi:nil="true" xmlns:xsi="http://www.w3.org/2001/XMLSchema-instance"/>
                          <ns1:executedCommand>@</ns1:executedCommand>
                          <ns1:instrumentStatus xsi:nil="true" xmlns:xsi="http://www.w3.org/2001/XMLSchema-instance"/>
                        </ns1:return>
                      </ns1:executeCommandResponse>
                    </part>
                  </switchConfigurationResponse>
                </messages>
    </sequence>
</scope>
```

Figure 12.53   Result of execution of computeLPTCommand.

```java
String sw2 = ((Element)getVariableData("inputVariable","payload",
"/client:TPoditProcessRequest/client:SW2")).getFirstChild().
getNodeValue();
value += sw2.equalsIgnoreCase("on") ? 4 : 0;
String sw3 = ((Element)getVariableData("inputVariable","payload",
"/client:TPoditProcessRequest/client:SW3")).getFirstChild().
getNodeValue();
value += sw3.equalsIgnoreCase("on") ? 8 : 0;
// test point configuration
String tp4 = ((Element)getVariableData("inputVariable","payload",
"/client:TPoditProcessRequest/client:TP4")).getFirstChild().
getNodeValue();
value += tp4.equalsIgnoreCase("on") ? 16 : 0;
String tp5 = ((Element)getVariableData("inputVariable","payload",
```

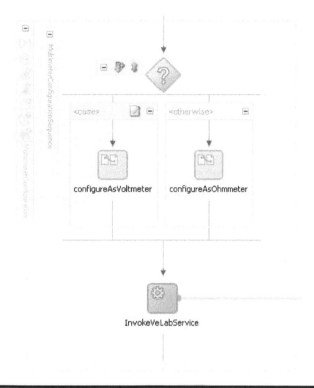

Figure 12.54 The multimeter configuration block.

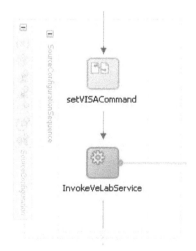

Figure 12.55 Source configuration block.

```
⊟ <scope name="SourceConfiguration">
  ⊟ <sequence>
    📇 setVISACommand
      [2009/02/03 18:43:35] Updated variable "sourceConfigurationCommand" More...
      [2009/02/03 18:43:35] Updated variable "sourceConfigurationCommand" less
        - <sourceConfigurationCommand>
          - <part xmlns:xsi="http://www.w3.org/2001/XMLSchema-instance" name="parameters">
            - <executeCommand xmlns="http://velab.service/">
                <instrumentDescriptor>visa://192.168.67.101/GPIB0::4::INSTR</instrumentDescriptor>
                <command>SU1 1</command>
              </executeCommand>
            </part>
        </sourceConfigurationCommand>
    ⊚ InvokeVeLabService
      [2009/02/03 18:43:35] Invoked 2-way operation "executeCommand" on partner "VeLabService". less
        - <messages>
          - <sourceConfigurationCommand>
            - <part xmlns:xsi="http://www.w3.org/2001/XMLSchema-instance" name="parameters">
              - <executeCommand xmlns="http://velab.service/">
                  <instrumentDescriptor>visa://192.168.67.101/GPIB0::4::INSTR</instrumentDescriptor>
                  <command>SU1 1</command>
                </executeCommand>
              </part>
          </sourceConfigurationCommand>
          - <sourceConfigurationResponse>
            - <part xmlns:xsi="http://www.w3.org/2001/XMLSchema-instance" name="parameters">
              - <ns1:executeCommandResponse xmlns:ns1="http://velabservice/">
                - <ns1:return>
                    <ns1:instrumentDescriptor>visa://192.168.67.101/GPIB0::4::INSTR</ns1:instrumentDescriptor>
                    <ns1:executionStatus>OK</ns1:executionStatus>
                    <ns1:errorMessage xsi:nil="true" xmlns:xsi="http://www.w3.org/2001/XMLSchema-instance"/>
                    <ns1:result>
                      ERROR: NO LEADING COMMAND!
                    </ns1:result>
                    <ns1:executedCommand>SU1 1</ns1:executedCommand>
                    <ns1:instrumentStatus xsi:nil="true" xmlns:xsi="http://www.w3.org/2001/XMLSchema-instance"/>
                  </ns1:return>
                </ns1:executeCommandResponse>
              </part>
            </sourceConfigurationResponse>
          </messages>
  </sequence>
</scope>
```

Figure 12.56 Result of execution of source configuration.

```
"/client:TPoditProcessRequest/client:TP5")).getFirstChild().
getNodeValue();
value += tp5.equalsIgnoreCase("on") ? 32 : 0;
String tp6 = ((Element)getVariableData("inputVariable","payload",
"/client:TPoditProcessRequest/client:TP6")).getFirstChild().
getNodeValue();
value += tp6.equalsIgnoreCase("on") ? 64 : 0;
String tp7 = ((Element)getVariableData("inputVariable","payload",
"/client:TPoditProcessRequest/client:TP7")).getFirstChild().
getNodeValue();
value += tp7.equalsIgnoreCase("on") ? 128 : 0;
// set value
setVariableData("switchConfigurationCommand","parameters","/
ns1:executeCommand/ns1:command", String.valueOf((char)
value));]] >
</bpelx:exec>
```

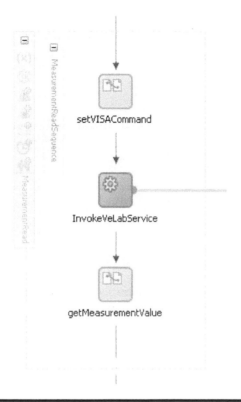

Figure 12.57   The measurement read block.

The execution of this block has the result as given in Figure 12.53.

The multimeter configuration block: Based on the value of the SW3 the multimeter can be set as ohmmeter or voltmeter (Figure 12.54). This is done by the BPEL decide activity which evaluates the SW3 input parameter and if this is set to "on" it will execute the "configureAsVoltmeter" assign activity, otherwise it will execute the "configureAsOhmmeter" assign activity.

The source configuration block is composed of two activities (Figure 12.55):

■ An assign activity which prepares the input variables for the VeLabService invocation.
■ The invoke activity, which calls the VeLabService.

The execution has the result as given in Figure 12.56.

The measurement read block: This block (Figure 12.57) is composed of three activities:

■ An assigned activity which sets the input variables for the invocation of the VelabService.

- The invoke activity which calls the VeLabService.
- An assigned activity which copies the returned value by the VeLabService invocation into the output variable of the BPEL process.

References

1. I. Szekely, F. Sandu, A.N. Balica, and D.N. Robu, Analysis of wireless measurement transmission performance. *Proceedings of the 15th IMEKO TC 4 Symposium on Novelties in Electrical Measurements and Instrumentation*, Iasi, Romania, September 19–21, 2007, ISSB/ISBN: 978-973-667-260-6, 978-973-667-262-0, selected for re-publication in extenso, as Performance measurement for mobile data streaming. F. Sandu, I. Szekely, D.N. Robu, A.N. Balica, *Computer Standards and Interfaces—The International Journal on the Development and Application of Standards for Computers, Software Quality, Data Communications, Interfaces and Measurement*, Elsevier Publications, ISSN: 0920-5489.
2. http://www.3gpp.org, *3gpp Specifications—23.107 Quality of Service (QoS) Concept and Architecture*.
3. J. Eberspächer, H.J. Vögel, and C. Bettstetter, *GSM Switching, Services, and Protocols*, 2nd Edition, John Wiley & Sons, New York, NY, 2001, ISBN 047149903X.
4. J. Bannister, P. Mather, and S. Coope, *Convergence Technologies for 3G Networks: IP, UMTS, EGPRS and ATM*, John Wiley & Sons, New York, NY, 2004, ISBN 047086091X .
5. www.nokiasiemensnetworks.com
6. www.cisco.com
7. Z. Wu, J. Wu, D. Sun, and X. Wu, Remote measurement platform based on DataSocket and .NET framework, *Proceedings of SPIE (the International Society for Optical Engineering)*, Vol. 6358, dedicated to the *6th International Symposium on Instrumentation and Control Technology: Sensors, Automatic Measurement, Control, and Computer Simulation*, Beijing, China, 2006, http://www.spiedl.org/
8. National Instruments, Tutorial on *DataSocket Transfer Protocol*, http://zone.ni.com/devzone/cda/tut/p/id/3223.
9. National Instruments, Tutorial on *Development of Wireless Measurement Systems*, http://zone.ni.com/devzone/cda/tut/p/id/3247.
10. R. Ruchala and K. Zielinski, Measurements of data streaming QoS during handover process in Mobile IPv6 testbed, *Proceedings of the 2nd International Conference on Testbeds and Research Infrastructures for the Development of Networks and Communities— "TRIDENTCOM 2006,"* Barcelona, Spain, ISBN: 1-4244-0106-2.
11. C.S. Wang, J.H. Lee, and Y.T. Chu, Mobile telemedicine application and technologies on GSM, *Proceedings of the 1st International Conference on Bioinformatics and Biomedical Engineering, ICBBE 2007*, Wuhan, China, pp. 1125–1128, ISBN: 1-4244-1120-3.
12. A.S. Tanenbaum, *Computer Networks*, 4th Edition, Prentice-Hall, New Jersey, USA, 2002, ISBN 978-0-13-066102-9.
13. A.A. Gokhale, *Introduction to Telecommunications*, 2nd Edition, Thomson Delmar Learning, New York, USA, 2004, ISBN 1401856489.
14. A. Hac, *Mobile Telecommunications Protocols for Data Networks*, John Wiley & Sons, 2003, New York, NY, ISBN 0-470-85056-6.
15. M. Sauter, *Communication Systems for the Mobile Information Society*, John Wiley, 2006, New York, NY, ISBN 0-470-02676-6.

16. http://www.3gpp.org, 3GPP Specifications—2X Series.

17. www.velleman.be

18. D.E. Comer, *Internetworking with TCP/IP: Principles, Protocols, and Architecture*, 5th Edition, Prentice-Hall, New Jersey, USA, 2006, ISBN 9780131876712

19. J. Postel, User Datagram Protocol, Request for Comments (RFC) no. 768, the Internet Engineering Task Force (IETF), 1980.

20. http://www.nlanr.net/, Iperf tool for measuring maximum TCP/UDP bandwidth, the U.S. National Laboratory for Applied Network Research, http://www.ncsa.uiuc.edu, University of Illinois at Urbana-Champaign, Jperf interface [for Iperf].

21. A. Orebaugh, G. Ramirez, and J. Beale, Wireshark and ethereal network protocol analyzer toolkit, *Syngress, 2006*, ISBN 1597490733.

22. www.ni.com/labview

23. F. Sandu, A. Piironen, P.N. Borza, and C. Gerigan, *Measurements and Automated Test Systems*, LUX LIBRIS Publishing House, Braşov, 2004, ISBN 973–9428–96–9.

24. F. Sandu, D.N. Robu, P.N. Borza, and W. Szabo, Twin-microcontroller GSM modem development system, *Proceedings of the 8th International Conference on Optimization of Electrical and Electronic Equipment "OPTIM 2002"*, Braşov, Romania, vol. 4, pp. 155–160, ISBN 973-635-004-5.

25. F. Sandu, P.N. Borza, A. Balica, and El. Kayafas, *Remote Access to Signal Generation and Acquisition*, 2003, Conspress Publishing House, Bucharest, ISBN 973-8165-44-X.

26. S. Soucek and T. Sauter, Quality of service concerns in IP-based control systems, *IEEE Transactions on Industrial Electronics*, 51(6), 1249–1258, 2004.

27. D. Cheij, Software Architecture for Building Interchangeable Test Systems, *IEEE Aerospace and Electronic Systems Magazine*, 17(1), 27–30, 2002, ISSN 0885-8985.

28. www.ivifoundation.org

29. National Instruments, Tutorial on LabVIEW DSC, http://www.ni.com/swf/presentation/us/labview/newdsc/

30. National Instruments, Tutorial on Using Shared Variables, http://zone.ni.com/devzone/cda/tut/p/id/4679

31. A. Bryan, M-Learning: Emerging pedagogical and campus issues in the mobile learning environment, *EDUCAUSE Center for Applied Research (ECAR) Bulletin*, 2004(16), (August 2004), a publication of ECAR (http://www.educause.edu/ecar).

32. A. Bryan, Going nomadic: Mobile learning in higher education, *EDUCAUSE Review*, 39(5) (September/October 2004), 28–35, http://www.educause.edu/pub/er/erm04/erm0451.asp.

33. E. D. Wagner and R. Robson, Education Unplugged: Mobile Learning Comes of Age, Annual Meeting of the National Learning Infrastructure Initiative, New Orleans, LA, January 24, 2005; Colleen Carmean, blog entry, January 24, 2005, http://blog.educause.edu/carmean

34. W. Hodgins, *The Future of Learning Objects*, Learning Objects Forum, Menlo Park, CA, September 3, 2002.

35. F. Sandu, W. Szabo, and P.N. Borza, Automated measurement laboratory accessed by Internet, in *Proceedings of the XVI IMEKO World Congress 2000*, Vienna, Austria.

36. F. Sandu, W. Szabo, and V. Cazacu, Automated test system with wireless access, *Romanian Magazine of Virtual Instrumentation*, nr. II, 2002.

37. F. Sandu, S. Lahti, S. Rantapuska, I. Tollet, J. Löytöläinen, and R. Kivelä, Remote and mobile control of multidisciplinary experimental systems, *6th IFAC Symposium on Advances in Control Education, "ACE 2003,"* Oulu, Finland.

38. E. Kayafas, F. Sandu, I. Patiniotakis, and P.N. Borza, Approaches to programming for tele-measurement, *Proceedings of the XVII IMEKO World Congress 2001*, Lisbon, Portugal.

39. A.C. Stanca, E. Kayafas, F. Sandu, and M. Demeter, Mobile access to real remote experiments by mini browsers based on content-to-terminal adaptation, *Proceedings of the 10th International Conference on Optimization of Electrical and Electronic Equipment, "OPTIM'06"*, May 18–20, 2006, Braşov, Romania, in coop with: IEEE (IAS, IES, PELS), IEE, VDE, Eds.: M. Cernat, A. Nicolaide, and I. Margineanu, + CDROM, Transylvania Univ. Press, Braşov, ISBN 973-635-705-8.

40. F. Sandu, P.N. Borza, E. Kayafas, and S.A. Moraru, Labview-based remote and mobile access to real and emulated experiments in electronics, Workshop on Pervasive Technologies in e/m-Learning and Internetbased Experiments ("PTLIE"), *Proceedings of "PETRA 2008," the 1st International Conference on Pervasive Technologies Related to Assistive Environments*, ACM (Association for Computer Machinery), Athens, Greece, May 15–19, 2008 (organized by the University of Texas at Arlington, USA).

41. S. Moraru, F. Sandu, and P. Borza, (Eds), Advanced Technologies for e-Learning, 420, Ed. Lux Libris, Braşov, Romania, 2008, ISBN 978-973-131-045-9.

Chapter 13

Comparative Study on the Mobile Data Service Development in the United States, China, and South Korea: Interaction among Performance, Strategies, and Policies*

Jing Zhang, Tugrul U. Daim, and Byung-Chul Choi

Contents

---

* This chapter expands the earlier work of the authors in Daim, T., Zhang, J., and Choi, C., Exploring diffusion of mobile data services, in *Encyclopedia of E-Business Development and Management*, Lee (Ed.), IGI Press, 2010, pp. 325–334.

13.1 Introduction

The mobile communications industry is experiencing the 3G transition globally, which represents a shift from voice-centric services to multimedia-oriented services. According to the statistics from Universal Mobile Telecommunications System (UMTS) forum and CDMA (code division multiple access) Development Group (CDG), 182 Wideband Code Division Multiple Access (WCDMA) networks, and 250 CDMA2000 networks have been deployed all over the world, and the total subscriber number of 3G (including WCDMA, High-Speed Downlink Packet Access (HSDPA), CDMA 1X, and CDMA Evolution-Data Optimized (EV-DO)) had reached 660 million by the end of March, 2008.

In the 3G era, it is surprising that the successful development of new mobile technologies and services has appeared in Asia, especially in Japan and South Korea, instead of Western Europe and North America. Japan is the first nation to have more than 50% of its subscribers using 3G, followed closely by South Korea [1]. Some developing countries have presented higher growth rate in mobile data services than developed countries. This phenomenon motivated this comparative study.

Mobile data services, for which contents and applications are indispensable, are quite different from voice services. The 3G transition is not only a technology upgrade but a big economic transformation requiring a reconstruction of the industry's value network [2,3]. It is not easy to complete this transformation. We can identify various developing paths in different countries, because the involved companies and governmental departments have taken different policies and strategies. What are the success factors and problems in those paths? Are there any common successful strategies that can be referenced by others? Aiming to answer the questions, we select three typical countries in terms of market revenue, market size, and technology—the United

States, China, and South Korea, respectively, to study their paths of developing mobile data services. The policies and strategies are largely different in these three countries. Although we narrow the range of analysis down to three specific countries, we believe that their unique characteristics provide us with a broad spectrum of governmental policy, market, level of technology development, and business strategy. Therefore, the results of comparison can cover important issues in mobile communication industry and have significant insights for most other countries.

13.2 Methodology

13.2.1 Terminology and Abbreviations

In this chapter, the term "mobile services" only refers to the commercially available mobile communication and information services based on terrestrial facilities. Mobile satellite services are excluded. And "mobile data services" are considered to be the delivery of nonvoice information to a mobile device, including broadcasting and two-way services. Besides, some abbreviations used in this chapter are defined as follows.

- Mobile network operators (MNOs) refer to the facilities-based providers that possess mobile communication infrastructure to provide mobile communication services.
- Mobile virtual network operators (MVNOs) do not have the necessary infrastructure, but they provide services to the public by purchasing airtime or leasing facilities from an MNO.
- Value-added services (VAS) is a certain term for noncore telecom services, or for all services beyond standard voice calls and fax transmissions. Mobile data services are the main part of mobile VAS. Unless specifically noted, VAS in this chapter only refers to mobile VAS.
- A content provider (CP) creates contents/applications for mobile subscribers. A pure CP does not have the channels, gateways, or access number for connecting with an MNO/MVNO.
- A service provider (SP) in this chapter only refers to a value-added service provider (VASP), which is a third-party provider rather than an MNO or MVNO. SPs have connection channels and access numbers with which they can work as a bridge that integrate contents/applications and access to the mobile network. Usually an SP also creates and provides contents/applications, working as a CP.

13.2.2 Comparison Model

There are many factors that influence the development of the new technologies and services, including market scale, customer behavior, regulations, business strategies, competition, and the interactions among them. To make the comparison logically clear, we build a comparison model. In this model, the comparison is made on two levels: the performance of mobile data services and the influencing factors in this case the corporate strategies and government policies.

For the performance, we care about the diversity and prevalence of mobile data services, as well as the profitability for the companies. As for the influencing factors, with the consideration of taking the characteristics of the telecom industry, market characteristics, regulatory policies, and corporate strategies are included in the model.

The regulatory policies influence the mobile service market mainly through entry regulation and competition policies to shape the industry structure and to restrict or promote the competition in the market. The corporate strategies involve multiple aspects including technology selection, product and service development, value chain construction, growth strategy, and marketing. To cope with customer trend changing rapidly and maintain competitiveness with various incompatible technologies is main issue for operators. Especially, in the service development, the value chain of providing mobile data services is much more complex than that of voice services. Therefore the value chain strategy which considers the provision of complementary products and services and the integration of multiple products and services into solutions for customers is a crucial part of the strategies.

13.2.3 Data Collection

The data for comparison are collected with a focus on the MNOs. Two Chinese MNOs, China Mobile and China Unicom which were the only two MNOs in China before October 2008, are studied. In the United States, the four nationwide MNOs, AT&T, Verizon Wireless, Sprint Nextel, and T-Mobile have about 87% market share by the volume of subscription. In South Korea, three MNOs—SKT, LGT and KTF— compete in the market. Since all the three countries have few major operators covering dominating market share, they are holding real power over the industry. With the infrastructure integrating the value from terminals, equipment, systems, and services, they are the most important hubs and have the strongest influence in the industry. The CPs and SPs are all connected to MNOs. Therefore, in terms of description of business strategy, we expect that operator-centric perspective can be a prism reflecting the general tendencies of various players in the mobile communication industry.

13.3 Performance of Developing Mobile Data Services

13.3.1 Diversity

Based on various technologies, mobile data services can support a large number of applications. It results in multiple service categories, and some categories overlap each other. Here we describe the diversity of mobile data services in the three countries from two levels: technological platforms and applications (see Figure 13.1). The former refers to the service providing means to carry nonvoice information, currently including short message service (SMS), multimedia message service (MMS), Wireless Application Protocol (WAP) (which is a protocol for mobile Internet access), JAVA, and binary runtime environment for wireless (BREW, a programming platform for CDMA system). The latter refers to the contents or applications that customers can enjoy based on the services, including ringtones, news, pictures, games, video and music downloading, location-based services (LBSs), email, stocks, banking, and so on. For example, SMS provides a point-to-point communication method as well as contents like weather forecasting, greetings, jokes, ads, and text-based games. Therefore, the classification will be confused if we regard platforms and applications as the same level.

Currently, the technological platforms are available and deployed in all the three countries. But the diversity of contents and applications provided based on them varies across countries. The walled-garden strategies of U.S. MNOs and low interest of customers made the applications limited. Although the U.S. MNOs have begun selectively to allow third-party CPs to market content directly to their subscribers, it is estimated that about three-quarters of all U.S. mobile content sales were still going through the operators' branded portals, or storefronts, in late 2006 [4]. The lack of contents and applications weaken the capabilities of some services. The SMS and MMS are just positioned as point-to-point communication methods instead of contents/applications carriers.

In contrast, in China, thanks to the keystone strategy of MNOs, thousands of VAS providers and CPs are developing and providing over 100,000 products of contents and applications through mobile networks [5,6]. These services set up close relations with many other businesses including media, entertainment, education, publication, travel, retailing, finance, banking, security, medical service, and so on, and changed both business models and lives of people greatly. For example, the interaction with audience by SMS has become an important part of all the TV

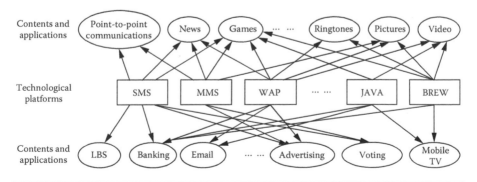

Figure 13.1   The categories of mobile data services.

shows and radio broadcasting programs, and mobile music downloading has become a major channel for digital music distribution.

In South Korea, multiple data service platforms technologies such as SMS, MAP (Mobile Application Part), WAP, and BREW are utilized for mobile data services. Based on the highest penetration rate of cell phones (over 90%) among three countries, the world-class wireless telecommunication infrastructure and the demanding and sensitive customers, over 250 kinds of application services were already available in the market in 2005 [7]. Although operators are pursuing almighty strategy, each of the CPs, operators, and handset companies, are still developing various applications such as music, game, multimedia, and information with their own technologies or cooperation among them. Currently, especially, South Korea has many CPs that have highly competitive and unique mobile data service, and these companies are trying to make inroads into the world market independently or under support of operators.

13.3.2 Prevalence

It is difficult to measure and compare the level of mobile data services development because of the high diversity of mobile data services. Different indicators are used in different countries. We try to find comparable data to show the prevalence of mobile data services in the three countries. The common data services as technological platforms in the three countries are SMS, MMS, and mobile Internet access. Therefore, we use the three types of mobile data services as main indicators.

According to the data from a survey of mobile subscribers ending at the beginning of 2007, it was estimated that in 2007 about 40% of U.S. mobile subscribers use SMS, 15% use MMS, and 11% access to the mobile Internet [4]. In 2006, the total SMS traffic in the United States market rose to more than 158 billion messages and the volume of MMS reached 2.7 billion messages [4]. The average monthly SMS and MMS volume per subscriber in 2006 were, respectively, 58 and 0.99.

In China, SMS is so popular that 92% of the subscribers are active SMS users covering all age groups [8]. At the end of 2006, about 15.7% were MMS users, and this percentage reached 28% in 2007. There are about 60 million active WAP users accounting for 11.0% of the mobile subscriber [8,9]. In 2006, 430 billion SMS messages were sent, and this number rose to 529 billion in 2007 [10]. The volume of MMS rose from 5.5 billion in 2006 to 14.5 billion in 2007 [5,6]. The average monthly SMS and MMS volume per subscriber in 2006 were 84 and 1.07. And these numbers reached 98 and 2.40 in 2007.

In South Korea, it was estimated that in 2007 about 70% of South Korea mobile subscribers use SMS and 30% use MMS. In 2007, approximately 23 million people accounting for 54% of total cell phone users experienced wireless Internet through mobile phone [11]. In 2007, the total SMS traffic in South Korea market came to more than 91 billion messages and the average monthly SMS volume per subscriber in 2007 was 190.

13.3.3 Profitability

The profit margins of all kinds of mobile data services have not been measured accurately. But it is commonly believed that mobile data services, especially the services that do not need broad bandwidth, such as SMS and ringtones download-ing, have much higher margin than voice services, as voice services becoming commoditized. They are powerful weapons to maintain and improve the average revenue per user (ARPU) of mobile services.

The contribution rate of data services in the U.S. mobile service revenue is rela-tively low at 13.5% in 2006 (see Figure 13.2). But the increasing data service ARPU compensated the declining voice service ARPU. In China, the revenues of data services have increased rapidly since 2001, and accounted for averagely 22.6% in 2006 and 25% in 2007 in MNOs' total revenue (see Figure 13.2). In addition to the mobile data service revenue of MNOs, mobile data services have created a new industry of VAS providers and CPs in China. The average margin of providing mobile VAS is around 60% [8], reflecting the high profitability of mobile data services in China. In South Korea, the revenue of data services has steadily increased, but become lower than China since 2006. However, it does not mean that popular-ity of mobile data service starts to decrease in South Korea. This phenomenon

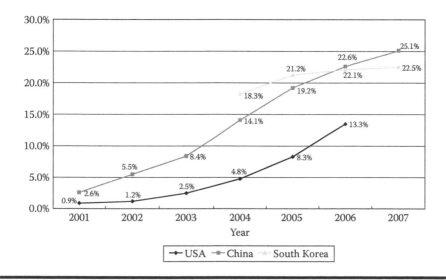

Figure 13.2 Mobile data revenue as percent of total mobile service revenue. (Adapted from FCC, Annual report and analysis of competitive market conditions with respect to commercial mobile services (12th report), February 2008, Federal Communications Commission: Washington, DC; China Mobile Limited Annual Report 2001–2007; China Unicom Limited Annual Report 2001–2007; MINTEL, Mobile phone services—U.S. November 2007.)

might be for a temporary period originated from adjusting the current extremely expensive bill system for multimedia service based on 3G network.

13.4 Regulatory Policies and Corporate Strategies

13.4.1 Regulatory Policies

Different regulatory policies are observed in the three countries. But basically the telecommunication industry was designed by operators putting first because they possess oligopolistic power based on infrastructure. As a result, an operator-centric structure is found in all three countries.

13.4.1.1 United States

In the United States, the promotion of competition was established by the Congress as a fundamental goal for policy formation and regulation on mobile services in 1993. Toward this goal, the major policies from Federal Communications Commission (FCC, the telecom regulatory institution in the United States) include

- A market-oriented approach is taken to managing spectrum helps to lower the entry barriers.
- The spectrum license is related to certain geographical area. Licensees can disaggregate (divide the spectrum into smaller amounts of bandwidth) or partition (divide the license into smaller geographical areas) their licenses, or both, to other entities.
- The mobile SPs have the freedom to choose technologies and develop services.
- Competition is promoted further by policies such as requiring automatic roaming among SPs and the implementation of both local number portability (LNP) and mobile number portability (MNP).

As a result, there is effective competition in the U.S. mobile service marketplace. The report by FCC showed that by the end of 2006, more than 150 companies identified themselves as mobile SPs [4]. However, they do not all compete head to head in each and every region because they have different geographic service areas. A high number of them are regional players and MVNOs providing services for certain groups of subscribers. Four nationwide MNOs—AT&T, Verizon Wireless, Sprint Nextel, and T-Mobile—are holding the spectrum licenses covering the entire land area, and thus population, of the United States. Their market shares by volume of subscription in 2006 are, respectively, 26.8%, 26.0%, 23.0%, and 11%. No single company has a dominant share of the market. Different providers have chosen to implement a variety of different technologies, including CDMA,

GSM, TDMA, and iDEN, with divergent technology migration paths. Competition among multiple incompatible standards emerged as a distinctive feature of the U.S. mobile industry model.

13.4.1.2 China

In China, the communication services are separated into basic services and VAS. Different principles and regulations are used. The main problem is how to set up orderly competition in the telecom service market.

- VAS are open for competition. Qualified companies can apply and gain the license from regulatory agencies.
- Basic service operation is under strict control of the government. For any basic service operator, the state-owned equity interest or shares shall not constitute less than 51%. Mobile voice and infrastructure services are largely classified as basic services. The number of MNOs in the market is determined by the government through licensing system.
- As the representative of the most important stakeholder, the government can interfere directly into the companies' operations, when it is necessary.

Consequently, the competition in basic service market is controlled by the government. All the basic operators are mainly state-owned companies. The two MNOs—China Mobile and China Unicom—were started in the 1990s during the telecom reform. China Mobile was in the absolute leading position with 69.4% market share at the end of 2007 [10]. The two strong nationwide MNOs left no room for MVNOs to survive in the market. In May 2008, the Chinese government started a new round reconstruction of the telecom industry for granting 3G licenses. It is expected that three MNOs will still be functioning after the reconstruction. In contrast, more effective competition system would have been established in VAS field. By the end of 2007, over 22,000 VAS licenses were granted by the regulatory agencies in China [10]. To access customers and necessary infrastructure, these VAS companies must cooperate with basic operators and provide various information services through the networks.

13.4.1.3 South Korea

The policy of South Korea government has gradually changed for selecting operator. When South Korea government announced the first bid for selecting national operators in 1992, they also set the rules themselves. However, later in 2000 and in 2001, South Korea government implemented a transparent bid procedure.

- VAS are open for competition and market. M&A between operator and CP including even broadcasting station is legally possible.

- Basically, the government determines MNOs and controls the number of MNOs. However, government does not interfere in the management of operators since all of them are private companies.
- In 2008 June, a bill for MVNO was passed. Accordingly, though there was no MVNO launching business till then, more effective competition was to be expected in the future.

The major role of the government changed into being a strategic player from a ruler. Accordingly, operator companies were recognized as partners of government for more efficient competition and contribution to the national economy. This change was due to the change of strategic cognition of government of the telecommunication industry. There are two reasons for this change in the attitude of government toward the mobile communication industry. First, South Korea government realized that the mobile communications industry can greatly contribute to the national economy. Second, the markets were not limited to domestic only. South Korea currently has three private operators, SKT, LGT, and KTF, which have dominative influence on industry with taking approximately 50%, 20%, and 30% market share, respectively.

13.4.2 Corporate Strategies

13.4.2.1 United States

The MNOs have deployed mobile broadband networks based on the various incompatible 2G technologies, following different paths since 2001. GPRS (General Packet Radio Service), EDGE (Enhanced Data GSM Environment), CDMA 1X, CDMA EV-DO, CDMA EV-DO Rev. A, WCDMA, and HSDPA were adopted. The CDMA EV-DO and EV-DO Revision A networks of CDMA operators (Verizon Wireless and Sprint Nextel) covered 82% of the U.S. population, and WCDMA/HSDPA networks of GSM/TDMA operators (AT&T and T-Mobile) covered 43% [4]. Some SPs committed to deploying the WiMAX networks.

While the basic data services like SMS began to be available at the end of the 1990s, the SPs paid more attention to voice services and marketed data services primarily as and add-on to voice services in the following years. The quality and price of voice service have always been key in the competition. The rapid decline of the voice service price greatly increased the usage of mobile calls. However, the well-developed voice services and relatively high prices of data services have restricted the growth of data service usage. Currently in the U.S. mobile services market, mobile service for business activities such as e-commerce and internet business are more developed as an auxiliary service [12].

The U.S. MNOs commonly take a walled-garden strategy, dominating the value chain with the control of the provision of handsets, contents, and applications.

MNOs sell low-price locked handsets to customers together with a mobile service contract and customers generally have to pay the early termination fee if they choose to terminate the services before the end of the minimum term. Meanwhile, they keep tight control on what applications are available and what services consumers can access on handsets by selling content through their own branded portals. The Apple iPhone launched by AT&T in June 2007 represents a fundamental change in this walled-garden business model. However, Apple has itself also created a walled garden on the iPhone in terms of branding and applications and the exclusive agreement between Apple and AT&T makes iPhone unavailable for subscribers of any other SPs. The providers have begun selectively to allow third-party CPs to market the multimedia content directly to their subscribers. However, the walled-garden business model still dominates the mobile data service operation [4].

Those strategies are largely shaped by the tough competition. The strategies of incompatible technologies and services, high subsidy on handsets, and exclusive access to contents and applications increase the switching costs of consumers and accordingly help to retain customers. In addition, the pressure of competition makes the providers cautious to promote new services which usually go with high risks. The walled-garden strategy lets them to keep the potential value for themselves as much as possible and helps them to reduce the risks. The objectives of those strategies are for rivalry rather than for developing a new market.

13.4.2.2 China

Although China granted 3G licenses much late (in January 2009) and fell behind other countries as a result, the MNOs and related VAS providers have prepared for mobile data services for years [13]. China Mobile has been upgrading its GSM network since 2001, and deployed EDGE technologies in 2007. China Unicom has focused on the data services mainly based on its CDMA network, which has been upgraded to EV-DO. Based on the infrastructure technologies, multiple data service platforms such as SMS, MMS, WAP, BREW, and JAVA were built to carry diverse contents and applications.

China Mobile introduced its "Monternet" program in November 2000 to promote mobile data services, and adopted the similar way of NTT DoCoMo's i-mode in Japan [14,15]. The objective was the cooperation and revenue-sharing schemes between China Mobile and its VAS provider partners which are called SPs in the Monternet program. The Monternet program enables SPs to access China Mobile's network at any place, provide nationwide VAS with contents/applications to the end customers, and charge the "information fee." By providing network transmission channels and sharing its billing system or other resources with SPs, China Mobile is able to charge the "traffic fee" and shares a part of the information fee. The percentage that China Mobile can share with an SP in the information fee changes according to their cooperation mode. In the cases where China Mobile takes more responsibilities and contributes more in the cooperation, it may keep a higher percentage of the information fee.

In China, mobile data services are considered as a new market with great potential. The cooperation among MNOs, SPs, and CPs has expanded the contents and applications of mobile data services greatly. Cell phones and mobile services are playing an increasingly significant role in the lives of people as a cultural element in China.

13.4.2.3 South Korea

South Korea is the first country which commercialized CDMA technology in 1996. Due to the advantage of early starter, 3G technology already became popular in market and various cutting-edge technologies are currently competing for the future market. Currently, government institutes and companies are preparing for 4G era and trying to develop standard technology for it. Convergence between broadcasting and mobile data service is lively in South Korea. South Korean operators have been utilizing various marketing strategies based on diverse applications and various customized data services. They produce various customized cellular phone including diabetes-phone for a diabetic person and Mecca phone for Muslims.

The operator's position in industry in South Korea is very similar to both the United States and China. In South Korea, few operators have the whole market share such as two operators are doing in China. They have national distribution channels like the four major national operators in the United States and exercise their power to handset companies as a retail dealer of handset. However, Korean operators' growth strategy to cope with the change of the value chain is quite different from the United States and China. Three operators in South Korea are trying a new approach no one had tried before. Direct investment in CP and M&A with various CP and broadcasting station based on their abundant resource are the initial activities for this goal of becoming the almighty managing from producing contents to distributing those. There are two major motivations for this: diversification of distribution media channel (TV, computer, phone, etc.) and emergence of One Source Multiuse (OSMU) move the source of value creation from distribution capacity to contents power.

13.5 Conclusions

The comparison in the prior sections shows that among the three countries, the United States is behind the other two countries in developing mobile data services. Although the technologies have not been well developed in China, the mobile data services have gained a high prevalence among Chinese mobile subscribers. In South Korea, with strong manufacturing and mobile service technologies, diverse mobile services are provided for customers.

Compared to the operators in China and South Korea, the U.S. MNOs kind of ignored mobile data services in the early stage. In the depression period of the U.S. telecom industry they chose to put more attention on the low-risk voice services under the pressure of competition, while China and South Korea put mobile data services into an important strategic position and began to prepare from technology to market to expand this new market. Especially in South Korea, mobile technology and service industry has important position in the national strategy. The strong support from the government made it leading in the world both in terms of technology and service development. Mobile data services can be a strong element in improving the international competitiveness.

The value chain strategy is very important in developing mobile data services. The experience from China showed that a keystone ecosystem strategy to foster a healthy ecosystem has gained considerable achievement in developing mobile data services. Effective partner relationship management and active innovations are major success factors in it. The cooperation among MNOs, SPs, and CPs has expanded the contents and applications of mobile data services greatly. Mobile data service could become a big industry and play an extremely important role in the economy. Some traditional industries' business model may change with the usage of mobile data services. For a large market, to be open for the development of contents and applications will promote growth greatly. Compared to the keystone strategy, the U.S. operators' walled-garden strategy is a restricting factor for mobile data service development. Fortunately, the FCC has begun to require the openness to third-party CPs in the new round of licensing. The almighty strategy that South Korean operators are implementing is effective to integrate competitiveness among diverse players in the industry and developing high-quality contents based on operators' abundant resource. In South Korea, which has both leading technologies and relatively small market size, almighty integrator controlling the entire value chain might be the best way or at least necessary evil to make inroads into the world market and compete with other global companies. However, it internally contains the risk which is able to damage diversity of business ecosystem in the long term and hinder in fair competition for customer.

According to the previous analysis, we say that mobile data contents are more prosperous in China and South Korea than the United States, and this difference basically is originated from operators' business strategy and market situation. However, there are some facts we have to notice when we consider implementing the keystone strategy of China and almighty strategy of South Korea.

In the implementation of a keystone strategy, the regulators and operators should pay much attention on the management of value creation process. The experience from China shows that the lack of effective regulations on this new market may result in frauds in service provision, which will degrade the reputation of the whole industry. And the information asymmetry among the players can bring badly negative effects on the value-sharing schemes. That means the players with the control of more information can get improper value by cheating others.

It is extremely important to improve data transparency among the participants in the ecosystem.

Technology is only a beginning in developing mobile data services; marketing is the more important part. In marketing, to promote the highly diversified mobile data services, accurate market segmentation is quite important. We can see that in South Korea, the in-depth market research resulted in abundant applications for mobile data services. Emotional marketing is also a powerful weapon, which has been proved in China. In South Korea and China, we can see the importance of the trend creation for mobile data services. The operators have been making great efforts in preparing the market for several years. Market segments, customer differentiation, and special branding are effective approaches in opening a new market. Particularly, for the possible pioneer users of new mobile services—the young generation—the operators both in South Korea and China have designed particular service packages and brands and made them become an identity for the young generation. As a result, Cell phones and mobile services are playing an increasingly significant role in the lives of people as a cultural element in China and South Korean.

The experience from China and South Korea has shown that customers' preference and tendency can seriously be influenced by the advertisement of MNOs. The situation that U.S. consumers present little interests in data services can be largely caused by the providers' marketing strategies. In the advertisement of the U.S. providers, there have been a few highlights for mobile data services. Customers have been led to focusing on voice services plans and free or low-price cell phones so that they do not have deep impression of any mobile data service. But it should be noticed that the process of trend creation needs investment on advertising and market research. At the beginning years, the return on that investment could not be huge. There might be a high risk to explore a new market. In the United States, each MNO is facing tough competition from others. It is dangerous for one company to try changing the preference of the consumers. It might lose its customers because when it is doing this, its rivals are satisfying the current preference. Therefore, operators should pay attention to the balance between voice services and data services. The United States should have made more efforts on developing mobile data services. If subscribers are only led to use more voice services, there might be less room for data services as we have seen in the United States.

References

1. RNCOS, 3G market outlook, 2007 (2007–2010). http://www.rncos.com/Report/IM506.htm
2. Kuo, Y.F. and C.W. Yu, 3G telecommunication operators' challenges and roles: A perspective of mobile commerce value chain. *Technovation*, 26(12), 1347–1356, 2006.
3. Tilson, D. and K. Lyytinen, The 3G transition: Changes in the US wireless industry. *Telecommunications Policy*, 30, 569–586, 2006.

4. FCC, Annual report and analysis of competitive market conditions with respect to commercial mobile services (12th report), February 2008, Federal Communications Commission: Washington, DC.
5. China Mobile Limited Annual Report 2001–2007.
6. China Unicom Limited Annual Report 2001–2007.
7. Choi, B.-S., The competition of (value-added services) in Mobile Telecommunication Industry (Economy Focus report no. 24), 2005. Samsung Economy Research Institute. South Korea.
8. iResearch, *Mobile VAS Market Development in China 2007–2008*. 2008, iResearch Consulting Group. China.
9. CNNIC, *WAP Development in China*. May 2007, China Internet Network Information Center. China.
10. MII, *The Development Report of the Communications Industry 2007* (in Chinese). 2008, Ministry of Information Industry of People's Republic of China: Beijing.
11. NIDA, Survey for wireless internet usage of wireless service type. September 2007.
12. Ko, J.-m., *Contents Business in Mobile Era* 2003, Samsung Economy Research Institute.
13. Zhang, J. and X.J. Liang, 3G in China: Environment and Prospect, in Portland International Conference on Management of Engineering and Technology (PICMET). 2007: Portland, OR.
14. Xu, Y., Mobile data communications in China. *Communications of the ACM*, **46**(12), 81–85, 2003.
15. Preez, G.T.d. and C.W.I. Pistorius, Analyzing technological threats and opportunities in wireless data services. *Technological Forecasting and Social Change*, **70**(1), 1–20, 2003.
16. Daim, T., Zhang, J., and Choi, C. Exploring diffusion of mobile data services, In *Encyclopedia of E-Business Development and Management*, Lee (Ed.), IGI Press, 2010, pp. 325–334.

Chapter 14

Participatory Immigration Policy-Making and Harmonization Services Based on Collaborative Web2.0 Technologies

Athanasios Karantjias, Nineta Polemi, and George Pentafronimos

Contents

14.1 Introduction

Nowadays, mobility constitutes a basic principle when considering people, infrastructures, and services under the light of globalization and free movement. Also, the right of free movement represents one of the fundamental civil liberties and it is one of the main objectives of the European Commission (EC) Treaty playing a vital role for the achievement of a single European market. Correspondingly, European Union (EU) governmental organizations facilitate legal immigrants to move freely or seek employment within the EU [1], while at the same time are seeking effective ways of monitor, trace, and audit movement and employment of illegal immigrants. Furthermore, the European Council of December 15–16, 2005 underlined the need for a global approach based on concrete actions and decided to allocate up to 3% of the relevant financial instruments to intensify financial assistance in areas concerning or related to migration.

On the other hand, existing organizational, legal, technological, societal, and political differences in the way EU Member States adopt migration policies, is a serious obstacle to the effective monitoring and management of immigration flows [2–4]. Fortunately, the entry in the ICT Society constitutes a basic strategic choice both for all the members of EU and the Balkan countries as well. Existing ICT advancements can be considered mature enough to enable the provision of innovative national and cross-border electronic services. Thus, progress in harmonization efforts in accordance with the potential advancement in migration services and tools will open paths for new types of Migration Information Systems to the benefits of European citizens.

This chapter proposes a range of migration-oriented services, which facilitate the participation of citizens in the migration policy development process and at the same time boost the harmonization of migration policies and existing efforts across EU Member States [5]. All integrated migration awareness services take advantage of the state-of-the-art ICT technologies, notably Service Oriented Architecture (SOA) [6], Web2.0 [7], social networking [8], the semantics Web [9–11], and policy models and languages, in order to offer a motivating and user-friendly environment, empowering open participation in the policy-making process by targeting both citizens and decision makers. The main objectives are to

- Enable public administrators, politicians, and decision makers to work and collaborate toward a harmonized EU migration policy. The offered services will support these user groups in drafting immigration policy text, developing and testing policy models as well as to perform evaluation scenarios ("what-if" process).
- Facilitate citizens (including societal groups of immigrants) in order to be able to get informed and evaluate various migration-related proposals and policies, while at the same time expressing their feelings and opinions about

existing or developing policies. The process will involve the adoption of new models for citizens' participation, as well as the collection and analysis of collaborative input from various communities.

Specifically, the ImmigrationPolicy2.0 integrated services are offered by a centralized, collaborative, and trustful migration platform (conveniently called ImmigrationPolicy2.0 platform), enabling users (i.e., policy makers, politicians, decision makers, citizens) to identify, model, visualize, analyze, and evaluate national migration policies and practices, as well as to monitor and accordingly harmonize their procedures and data formats related to civil status documents, such as residence permits/certificates, work permits, civil status certificates, and family unification certificates, among others.

The core platform provides a single point of access to all aforementioned services and its integration is based on advanced Web2.0 awareness components (e.g., mash-ups, portlets), SOA artifacts (e.g., Web Services, Enterprise Service Bus), semantic components [12], and rule/inference engines.

The rest of this chapter is organized as follows: Section 14.2 provides background for the motivation and outcomes of this work. Section 14.3 presents the proposed immigration user-centric collaborative services (ImmigrationPolicy2.0 services) as well as the required core design principles that have been properly addressed. Section 14.4 describes the ImmigrationPolicy2.0 platform and its basic components. Finally, Section 14.5 draws conclusions.

14.2 Rationale

For more than 20 years (from the Maastricht Treaty, 1986, to the Treaty of Amsterdam in force since 1999), the Member States have joined forces to combat international phenomena, such as illegal immigration, while promoting the idea of a single European region. On the other hand, legal, political, technological, and societal differences in the Member States as well as the isolation of legal immigrants from the decision-making processes and the formulation of balanced migration proposals, are the main causes blocking the adoption of a common, acceptable, and applicable pan-European migration policy.

Recently, several projects have been launched to research and elaborate on the existing problematic situation, while the implementation and adoption of Migration Information Systems is emerging slowly. Action and development programs funded or cofunded by the EC, such as the EC-UN Joint Migration and ARGO, as well as research projects like Euromed Migration II and those under the MARRI Initiative, are mainly aiming at promoting administrative cooperation and supporting civil society organizations and local authorities [13]. In general, their main objectives are to bring together practitioners, and further develop networking and partnerships by conducting training actions and staff exchanges, studies, conferences, or seminars. Among them, the PROMNISTAT and DCIM-EU projects focus on the establishment of comparable indicators for data collection on migration issues, while the objectives of the ERLAIM project are policy oriented, providing awareness for the integration of immigrants by exchanging good practices, policies, and experiences. However, all aforementioned activities lack in advanced technological adoption and open collaboration approaches, rendering them merely ineffective due to complexity and constant change awareness problems [13].

The identification of the continuously evolving nature of migration issues and the need for adopting automated procedures that integrate widely available technological tools for facilitating in a more holistic way the development and future advancement of common migration policies and more sophisticated initiatives, have introduced a number of software tools in order to enhance traditional ways of migration information diffusion and exchange. Representative examples is the MIPEX project that provides informative national reports and comparative statistical data, as well as the Web-based services offered by the European Civil Registry Network and the Information Exchange System of European Migration Network (EMN), whose main objective is the exchange of Civil Acts documents and other relative data. Additionally, more generic approaches such as the Policy Mix Web Portal and the Common Assessment Framework (CAF) constitute significant sources of information and experience for further analysis and possible interrelation with migration issues [13].

Consequently, while there are various national migration-related initiatives and services, the problems arising when these services/systems move to European level or when non-EU low ICT-based migration authorities are willing to participate and use them as well, are not considered. Most of these tools are isolated and none of them are interoperable with other ICT migration systems and collaborative/interactive with all relative stakeholders for using them persistently in order to monitor, analyze, reengineer, and harmonize their migration procedures and processes, supporting their national policies. Explicitly, they do not aim at the development of collaborative, cross-border citizen-centric migration services.

Subsequently, there is an urgent acknowledged need for innovative and effective collaborative actions for harmonizing national migration policies and specific migration procedures, and for providing advanced migration services, adopting more sophisticated methods by fully exploiting the existing technological advancement. Latest ICT developments based on Web2.0 technologies, collaborative knowledge management systems, and semantic and syntactic analysis technologies, provide numerous opportunities for participatory governance, and citizen-generated policy making. Furthermore, they can boost innovation in terms of both policy-making models and decision support tools. In view of these developments, the following sectors propose a pilot tool, techniques, solutions, and overall services for collaborative development of immigration policies, as well as for supporting related decisions.

14.3 Collaborative Immigration e-Services

The i2010 Government Action Plan under the name "Accelerating e-Government in Europe for the Benefit of All" has already been developed, comprising five framing objectives that can be considered as key methods:

- *All citizens must be taken into consideration:* e-Government will really make a difference if everyone can use and be able to benefit from trusted, innovative services, regardless of gender, age, nationality, income, and disability.
- *Efficiency and effectiveness must be a reality:* Significantly contributing to high user satisfaction, transparency, and accountability, a lighter administrative burden and efficiency gain.
- *High-impact key services for citizens and business have to be implemented:* A road map is necessary for large-scale, cross-border procurement with agreement on cooperation of further high-impact online citizens.
- *Safe access to services has to be ensured:* Governments must enable citizens and businesses to benefit from convenient, secure, and interoperable access to public services.
- *Participation and democratic decision making have to be strengthened:* The demonstration of tools and new propositions for effective public debate and participation in democratic decision making are required.

Taking these five main objectives into account and focusing on the design and implementation of an innovative Migration Information System, the following paragraphs identify several issues and basic requirements, arising from the development and further adoption of collaborative, citizen-centric migration awareness e-services.

14.3.1 Issues and Requirements

The provision of a complete and informative knowledge base for the potential advancement of migration oriented electronic services necessitates the fulfillment of basic requirements, posed from the integration of robust and effective Migration Information Systems, able to provide European cross-border immigration services.

Scalability issues: The increased number of decision makers and stakeholders involved in migration policies, increased and diverged legislations, and implemented migration procedures, different organizational structures, large and inhomogeneous legacy systems cause a "chaos" in the monitoring of national policies and the harmonization of procedures, processes and data formats. Hence, Migration Information Systems demand the creation of a dependency between business and information technologies in order for organizations to be able to efficiently maintain and further support their migration awareness activities. They should allow the abstraction of proprietary applications through the use of adapters, brokers, and orchestration engines. Thus, the resulting integration architecture will be more robust and extensible, especially with the advent of the open Web Services framework and its ability to fully abstract proprietary technology.

Research and Standards: The development of synchronous Migration Information Systems is an emerging market and needs close collaboration with relevant standardization bodies and so they will not only ensure best of breed technologies with a strong focus on European needs, but also that these requirements are translated into globally accepted industry standards. However, private companies are typically compelled by the realities of profit making, while the public sector's organizations do not have such profit-oriented motivations. Essentially, such implementations should adopt peak technologies and worldwide-accepted and mature standards in order to build Enterprise Application Integration (EAI) and technology frameworks, providing advanced added-value electronic migration awareness services according to Software-as-a-Service (SaaS) model. This approach ensures that new technologies come with sustainability insurance for potential investors and commercial exploitation.

Interoperability at all levels: Achieving interoperability at technological (semantically defined content, interoperable systems and applications), legal (migration-related directives, decisions), and organizational (migration stakeholders) level, among many distributed and heterogeneous enterprise systems, is a difficult task, requiring easily identifiable and publishable services, as well as interfaces for the establishment of secure and reliable connection points. A main difficulty being dealt with when designing and implementing Migration Information Systems is to find a universal and standardized way to interoperate with the different kinds of applications and tools, which adopt proprietary interfaces, limited communication protocols, and lack of scalability. The fulfillment of this requirement necessitates the provision of collaborative tools to national stakeholders, decision makers and migration organizations in order for them to be able to continuously and collaboratively identify differences, gaps, and barriers, taking into account that they have better knowledge than anyone on their systems, procedures, and migration policies.

Reusability: The goals behind service reusability are tied directly to some of the most strategic objectives of service-oriented computing, which should be strongly supported by every synchronous Migration Information System's integration. These objectives are as follows:

■ Allow for service logic to be repeatedly leveraged overtime so as to achieve an increasingly high return on the initial investment of delivering the service.
■ Increase business agility on a migration-based level by enabling the rapid fulfillment of future business automation requirements through wide-scale service composition.
■ Enable the realization of agnostic service models.
■ Enable the creation of service inventories with a high percentage of these agnostic services.

Rather than embedding functionality that should be deployed across every specific migration awareness service, an advanced Migration Information System should offer advanced and reusable service interfaces in order to easily expand its functionality and build upon it. The main goal is to offer an innovative implementation framework, in which all essential migration awareness functions could be easily reused, configured, and customized in every e-service provided.

Societal aspects: The involvement of various groups of EU and non-EU decision makers and stakeholders with different societal, ethical, and religious habits, speaking different languages, imposes the need for interoperable, multiaccess modules with multilanguage capabilities. This problem necessitates to be addressed at the European level, gathering societal requirements for the design of immigration services with multimodal multilingual access.

Cost: The funding in many countries in Europe is limited in the public sector for the introduction, development, training and support of value-added services as well as for hiring IT experts. Therefore, the standard-based solutions need to consider the cost as a main issue. It necessitates the use of low-cost and/or royalty-free software components, while also exploit tools and techniques that minimize service deployment costs.

14.3.2 *ImmigrationPolicy2.0 Services*

The ImmigrationPolicy2.0 platform (presented in Section 14.4) provides a single point of access to various services (ImmigrationPolicy2.0 services) enabling and motivating citizens' participation in the process of migration policy making, as well as enhancing existing harmonization actions by supporting politicians and policy, or decision makers in drafting and structuring immigration policy documents. ImmigrationPolicy2.0 ontologies, data structures, and repositories allow users to access a single consolidated view of migration policy models, processes, and procedures.

The integrated and adopted data structures are properly implemented allowing the effective and standard-based hosting of information (fed as raw material) that is relevant to migration-related legislation (e.g., see [14]), decision making, and policy design. Primary supported and efficiently exploited information assets include

- Council Directives and Decisions
- EU Directives and Decisions
- EU Migration Policies
- National Migration Policies
- National Migration Procedures and Processes
- Civil Status Documents
- National and EU Mandates
- National and EU Related Legislation and Policies
- Best Practices
- Migration Studies

The proposed ImmigrationPolicy2.0 services, as depicted in Figure 14.1, are classified into the following concrete categories:

- *Search services* that provide convenient access to sustainable, scalable, structural, searchable inventories, focusing on the following topics:
  - *National migration policies and related legislation*: national migration policies, relevant to EU and national decisions, directives, policies, mandates, EU conventions, joint actions, recommendations, reports, and best practices.
  - *National practices, processes, organizational structures and data formats*: national practices, processes, procedures, organizational structures of involved public administrators, and data formats of civil status documents (residence permits/certificates, work permits, family unification, civil status certificates).
- *Knowledge Harvesting and Content Extraction Services (KHCES)*, from multiple distributed information sources/repositories, based on a set of front end technologies for data collection and collaboration. In particular, the

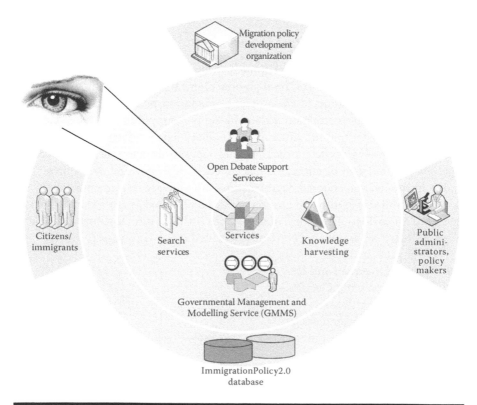

Figure 14.1   Anatomy of the Immigration2.0 services.

ImmigrationPolicy2.0 platform integrates several existing front end data collection technologies and tools (e.g., based on tools and techniques similar to those in [6,15,16]. Moreover, it provides the means for extracting and gathering existing procedures and data formats related to civil status documents in a collaborative environment integrating various Web2.0 services/applications, for example, Wikis, Blogs, Forums, and other.

■ *Governmental Management and Modeling Services (GMMS),* enabling the development and structuring of migration policy models. The GMMS service is offered along with a range of tools enabling policy makers to develop models comprising sets of both structured and unstructured information. The tools offer the capability to transform any type of unstructured document to a structured one according to the target migration policy model. Furthermore, for a given migration policy model it allows the disciplined development and illustration of procedures, processes and workflows associated with the model. The service allows the fine-tuning of models based on building, running, and evaluating "what-if" scenarios associated with the models.

■ *Migration Policy Synchronization and Harmonization Services (MPSHS),* that provide an efficient synchronization framework, which succeeds in managing data and information flows between the various public authorities involved in the handling of civil status documents. These are supported by an interactive system for collaborative synchronization of policies, satisfying the requirement for coordination and data synchronization, which is among the primary reasons of wrong decisions during harmonization efforts. As part of the synchronization and harmonization services, ImmigrationPolicy2.0 platform provides the involved actors with common, uniform, and ubiquitous access to properly customized graphical interfaces for collecting, analyzing, and sharing real-time data as well as for supporting management decisions.

■ *Open Debate Support Services (ODSS),* enabling user communities to work together in building open debates, adding diverse opinions, thoughts, and relevant contributions to each debate, while also collaboratively elaborating in order to either reach a final consensus or formulate an idea about possible best future actions. These services are supported by debate tools enabling complex aggregation of data. Data aggregation focuses on conflicting demands, positive and negative opinions, similar expressions, as well as open and structured reasoning.

Even though the aforementioned services are all deployed in the ImmigrationPolicy2.0 platform, they are not available to all users, since they have varying scope for accommodating the needs of the stakeholders. Hence, some services are offered at a local/regional level (e.g., to municipalities only), whereas other services have a national, transnational, or even EU scope. Table 14.1 illustrates the scope of each type of service.

**Table 14.1   Scope (Regional, National, EU) of the ImmigrationPolicy2.0
Services**

ImmigrationPolicy2.0 Services	*Scope*	*Remarks*
Search services	Regional, National, EU	Search of information assets is offered at all levels. ImmigrationPolicy2.0 aims at providing an infrastructure for the future pan-European integration of information assets about migration policies.
KHCS	Regional, National	Information aggregation in ImmigrationPolicy2.0 is performed at each individual site in order to avoid the localization problems/complexity, which incurs an EU level knowledge harvesting case.
GMMS	Regional, National	ImmigrationPolicy2.0 offers tools and techniques for policy development in each of the pilot sites.
MPSHS	Transnational EU	The problem of policies interoperability and harmonization is by its nature a transnational (cross-border) and EU level problem
ODSS	Regional	ODSS services are offered at local/regional level

Envisaged typical usage of the ImmigrationPolicy2.0 services per user group
has as follows:

■ *Citizens' usage:* Individual immigrants, their representative organizations and
groups, may use the ImmigrationPolicy2.0 collaborative platform in order to
provide the base initial information sources to be fed into the systems as raw
material for mass cooperation, decision making and policy design. Moreover,
immigrants are able to search, browse, and comment on migration-based
information assets, procedures, directives, and legislations. In terms of the
aforementioned outlined services, citizens benefit from search, content management, and open-debate services. Specifically, the content management
services allow citizens to provide their information to the platform, regardless
of any legal, cultural, territorial, and social peculiarities, associated with the
provision of this information. Search services empower citizens to access content assets, procedures, and directives of their interest.

■ *Immigration Stakeholders'* (e.g., *Decision/Policy Makers and Public administrators) usage:* Decision/Policy Makers, such as European Agencies of Migration Policies and Public Administrators, such as Ministries of Interior, and Municipalities may use the ImmigrationPolicy2.0 platform for monitoring and benchmarking national migration policies, as well as building harmonized EU procedures, processes and data formats of civil status documents, as depicted in Figure 14.2 that follows. In addition, they are able to easily identify legal, organizational, security, privacy, and political gaps and barriers at the EU level for defining common-unified policies. In the same context, the ImmegrationPolicy2.0 platform enables policy makers and public administrators to report, search, browse, model, and analyze their national migration policies, specific civil status documents' handling procedures, data formats, and organizational structures. Decision/Policy makers leverage the GMMS services, the ODSS services, as well as the search services in order to be able to form a holistic view of policies, best practices, regulations, and migration awareness procedures. Furthermore, policy makers and administrators may have access to harmonized services [17] for identifying differences in message formats and building EU-wide standards for cross-border service interaction.

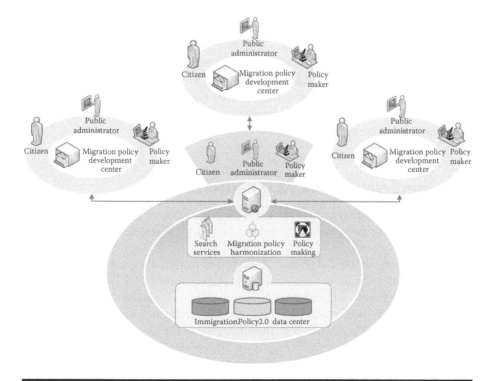

Figure 14.2   Interoperation across ImmigrationPolicy2.0 systems.

The provision of the integrated migration awareness services could be performed in various ways:

- *Information Asset Management:* Public administrators and migration-related organizations are able to use ubiquitous and user-friendly Web interfaces to create and insert into the system data, migration policies, directives, practices, procedures, structures, processes, and studies. Both citizens and public administrators are able to search and browse these information assets according to their access privileges (which will be different for the specified user groups and roles). In the scope of information asset management, end-users are enabled to monitor and track changes in a controlled way, while also participating in open debates.
- *Governmental Process Modeling:* Public administrators and policy makers may model, further analyze and appropriately manage the information assets contained in the ImmigrationPolicy2.0 platform.
- *Benchmarking and Assessment:* End-users (notably decision and policy makers) are able to benchmark national practices, policies, procedures, and studies as well as perform assessment on these in terms of legislation, organizational, political, security characteristics, penetration/acceptance rates, forecasting in order to identify gaps with respect to EU directives and best practices. This enables end-users to reengineer their current internal affairs.
- *Semantic Interoperability Guidance:* Public administrators and policy makers are guided by the platform on how to use semantically awareness schema definitions in order to build the basis for an EU-wide standardization of messages and policies in migration area. This semantic knowledge is shareable among systems to enable collaborative evolution of National and EU ontologies/systems on migration assets.

14.4 ImmigrationPolicy2.0 Platform

The following paragraphs present the main entities and analyze their fundamental characteristics that compose the ImmigrationPolicy2.0 platform. These are the following:

The ImmigrationPolicy2.0 Portal: A Web Interactive System which provides electronic-based point of secure access to ImmigrationPolicy2.0 information and content, retrieved and processed from diverse sources, in a unified and user-friendly way. This system is based on collaborative Web2.0 technologies (such as collaborative forums, blogs, Wikis, etc.). The Immigration2.0 portal serves a single secure access point to the full range of Immigration2.0 services, including search services, harvesting knowledge and migration intelligence services, policy modeling services, as well as harmonization services. Different portal instances may be deployed in order to accommodate the peculiar needs of the various sites. The portal infrastructure incorporates security mechanisms and policies that enhance the ImmigrationPolicy2.0 platform and tools with proper authentication and authorization properties and Single-Sign-On (SSO) procedures, enclosing end-user's preferences and requirements.

The ImmigrationPolicy2.0 CMS: A Content Management System (CMS), accompanies the Portal and undertakes the creation, editing, management, and publishing of all the primary and processed content in a consistently organized fashion. Specifically, the CMS mainly provides

- ■ Advanced content management tools, such as rich text editors, live page editing and scheduling, and advanced document management components.
- ■ Web2.0 awareness technologies with their own set of authorizations, message boards for facilitating conversations around migration awareness topics, blogs for allowing users to convey information and RSS feeds from the last-mentioned message boards and blogs within the system.
- ■ A multitier search engine so that end-users are able to search relevant migration awareness information throughout the entire Web interactive system, within specific portlets such as Wikis, Message Boards, other Web2.0 awareness technologies and even in external integrated applications through its advanced multiple interfacing module.
- ■ Intuitive front end-user interfaces share a set of common characteristics to promote user friendliness and accessibility. These are multilingual, in order for users to easily toggle between different language settings and they will follow standardized best practices for accessibility (especially for special groups).
- ■ Web publishing tools, to easily create and manage content, from a simple article of text and images to fully functional Web sources.

The ImmigrationPolicy2.0 Ontologies and Semantic Structures: A collection of semantic structures, notably ontologies/taxonomies, modeling migration policies, migration information assets, as well as their semantic relationships have been designed, implemented, and integrated within the ImmigrationPolicy2.0 portal. Thematic, service, governmental, and migration-related ontologies/taxonomies are defined to better organize the various quantities of migration-based information assets in the databases of ImmigrationPolicy2.0. These bring context to words, topic areas, and search results, providing a hierarchical structure of asset categories, from general to specific. It conveniently calls the set of semantic structures of the ImmigrationPolicy2.0 platform, as the ImmigrationPolicy2.0 Knowledge Base. Note also that as part of the ImmigrationPolicy2.0 data and semantics modeling, a set of service and migration-based Content Objects in XML format are fed to the CMS engine so that it may appropriately generate the required electronic forms to be used by end-users in order to insert the primary information assets. Each of these objects is based on e-GIF policies and standards (http://www.govtalk.gov.uk) [18] describing specific attributes that define migration-based procedures, directives, structures, civil status documents, and best practices.

The ImmigrationPolicy2.0 SOA Environment and BPM Tools: An SOA infrastructure (e.g., application servers, Web services containers) has been established to enable integration of diverse data (i.e., access to data sources and repositories) and business logic (i.e., policy development services) in the scope of complex distributed workflows for policy modeling, decision support, and migration policy harmonization. In the scope of ImmigrationPolicy2.0 SOA environment, standards-based BPM tools (utilizing the BPMN (Business Process Modeling Notation) and/or BPEL (Business Process Execution Language) standards are used in order to enable modeling, composition, and deployment of service workflows.

The ImmigrationPolicy2.0 Decision Support and Business Intelligence System: On top of the ImmigrationPolicy2.0 knowledge base, the ImmigrationPolicy2.0 provides tools and applications for "Immigration Intelligence." In particular, a properly designed Decision Support System (DSS) is provided to facilitate decision makers in compiling useful information from primary raw data, documents, national and EU knowledge, and migration models to identify, solve problems, and reach decisions. The DSS system and accompanying rules and tools are applicable to both—local/regional and EU/transnational level policies based on different runtime parameters and deployment configurations. The DSS system capitalizes on the BPM tools.

The ImmigrationPolicy2.0 Governmental/Policy Modeling Toolkit: The platform also offers a Governmental Process Modeling Management Tool, which enables experts to design and model complex workflows together in a graphical environment. The BPM tools of the platform are used to compose workflows on the basis of distributed services.

The ImmigrationPolicy2.0 Migration Policy Harmonization: The migration policy harmonization tools also capitalize on the BPM tools and infrastructure of the ImmigrationPolicy2.0 platform. The respective services and workflows are based on the common semantic infrastructure/structures developed for the various policies as part of the ImmigrationPolicy2.0 knowledge base.

The ImmigrationPolicy2.0 Open Debate Support Applications: A set of open debate applications are provided over the ImmigrationPolicy2.0 portal infrastructure integrating a collection of Web2.0 components (wikis, forums and blogs, collaborative domain ontologies, and collaborative documents). These applications act as collaboration tools among stakeholders in order to report their current practices, provide their opinions, suggestions, and comments on any migration-related content established in the system with the goal of specifying the future migration policy European framework. The same Web2.0 technologies are used to setup online Questionnaires, interviews, workshops, surveys, collaborative documents, and open debates with key stakeholders and decision makers.

The development and integration of the aforementioned ImmigrationPolicy2.0 platform's elements are based on existing components and tools, as illustrated in Table 14.2.

Table 14.2   Open Source Components of the ImmigrationPolicy2.0 Platform

ImmigrationPolicy2.0 Infrastructure or Service Element	*Components*
Portal CMS and Information Delivery	LifeRay Portal Infrastructure
eb2.0 Technologies and Elements (Wiki, Blogs, Forums)	LifeRay Social Networks Infrastructure
SOA Infrastructure	Open Source JavaEE Server (e.g., JBoss, JonAS)
Business Process Modeling	ADONIS Toolkit
Government Policy Modeling and Migration Policy Harmonization Workflows	ADONIS Toolkit, PolicyMix Project
DSS Workflows	ADONIS Toolkit
Open Debate Support Services	LifeRay Social Networks Infrastructure

The JavaEE compliant JBoss Application Server (http://www.jboss.org) is used as the base middleware infrastructure of the entire ImmigrationPolicy2.0 platform. It provides the required containers for deploying Web-based components and migration awareness services in an SOA-oriented environment. All processes deployed in it organize their embedded logic into separate and easily changed "state machines," establishing a highly agile automation environment, fully capable of adapting to change, which is realized by abstracting the process logic into its own tier. Thereby, the integrators are able to alleviate other services from the need to repeatedly embed process logic, and support process optimization as a primary source of change for which migration awareness services can be recomposed. The generation of all Business Process Management Notations is achieved through ADONIS (www.adonis-community.com) system, which is properly integrated in policy makers' and administrators' Web interfaces.

The generation of all required Web interfaces, as well as the proper management of the migration awareness content of the platform is performed through the integration of the Liferay (http://www.liferay.com) open source platform. It actually integrates core migration awareness taxonomies, and the required XML-based content objects. The ImmigrationPolicy2.0 taxonomies consider worldwide standards, such as the ISO-2788:1986 (www.iso.org) [19] and BS-8723 (schemas.bs8723.org) [20]. Among these are the *Integrated Public Sector Vocabulary (IPSV)* (www.govtalk. gov.uk) [21], published from the British Government, and the *Taxonomy of Human Services*, the *European Communities Glossary*. Correspondingly, the content objects' structure is based on the UK e-Government Interoperability framework (www. govtalk.gov.uk) in order for the system to enable the seamless flow of information assets and to provide a long-term strategy that will be able to accommodate and adapt in the future. Several Web2.0 tools and functionalities (including collaboration wikis, blogs, and forums) come as readily available components from the Liferay platform.

The use of the above tools boosts the openness and the cost efficiency of the overall ImmigrationPolicy2.0 platform and its offered services.

14.5 Conclusions

Enabling EU Member States and other European governmental organizations to systematically collaborate on migration issues and provide migration e-services holds a lot of future prospects. Specifically, the strengthening of existing harmonization efforts and the enhancement of European cooperation and effective collaboration constitutes the solution for difficult pan-European problems such as immigration. To this end, this chapter contributes toward participatory governance in the immigration domain, based on a wide range of state-of-the-art technologies. Immigration2.0 services are capable to offer various benefits since they improve citizens' ability to actively participate in public debates about immigration policies and other relative activities. They also benefit migration stakeholders (decision makers, policy makers, and public administrators) since

- They will have at their disposal, a centralized system for accessing, analyzing, and processing citizen data, as well as policies, directives, and so on. This system will facilitate data aggregation and analysis, while also facilitating the process of deriving information from bulk immigration data.
- They will be offered tools and services for compromising the diversity of heterogeneous migration policies established in different EU countries, as well as in neighboring EU countries. This harmonization will allow different counties to join forces toward more effective migration policies.
- The presented Immigration2.0 services capitalize on a set of innovative technologies paving the way for new research activities including
  - The blending of Web2.0 components (used for visualization of policy components in government modeling and public debate services) with semantic structures (i.e., migration policies ontologies and taxonomies) is envisaged to be the next generation of services.
  - The employment of business process management (BPM) [22] techniques for the composition of complex e-government services spanning multiple elementary services that are dispersed both geographically and administratively fall in the scope of the next generation of composite business Web Services (complex fully fledged SOA environments).
- The development and integration of an ontology framework and related taxonomies toward establishing a common understanding of terms and a sharable

vocabulary of information associated with migration policies will support complex aggregation of data, including conflicting demands, positive and negative opinions, similar expressions, and both open and structured reasoning.

■ The integration of algorithms, patterns, and methodologies for policy decision making and monitoring composes a higher level of intelligence for policy makers, surpassing in effectiveness existing traditional monitoring and harmonization methods.

Integrating and properly orchestrating a number of advanced technologies and open-source software components, ImmigrationPolicy2.0 constitutes an innovative revolutionary Migration Information System, able to strengthen operational pan-European cooperation in fight against illegal immigration and the harmonization of policies, taking advantage of the interactive participation both of citizens and immigration stakeholders.

Acknowledgments

This chapter outlines the EU FP7 e-Participation proposal titled Immigration Policy 2.0 and the authors would like to thank all partners (SingularLogic S.A, ATOS ORIGIN, BOC Asset Management GmbH, European Forum for Migration Studies, Municipality Subotica, Hellenic Migration Policy Institute, TIPS Albanian National Bureau, and Research Centre of University of Piraeus) for their review and inputs. Special thanks to Dr. Stelios Pantelopoulos and the SLG team.

References

1. Favell, A., The Europeanisation of immigration politics, European Integration online Papers (Eiop), 1998, 2(10).
2. Geddes, A., Still Beyond Fortress Europe? Patterns and Pathways in EU Migration Policy, Queen's Papers on Europeanisation, No. 4/2003.
3. Kostakopoulou, T., Invisible citizens: Long-term resident third resident third country nationals in the EU and their struggle for recognition, in R. Bellamy and A. Warleigh (eds.) *Citizenship and Governance in the EU*, London: Continuum, 2001, pp. 180–205.
4. Niessen, J., Overlapping interests and conflicting agendas: The knocking into shape of EU immigration policies, *European Journal of Migration and Law*, 2001, 3, 419–434.
5. Guiraudon, V., The constitution of a European immigration policy domain: A political sociology approach, *Journal of European Public Policy*, 2003, 10(2), 263–282.
6. Yan, Z., Enterprise Service Oriented Architecture (ESOA) Adoption Reference, in the *Proc. of the IEEE International Conference on Services Computing, 2006 (SCC '06)*, September 18–22, 2006, ISBN: 0-7695-2670-5, p. 512.
7. Juffinger, A., Neidhart, T.H., Weichselbraun, A., Wohlgenannt, G., Granitzer, M., Kern, R., and Scharl, R., Distributed Web2.0 crawling for ontology evolution, in the

Proc. of the 2nd International Conference on Digital Information Management, 2007 (ICDIM '07), October 28–31, 2007, Vol. 2, pp. 615–620, ISBN: 978-1-4244-1475-8.

8. Peers, S., Key legislative developments on migration in the European Union, *European Journal of Migration and Law*, 2002, 4, 339–367.

9. van Harmelen, F., The semantic web: What, why, how, and when, *IEEE Distributed Systems Online* 1541–4922 copyright 2004, IEEE Computer Society, 2004, 5(3), 4.

10. Hendler, J., Web 3.0 emerging, *IEEE Computer*, 2009, 42(1), 111–113.

11. McFedries, P., The cloud is the computer, *IEEE Spectrum*, August 2008.

12. Ding, L., Finin, T., Joshi, A., Peng, Y., Pan, R., and Reddivari, P. Search on the semantic web, *IEEE Computer*, October 2005, 38(10), 62–69.

13. Pentafronimos, G., Karantjias, A., and Polemi, N., OPSIS: An open, preventive and scalable migration information system, Springer Lecture Notes Proceedings, Hellenic Scientific Council for the Information Society (HSCIS), *The Third International Conference on e-Democracy*, September, 2009, Athens, Greece.

14. Annual Information Society Report, Benchmarking i2010: Progress and fragmentation in Europe's information society (Staff Working Paper Vol. 1) and Country profiles (Staff Working Paper Vol. 3), 2008.

15. Ghita S., Nejdl W., and Paiu R., Semantically rich recommendations in social networks for sharing, exchanging and ranking semantic context. ISWC 2005.

16. Adomavicius, G. and Tuzhilin, A., Toward the next generation of recommender systems: A survey of the state-of-the-art and possible extensions. In *IEEE Transactions on Knowledge and Data Engineering*, 2005, 17(6), 734–749.

17. Papademetriou, D., *Europe and Its Immigrants in the 21st Century: A New Deal or a Continuing Dialogue of the Deaf*? Migration Policy Institute Publications, MPI and Luso-American Foundation, March 2006.

18. UK GovTalk, e-Government Interoperability Framework, CabinetOffice, Available at http://www.govtalk.gov.uk/schemasstandards/egif.asp

19. ISO 2788:1986, *Documentation—Guidelines for the Establishment and Development of Monolingual Thesauri*, International Organization for Standardization, Available at http://www.iso.org

20. BS 8723, *Structured Vocabularies for Information Retrieval, Exchange Formats and Protocols for Interoperability*, Available at http://schemas.bs8723.org/

21. UK GovTalk, *Integrated Public Sector Vocabulary (IPSV), CabineOffice*, Available at http://www.govtalk.gov.uk

22. Jim, S. and Janelle, H.B., *Gartner Predicts 2007: Align BPM and SOA Initiative Now to Increase Chances of Becoming a Leader in 2010*, Gartner Report, November 2006, pp. 1–4.

Index

Note: n = footnote

Milton Keynes UK
Ingram Content Group UK Ltd.
UKHW031125141024
449569UK00006B/423

9 781439 800898